Springer
Tokyo
Berlin
Heidelberg
New York
Barcelona
Budapest
Hong Kong
London
Milan
Paris
Santa Clara
Singapore

J.-M. Bony, M. Morimoto (Eds.)

New Trends in Microlocal Analysis

With 7 Figures

Springer

Jean-Michel Bony
Professor, Centre de Mathematiques, Ecole Polytechnique
91128 Palaiseau Cedex, France

Mitsuo Morimoto
Professor, Sophia University
7-1 kıoı-cho, Chıyoda-ku, Tokyo, 102 Japan

ISBN-13: 978-4-431-68415-2 e-ISBN-13: 978-4-431-68413-8
DOI: 10.1007/978-4-431-68413-8

Printed on acid-free paper

Softcover reprint of the hardcover 1st edition 1997

Typesetting: Camera-ready by editors and authors using Springer TeX macropackage.

Preface

This is a collection of articles on microlocal analysis and related topics in partial differential equations and mathematical physics; most of the articles were read at the JSPS–CNRS joint seminar "New Trends in Microlocal Analysis" organized by Hikosaburo Komatsu and Jean-Michel Bony, sponsored by Société Franco-Japonaise des Sciences Pures et Appliquées, and held at La Maison Franco-Japonaise, Tokyo, from September 11 through 16, 1995. The seminar was supported by the Japan Society for the Promotion of Science (JSPS) and Centre National des Recherches Scientifiques (CNRS). We are grateful for their generosity.

Microlocal analysis was initiated by Mikio Sato around 1970. He wrote, along with his coauthors Masaki Kashiwara and Takahiro Kawai, a decisive article on the structure of pseudodifferential equations, thus laying the foundation of $\mathcal{D}$-modules and the singular spectrums of hyperfunctions. Microlocal analysis is still an active area of mathematical research in itself. The key idea is to analyze problems on the phase space, i.e., the cotangent bundle of the base space, and this approach has influenced many fields of mathematics such as real and complex analysis, representation theory, topology, number theory, and mathematical physics.

The book is divided into three parts: partial differential equations and mathematical analysis, mathematical physics, and algebraic analysis — $\mathcal{D}$-modules and sheave theory. It covers a large variety of the latest research, in which readers will find new ideas, problems, and results.

This volume is dedicated to Professor Hikosaburo Komatsu on his 60th birthday. He has been one of the leaders of microlocal analysis, in particular, its application to partial differential equations.

We sincerely thank Professor Nobuyuki Tose for his cooperation in editing and typesetting this volume using TEX.

September 7, 1996

Jean-Michel Bony
Mitsuo Morimoto

Contents

Part I
Partial Differential Equations and Mathematical Analysis

Fourier integral operators and Weyl-Hörmander calculus

Dedicated to H. Komatsu on his sixtieth birthday

Jean-Michel Bony

Centre de Mathématiques URA 169 CNRS, Ecole Polytechnique, 91128 Palaiseau Cedex, France (e-mail: bony@math.polytechnique.fr)

Introduction

It is well known that the space of *classical* pseudo-differential operators is invariant under conjugation by *classical* Fourier integral operators. However, the Weyl-Hörmander calculus [Hö1] [Hö2] provides a much larger framework for the theory of pseudo-differential operators. Any riemannian metric g on the phase space $\mathbf{R}^n_x \times \mathbf{R}^n_\xi$, satisfying the conditions of definition 1.1, defines a graded algebra of pseudo-differential operators. The classical theory corresponds to a particular metric, namely $g(dx, d\xi) = dx^2 + d\xi^2/\langle\xi\rangle^2$.

Given a diffeomorphism F of the phase space well behaved with respect to two metrics g_1 and g_2, our program is to construct classes of operators P (to be called Fourier integral) such that conjugates $P^\star AP$ of pseudo-differential operators A associated to g_2 are pseudo-differential operators associated to g_1. Except for the case of metrics not too far from the standard metric, which is studied in [Hö1], very few was known about this problem. In [Bo1] and [Bo2], we gave an answer in the particular case of symplectic metrics. In the present paper, we shall study (almost) general metrics.

It turns out that the problem is closely related to the characterization of pseudo-differential operators. The prototype of such results is the theorem of R. Beals [Be] for constant metrics, using commutators with operators whose symbol is a linear form. We gave in [B&C] a characterization valid for any metric in terms of localized commutators (see theorem 3.3 below). The main result of section 3 is a characterization involving only the boundedness in L^2 of commutators, under an extra assumption (geodesic temperance) on the metric. The linear forms should be replaced here by symbols belonging to a space, called $S^+(1, g)$, which plays a crucial role.

Now, given a diffeomorphism F, an operator $P \in \mathcal{L}(L^2)$ and $b \in S^+(1, g_2)$, we introduce the *twisted commutator* : $BP - P\widetilde{B}$, where the symbols of B and $\widetilde{B}$ are respectively b and $b \circ F$. We define Fourier integral operators as operators P whose iterated twisted commutators are bounded on L^2. The algebraic properties : adjoints, composition, conjugation ... follow rather simply. Of course, some properties are required on the derivatives of F, but we have not to assume that

F is symplectic. However, except for the case of symplectic metrics, it should be *asymptotically symplectic* for having the existence of invertible Fourier integral operators.

In sections 5 and 6, when F is symplectic, we define the principal symbol of a Fourier integral operator and the corresponding symbolic calculus. As a byproduct, we get the existence of almost invertible Fourier integral operators. The symbol is a section of a line bundle over the (extended) graph of F. We do not use the Maslov index for constructing this bundle, using the affine metaplectic group instead. This is perhaps simpler, and certainly more coherent with the rest of our construction. Actually, the theory is quite easy when F is an affine symplectic map and, after localization, one has just to consider maps F tangent to the identity at the origin, for a fixed metric.

We shall not develop here the main application of our calculus : the study of operators solving an evolution problem : $\frac{d}{dt}P_t = iAP_t$. We refer to [Bo1] and [Bo2] for the case of symplectic metrics and to a forthcoming paper for the general case.

1 Weyl quantization, metrics and symbols

Given a temperate distribution a on $\mathbf{R}^{2n}$, its standard quantization is the operator $a(x, D)$ defined by

$$a(x,D)u(x) = \iint e^{i(x-y)\cdot\xi} a(x,\xi)u(y)\, dy\, d\xi/(2\pi)^n .$$

while its Weyl quantization is the operator a^w defined by

$$a^w u(x) = \iint e^{i(x-y)\cdot\xi} a\Big(\tfrac{x+y}{2},\xi\Big)u(y)\, dy\, d\xi/(2\pi)^n .$$

The application $a \mapsto a^w$ is an isomorphism of $\mathcal{S}'(\mathbf{R}^{2n})$ onto $\mathcal{L}(\mathcal{S}(\mathbf{R}^n), \mathcal{S}'(\mathbf{R}^n))$ and an isomorphism of $\mathcal{S}(\mathbf{R}^{2n})$ onto $\mathcal{L}(\mathcal{S}'(\mathbf{R}^n), \mathcal{S}(\mathbf{R}^n))$.

It will be convenient to use capital letters $X = (x, \xi)$ to denote points of the phase space $\mathbf{R}^{2n}_X = \mathbf{R}^n_x \times \mathbf{R}^n_\xi$. It is equipped with its canonical symplectic form

$$[X,Y] = y\cdot\xi - x\cdot\eta$$

and its normalized measure $đX = dx\, d\xi/\pi^n$.

When the composition $a^w \circ b^w$ is defined as an operator from $\mathcal{S}$ into $\mathcal{S}'$, one can define the composition of the symbols a and b by $(a\#b)^w = a^w \circ b^w$. An explicit formula is given by

$$(a\#b)(X) = \iint e^{-2i[X-S,X-T]} a(S)b(T)\, đS\, đT .$$

The composition of three symbols is given by a simple and useful formula

$$(a\#\alpha\#b)(X) = \iint e^{-2i[X-S,X-T]} a(S)\alpha(S+T-X)b(T)\, đS\, đT . \tag{1}$$

Fourier Transformation. The composition with $\boldsymbol{\delta} = (\pi)^n \delta$ gives the following definition

$$\mathcal{F}a(P) = \widehat{a}(P) = \boldsymbol{\delta}\#a(P) = \int e^{-2i[X,P]} a(X)\, đX\ . \tag{2}$$

The distribution $\widehat{a}$ is defined on $\mathbf{R}^{2n}$ and is actually the Fourier transform of a, if $\mathbf{R}^{2n}$ is identified with its dual space via the symplectic form. One can define also the conjugate Fourier transform $\overline{\mathcal{F}}a = a\#\boldsymbol{\delta}$, and one has $\check{a} = \boldsymbol{\delta}\#a\#\boldsymbol{\delta}$, with $\check{a}(X) = a(-X)$.

For each quadratic form γ on $\mathbf{R}^{2n}$, one denotes by γ^σ the inverse of the quadratic form γ, still using the identification of $\mathbf{R}^{2n}$ with its dual space :

$$\gamma^\sigma(T) = \sup_{W \neq 0} \frac{[T,W]^2}{\gamma(W)}\ .$$

Definition 1.1 *A Hörmander metric is a riemannian metric on $\mathbf{R}^{2n}$ satisfying the three following properties*

- Uncertainty Principle : $g_X(\cdot) \leq g_X^\sigma(\cdot)$.
- Slowness : *There exists $\overline{C} > 0$ such that*

$$g_Y(X-Y) \leq \overline{C}^{-1} \Rightarrow \Big(g_Y(\cdot)/g_X(\cdot)\Big)^{\pm 1} \leq \overline{C}\ . \tag{3}$$

- Temperance : *There exist $\overline{C} > 0$ et $\overline{N} \in \mathbf{N}$ such that*

$$\Big(g_Y(\cdot)/g_X(\cdot)\Big)^{\pm 1} \leq \overline{C}\Big(1 + g_Y^\sigma(Y-X)\Big)^{\overline{N}}\ . \tag{4}$$

Typical examples are the so-called *metrics of type* (ρ, δ), given by

$$g_Y(dx, d\xi) = \langle\eta\rangle^{2\delta} dx^2 + \langle\eta\rangle^{-2\rho} d\xi^2\ , \quad \langle\eta\rangle = (1 + |\eta|^2)^{1/2}.$$

They satisfy the above conditions for $\delta \leq \rho \leq 1$ and $\delta < 1$.

Definition 1.2 *A g-weight is a positive function on $\mathbf{R}^{2n}$ such that*

$$\Big(M(Y)/M(X)\Big)^{\pm 1} \leq \begin{cases} \overline{C} & \text{for } g_Y(X-Y) \leq 1/\overline{C} \\ \overline{C}\Big(1 + g_Y^\sigma(Y-X)\Big)^{\overline{N}} & \text{for } X, Y \in \mathbf{R}^{2n} \end{cases} \tag{5}$$

The class of symbols $S(M,g)$ is the space of $a \in C^\infty(\mathbf{R}^{2n})$ such that the following semi-norms are finite

$$\|a\|_{k\,;S(M,g)} = \sup_{\substack{l \leq k; X \in \mathbf{R}^{2n} \\ g_X(T_j) \leq 1}} M(X)^{-1} \left|\partial_{T_1} \ldots \partial_{T_l} a(X)\right|\ ,$$

where $\partial_T a$ means the directional derivative $\langle T, da\rangle$.

The space $\Psi(M,g)$ of pseudo-differential operators of weight M is the space of operators a^w, with $a \in S(M,g)$.

The fundamental results of the theory (see [Hö2] section 18.5) are the following :
— Operators belonging to $\Psi(M,g)$ map $\mathcal{S}$ into itself and $\mathcal{S}'$ into itself. For $M=1$, they are bounded on L^2.
— The set of weight functions is a multiplicative group, and $\Psi(\cdot,g)$ is an algebra graded by this group :

$$S(M_1,g)\#S(M_2,g)\subset S(M_1M_2,g)\ ,\quad \Psi(M_1,g)\circ\Psi(M_2,g)\subset\Psi(M_1M_2,g)\ .$$

2 Confinement

Let g be a Hörmander metric and let us denote by $B_{Y,r}$ the g_Y-ball of radius r centered at Y. The concept of confined symbol [B&L] will allow to do the main part of the work locally in these balls, using the constant metric g_Y. Global results will then be obtained using the biconfinement theorem below.

Definition 2.1 *The space* $\mathrm{Conf}(g,Y,r)$ *is the Schwartz space* $\mathcal{S}(\mathbf{R}^{2n})$ *equipped with the following sequence of semi-norms*

$$\|a\|'_{k\,;\mathrm{Conf}(g,Y,r)}=\sup_{l,X,T_j}|\partial_{T_1}\dots\partial_{T_l}a(X)|\,(1+g_Y^\sigma(X-U_{Y,r}))^{k/2}$$

$$\text{for } l\le k,\ g_Y(T_j)\le 1$$

or preferably with the (uniformly) equivalent family

$$\|a\|_{k\,;\mathrm{Conf}(g,Y,r)}=\sup_X|a(X)|\,(1+g_Y^\sigma(X-B_{Y,r}))^{k/2}$$

$$+\sup_P|\widehat{a}(P)|\,(1+g_Y^\sigma(P))^{k/2}\,|g_Y|^{1/2}\ . \qquad (6)$$

Here, $|g_Y|$ is the determinant of g_Y and $g_Y(X-A)$ (resp. $g_Y(A-B)$) denotes the infimum of $g_Y(X-X')$ (resp. $g_Y(X'-X'')$) for $X'\in A$ (and $X''\in B$).

A family $(a_Y)_{Y\in\mathbf{R}^{2n}}$ is *uniformly confined in* $B_{Y,r}$ if $\|a_Y\|_{k\,;\mathrm{Conf}(g,Y,r)}$ is bounded by a constant depending on k but not on Y. A typical example is a family (a_Y) of functions whose support is contained in $B_{Y,r}$ and which is bounded in $S(1,g)$.

To prove the equivalence of the families of semi-norms, it is easy to estimate $\|a\|_{\mathrm{Conf}}$ by $\|a\|'_{\mathrm{Conf}}$. Conversely, one can control the uniform norm of $\partial_{T_1}\dots\partial_{T_l}a(X)$ by the $\|a\|_{\mathrm{Conf}}$ and one can interpolate with the decay of a itself using the following lemma, where g_Y will be the metric γ and $(1+g_Y^\sigma(X-B_{Y,r}))^{-N}$ will be the function Φ.

Lemma 2.2 *Let* Φ *be a positive function defined on* $\mathbf{R}^\nu$ *such that the level sets* $\Phi^{-1}(]\lambda,\infty[)$ *are convex. Let* γ *be a positive definite quadratic form on* $\mathbf{R}^\nu$ *and* $u\in C^2(\mathbf{R}^\nu)$ *such that* $|u(x)|\le\Phi(x)$ *and* $|\partial_{T_1}\partial_{T_2}u(x)|\le M$ *for* $\gamma(T_j)\le 1$. *Then*

$$|\partial_T u(x)|\le 2\sqrt{M\Phi(x)}\ \textit{for}\ \gamma(T)\le 1\ .$$

The most important result about confinement is the following theorem [B&L] which uses the function

$$\Delta_r(Y,Z) = 1 + g_Y^\sigma(B_{Y,r}-B_{Z,r}) + g_Z^\sigma(B_{Y,r}-B_{Z,r}) \ , \tag{7}$$

measuring the "distance" between Y and Z.

Theorem 2.3 (biconfinement) *For r sufficiently small, if (a_Y) and (b_Y) are uniformly confined in $B_{Y,r}$ then, for any N, the family $\Delta_r(Y,Z)^N(a_Y \# b_Z)$ is uniformly confined in $B_{Y,r}$ and in $B_{Z,r}$.*

More precisely, given k and N, there exist C and l which do not depend on a, b, Y, Z (and even on g as far as the constants $\overline{C}, \overline{N}$ in definition 1.1 are fixed), such that

$$\|a\#b\|_{k\,;\mathrm{Conf}(g,Y,r)} + \|a\#b\|_{k\,;\mathrm{Conf}(g,Z,r)}$$
$$\leq C\,\|a\|_{l\,;\mathrm{Conf}(g,Y,r)}\,\|b\|_{l\,;\mathrm{Conf}(g,Z,r)}\,\Delta_r(Y,Z)^{-N} \ .$$

A rather simple proof can be given using the *twisted convolution*

$$a \circledast b(X) = \{a\#\boldsymbol{\delta}\#b\}(X) = \int e^{-2i[X,S]}a(S)b(X-S)\,đS$$

Using $\boldsymbol{\delta}\#\boldsymbol{\delta} = 1$ and the definition (2), we get four formulas

$$a\#b = \begin{cases} a \circledast (\mathcal{F}b) \\ (\overline{\mathcal{F}}a) \circledast b \end{cases} \ ; \ \mathcal{F}(a\#b) = \begin{cases} (\mathcal{F}a) \circledast (\mathcal{F}b) \\ \check{a} \circledast b \end{cases} \ . \tag{8}$$

The result follows just from estimating the module of these twisted convolutions by the usual convolutions of $|a|, \left|\widehat{b}\right|, \ldots$ whose decay is controlled by the semi-norms of confinement.

Other important properties are the Schur property for the kernel Δ^{-N} :

$$\exists N \ , \ \sup_Y \int \Delta_r(Y,Z)^{-N}\,đZ < \infty \ , \tag{9}$$

and the control of the L^2 norm

$$\exists C \ , \ \|a^w\|_{\mathcal{L}(L^2)} \leq C\,\|a\|_{2n+1\,;\mathrm{Conf}(g,Y,r)} \ . \tag{10}$$

There exist partitions of unity $1 = \int \varphi_Y\,đY$, where the family (φ_Y) is uniformly confined. As a consequence, any $a \in S(M,g)$ can be written

$$a = \int M(Y)a_Y\,đY \ , \quad (a_Y) \text{ uniformly confined} \ ,$$

it suffices to write $a_Y = M(Y)^{-1}\varphi_Y a$. Conversely, any a given by such an integral belongs to $S(M,g)$.

These decompositions make possible to recover very simply the fundamental properties given at the end of the last section. For instance, for $a = \int a_Y\,đY \in S(1,g)$, one has

$$\|a_Y^w \overline{a}_Z^w\|_{\mathcal{L}(L^2)} \leq C\Delta(Y,Z)^{-N}$$

in view of (10) using the biconfinement theorem. The boundedness of a^w in L^2 is then a direct consequence of (9) and of Cotlar lemma.

3 Characterization of pseudo-differential operators

Given a Hörmander metric g, the function λ (the inverse of the function h in [Hö2] section 18.5) is defined by $\lambda(X)^2 = \inf_T g_X^\sigma(T)/g_X(T)$. The uncertainty principle says precisely that $\lambda \geq 1$. We will say that g is a *symplectic metric* if $\lambda = 1$, that is $g^\sigma = g$.

There is an asymptotic expansion in λ of $a\#b$, for which we just give the first two terms

$$a\#b = ab + (2i)^{-1}\{a,b\} + r_2(a,b)$$

where $\{a,b\}$ is the Poisson bracket.

If a_Y and b_Z are uniformly confined, then one has $\lambda(Y)\Delta_r(Y,Z)^N\{a_Y, b_Z\}$ and $\lambda(Y)^2\Delta_r(Y,Z)^N r_2(a_Y,b_Z)$ uniformly confined.

If $a \in S(M_1,g)$ and $b \in S(M_2,g)$, then $\{a,b\} \in S(M_1M_2\lambda^{-1},g)$ and $r_2(a,b) \in S(M_1M_2\lambda^{-2},g)$.

However, the integral formula giving the value of $r_2(a,b)$ depends only on the second derivatives of a and b, and the results above are valid under weaker assumptions.

Definition 3.1 *A function a belongs to $S^+(M,g)$ if*

$$|\partial_S \partial_{T_1} \dots \partial_{T_k} a(X)| \leq C_k M(X)\ , \text{ for } g_X^\sigma(S) \leq 1\ ,\ g_X(T_j) \leq 1\ .$$

The best constants C_k, $k = 0,1,\dots$ will be the semi-norms of a in this space.

Proposition 3.2 *Let $a \in S^+(M,g)$, $b \in S(M',g)$ and $c \in S^+(M',g)$. Then*

$$a\#b - b\#a \in S(MM',g)$$

$$a\#c - c\#a = -i\{a,c\} + r, \quad \{a,c\} \in S(MM'\lambda,g) \subset S^+(MM',g),$$
$$r \in S(MM',g)\ .$$

As a consequence, if $c \in S(M^{-1},g)$ and $b \in S^+(M,g)$, then $c^w \operatorname{ad} b^w$, i.e. the application $A \mapsto c^w(b^wA - Ab^w)$ maps $\Psi(1,g)$ into itself. Conversely, pseudo-differential operators can be characterized by the action of such operators $c^w \operatorname{ad} b^w$. This is a simple corollary of the following characterization (see [B&C], theorem 5.5) which we recall first.

Theorem 3.3 *An operator A belongs to $\Psi(1,g)$ if and only if, for any uniformly confined family (θ_Y), one has*

$$\|(\operatorname{ad} L_1)\cdots(\operatorname{ad} L_k)\cdot(\theta_Y^w A)\|_{\mathcal{L}(L^2)} \leq C_k$$

when L_j are linear forms $X \mapsto [X,T_j]$ with $g_Y(T_j) \leq 1$, the constants C_k being independent on Y and L_j.

Corollary 3.4 *An operator $A \in \mathcal{L}(L^2)$ belongs to $\Psi(1,g)$ if and only if one has, for $c_j \in S(M_j^{-1},g)$ and $b_j \in S^+(M_j,g)$,*

$$\left\| \prod_1^N (c_j^w \operatorname{ad} b_j^w) A \right\|_{\mathcal{L}(L^2)} \leq C \prod \|c_j\|_{k\,;S(M_j^{-1},g)} \prod \|b_j\|_{k\,;S^+(M_j,g)} \tag{11}$$

where $C = C(N)$ and $k = k(N)$ can be choosen independently of the weights M_j as far as their structural constants $\overline{C}$ and $\overline{N}$ in (5) are bounded.

Let us remark first that linear forms $L_j(X) = [X, T_j]$ belong to $S^+(M_j,g)$, with $M_j(X) = g_X(T_j)^{1/2}$. Moreover, the weights $g_X(T_j)^{1/2}$ have the same structural constants, which is an immediate consequence of (3) and (4). It suffices now to use (11) with $b_j = L_j$ and $c_j = M_j(Y)^{-1}\theta_Y$ and we are reduced to theorem 3.3.

We shall now examine whether it is possible to characterize pseudo-differential operators using just commutators with elements of $\Psi^+(1,g)$. For the sake of simplicity, we shall assume that g^σ is conformal to g, i.e. that $g^\sigma = \lambda^2 g$. The following condition will play un important role, which is not surprising if one thinks that elements of $S^+(1,g)$ are Lipschitz continuous for g^σ, and thus that their variation between two points is bounded by the g^σ-geodesic distance of these two points.

Definition 3.5 *We shall say that a Hörmander metric g is geodesically tempered if one has*

$$\Big(g_Y(\cdot)/g_Z(\cdot)\Big)^{\pm 1} \leq \overline{C}\Big(1 + d^\sigma(Y,Z)\Big)^{\overline{N}},$$

where $d^\sigma(Y,Z)$ denotes the g^σ-geodesic distance of Y and Z

This condition is satisfied for all metrics of type (ρ,δ). We have no example of a Hörmander metric which is not geodesically tempered.

Theorem 3.6 *Let g be geodesically tempered with $g^\sigma = \lambda^2 g$. An operator $A \in \mathcal{L}(L^2)$ belongs to $\Psi(1,g)$ if and only if one has*

$$\prod_1^N (\operatorname{ad} b_j^w) \cdot A \in \mathcal{L}(L^2) \quad \text{for } b_j \in S^+(1,g)\,. \tag{12}$$

We shall just sketch the proof in order to show the main argument. We shall denote by $\delta_r(Y,Z)$ the geodesic distance of $B_{Y,r}$ and $B_{Z,r}$. The first point is to construct a family p_{YZ} which is bounded in $S^+(1,g)$ such that, for a convenient $C > 1$, the function p_{YZ} vanishes in $B_{Y,r}$ and is equal to the constant $\delta_{Cr}(Y,Z)$ in $B_{Z,r}$. This can be done by considering first the function

$$q_{YZ}(X) = \min(d^\sigma(X - B_{Y,Cr})\,,\ \delta_{Cr}(Y,Z))\,,$$

which is Lipschitz continuous for g^σ, and then by regularizing it. The second point is to prove the following lemma.

Lemma 3.7 *Let (α_Y) and (β_Y) be uniformly confined. Then one has*

$$\delta_{Cr}(Y,Z)\alpha_Y^w A\beta_Z^w = \sum_{j=1}^{3} \alpha_{jY}^w A_j \beta_{jZ}^w$$

where α_{jY} and β_{jZ} are uniformly confined and where A_j satisfy (12) with uniform constants.

One has

$$\alpha_Y^w A p_{YZ}^w \beta_Z^w = \alpha_Y^w p_{YZ}^w A \beta_Z^w + \alpha_Y^w A_1 \beta_Z^w \ , \ A_1 = -\operatorname{ad} p_{YZ} \cdot A \, .$$

Thanks to the fact that α_Y vanishes in B_{Yr}, it is not difficult to prove the uniform confinement of $\alpha_{2Y} = \alpha_Y \# p_{YZ}$. For the same reason, one has $p_{YZ}\#\beta_Z = \delta_{Cr}(Y,Z)\beta_Z + \beta_{3Z}$ with β_{3Z} uniformly confined. The lemma follows.

An easy induction shows that the property (12) implies

$$\|\alpha_Y^w A \beta_Z^w\|_{\mathcal{L}(L^2)} \leq C_N (1 + \delta_{Cr}(Y,Z))^{-N} \tag{13}$$

for α_Y and β_Y uniformly confined. Using the geodesic temperance, one gets the same estimate with $\Delta_{Cr}(Y,Z)^{-N}$ in the right hand side.

To end the proof of theorem 3.6, we have just to show that the assumptions of theorem 3.3 are consequences of (12). Given a linear form $L_j(X) = [X, T_j]$, with $g_Y(T_j) \leq 1$, set $p_j(X) = L_j(X-Y)\psi_Y(X)$, the functions ψ_Y being uniformly confined and equal to 1 in $B_{Y,r}$. These functions p_j belong to a bounded set of $S^+(1,g)$. It remains to prove that $\prod(\operatorname{ad} L_j - \operatorname{ad} p_j) \cdot A$ is bounded on L^2, which is not very difficult using (13).

Theorem 3.8 *Assume that g is geodesically tempered with $g^\sigma = \lambda^2 g$.*
(i) Let $A \in \Psi(1,g)$ be invertible in $\mathcal{L}(L^2)$. Then $A^{-1} \in \Psi(1,g)$.
(ii) Let $a \in S(1,g)$ real valued, and let $f \in C^\infty(\mathbf{R})$. Then the operator $f(a^w)$ (in the sense of the functional calculus on self-adjoint operators in L^2) belongs to $\Psi(1,g)$.
(iii) Let $M \geq 1$ be a g-weight and $a \in S(M,g)$ real valued such that a^w is elliptic. Let $f \in C^\infty(\mathbf{R})$ satisfying $\left|f^{(k)}(t)\right| \leq C_k(1+|t|)^{p-k}$. Then the operator $f(a^w)$ belongs to $\Psi(M^p,g)$.

In the last statement, *elliptic* means that there is an estimate

$$\forall u \in H(M,g) \, , \quad \|u\|_{H(M,g)} \leq C(\|a^w u\|_{L^2} + \|u\|_{L^2})$$

(see [B&C] for the definition of the Sobolev spaces $H(M,g)$), in which case a^w, with domain $H(M,g)$, is a self-adjoint unbounded operator on L^2.

The proof of the first part of the theorem is purely algebraic. One has $\operatorname{ad} b^w \cdot A^{-1} = -A^{-1}(\operatorname{ad} b^w \cdot A)A^{-1}$ and, by induction, $\prod \operatorname{ad} b_j^w \cdot A^{-1}$ is a sum of products of terms which are either A^{-1} or iterated commutators of A and b_j^w.

For $b_j \in S^+(1,g)$, they are thus bounded on L^2 under the assumption of (i), and the result follows from the characterization 3.6.

The same kind of arguments show that the resolvent operators $(z-A)^{-1}$ are pseudo-differential for $z \in \mathbf{C} \setminus \mathbf{R}$, and one can get estimates of their semi-norms under the assumptions of parts (ii) or (iii). The end of the proof relies on the formula of Helffer-Sjöstrand [H&S]

$$f(A) = -\pi^{-1} \iint_{\mathbf{C}} \overline{\partial} \widetilde{f}(z)(z-A)^{-1}\, dx\, dy\,, \tag{14}$$

where $\widetilde{f}$ is an almost analytic extension of f.

Remark 3.9 The conclusions of theorem 3.8 are valid for metrics which do not satisfy $g^\sigma = \lambda^2 g$, under the following assumptions which may look a little bit complicated : there exist two riemannian metrics g^- and g^+ such that

1. — $g^- \le g \le \lambda^2 g \le g^+ \le g^\sigma$.
2. — The slowness condition in definition 1.1 is valid for g^-.
3. — $\sup_S g^+(S)/g(S) \le \inf_T g^\sigma(T)/g^-(T)$.
4. — $g^-_Y(Y-Z) \le \overline{C}^{-1} \Longrightarrow (g_Y(\cdot)/g_Z(\cdot))^{\pm 1} \le \overline{C}$
5. — $(g_Y(\cdot)/g_Z(\cdot))^{\pm 1} \le \overline{C}\,(1+d^+(Y,Z))^{\overline{N}}$, where d^+ is the g^+-geodesic distance.

Let us give some explanations. If one takes $g^- = g$ and $g^+ = \lambda^2 g$, conditions 1 to 4 are satisfied, and thus the conclusions of theorem 3.8 are valid if g_Y/g_Z is controlled by the geodesic distance for $\lambda^2 g$. The full set of conditions says that it is possible to relax this last condition (control by the geodesic distance for g^+ which will be equal to g^σ in "good cases"), provided that g be *slower* than what is usually required (which is expressed by condition 4).

For instance, for the metric $dx_1^2 + dx_2^2 + d\xi_1^2 + d\xi_2^2/\langle\xi\rangle^2$ in dimension 2, one has $\lambda = 1$ and it is not true that g_Y/g_Z is controlled by the geodesic distance for g. However, one can take $g^- = (dx^2 + d\xi^2)/\langle\xi\rangle^2$ and $g^+ = g^\sigma = dx_1^2 + \langle\xi\rangle^2 dx_2^2 + d\xi_1^2 + d\xi_2^2$ and conditions 1 to 5 are satisfied.

Remark 3.10 For symplectic metrics, the geodesic temperance had been introduced in [Un] and the theorem 3.8 (i) had been proved in [Br]. In [B&C], we introduced the notion of a metric *dominated by* another one, and the argument given there shows that, for a metric dominated by a metric satisfying the conditions of the above remark, theorem 3.8 is still valid.

4 Definition of Fourier integral operators

This definition is directly suggested by the above characterization of pseudo-differential operators. For the sake of simplicity, we shall give the definition in a restricted framework, where we have the simple characterization given by theorem 3.6. For general Hörmander metrics, the definition should be parallel to the characterization of corollary 3.4.

In this section, g_1 and g_2 are two geodesically tempered Hörmander metrics, satisfying $g_i^\sigma = \lambda_i^2 g$, and F is a diffeomorphism from $\mathbf{R}^{2n}$ onto $\mathbf{R}^{2n}$ satisfying the following conditions :

$$(\lambda_2(F(Y))/\lambda_1(Y))^{\pm 1} \leq C \tag{15}$$

$$g_{2,F(Y)}\left(d^k F(Y\,;\,T_1,\ldots,T_k)\right)^{1/2} \leq C_k \prod g_{1,Y}(T_j)^{1/2} \tag{16}$$

$$g_{1,F^{-1}(Z)}\left(d^k F^{-1}(Z\,;\,T_1,\ldots,T_k)\right)^{1/2} \leq C_k \prod g_{2,Z}(T_j)^{1/2}\,. \tag{17}$$

Here, the kth differential $d^k F$ at point Y is a k-linear map $\left(\mathbf{R}^{2n}\right)^k \to \mathbf{R}^{2n}$ whose value on vectors $T_1,\ldots,T_k$ is noted as above. It is equivalent to say that locally, using $g_{1,Y}$ orthonormal coordinates centered at Y and $g_{2,F(Y)}$ orthonormal coordinates centered at $F(Y)$, the components of F and all their derivatives are bounded.

Remark 4.1 The application F is Lipschitz continuous from $(\mathbf{R}^{2n}, g_1)$ onto $(\mathbf{R}^{2n}, g_2)$ and, thanks to (15), it is also Lipschitz continuous from $(\mathbf{R}^{2n}, g_1^\sigma)$ onto $(\mathbf{R}^{2n}, g_2^\sigma)$. Moreover, one has $|g_{1Y}| = \lambda_1(Y)^{-2n}$ and thus, in a change of variable $Z = F(Y)$, the measure $|g_{1Y}|^{1/2}\, đY$ becomes $c(Z)\,|g_{2Z}|^{1/2}\, đZ$, where the functions c and c^{-1} are uniformly bounded.

Proposition 4.2 *If M is a g_2-weight, then $M \circ F$ is a g_1-weight. If a belongs to $S(M, g_2)$ (resp. $S^+(M, g_2)$) then $a \circ F$ belongs to $S(M \circ F, g_1)$ (resp. $S^+(M \circ F, g_1)$).*

Definition 4.3 (Fourier integral operators) *An operator $P \in \mathcal{L}(L^2)$ belongs to $\mathrm{FIO}(F, g_1, g_2)$ if*

$$K_{b_1} \ldots K_{b_l} \cdot P \in \mathcal{L}(L^2)$$

for $b_j \in S^+(1, g_2)$, where the "twisted commutators" are defined by

$$K_b \cdot P = b^w P - P(b \circ F)^w\,.$$

It is clear, in view of theorem 3.6, that $\mathrm{FIO}(\mathrm{Id}, g, g) = \Psi(1, g)$.

Remark 4.4 The application F is not uniquely determined by the class of Fourier integral operators. Let G be another diffeomorphism such that

$$g_{2,F(Y)}\left(d^k (F-G)(Y\,;\,T_1,\ldots,T_k)\right)^{1/2} \leq C_k \lambda_1(Y)^{-1} \prod g_{1,Y}(T_j)^{1/2}\,.$$

Then, for $b \in S^+(1, g_2)$ one has $b \circ F - b \circ G \in S(1, g_1)$. It follows easily that $P \in \mathrm{FIO}(F, g_1, g_2)$ imply $P \in \mathrm{FIO}(G, g_1, g_2)$. The condition depends heavily on g_1. For instance, if $g_1 = g_2$ is the metric of type $(0,0)$ and if F is the translation $X \mapsto X + X_0$, then it is equivalent to the identity and $\mathrm{FIO}(F, g_1, g_2) = \Psi(1, g_1)$. Of course, this property is not true if $g_1 = g_2$ is the metric of type $(1,0)$.

Theorem 4.5 *Let* $P \in \mathrm{FIO}(F, g_1, g_2)$.
(i) One has $P^\star \in \mathrm{FIO}(F^{-1}, g_2, g_1)$.
(ii) If $Q \in \mathrm{FIO}(G, g_2, g_3)$, *where* G, g_2, g_3 *satisfy the above geometric assumptions, then* $Q \circ P \in \mathrm{FIO}(G \circ F, g_1, g_3)$.
(iii) If $Q \in \mathrm{FIO}(F^{-1}, g_2, g_1)$ *and* $A \in \Psi(M, g_2)$, *then* $QAP \in \Psi(M \circ F, g_1)$.

To prove the first point, given $c \in S^+(1, g_1)$, one should prove that $c^w P^\star - P^\star b^w$ is bounded on L^2, where $b = c \circ F^{-1}$. It suffices to show that the adjoint of this operator, which is $P(\bar{b} \circ F)^w - \bar{b}^w P$, is bounded on L^2. This is clear from the definition of Fourier integral operators, and the same proof works for iterated twisted commutators.

Under assumption (ii), let $b_3 \in S^+(1, g_3)$ and set $b_2 = b_3 \circ G$ and $b_1 = b_2 \circ F = b_3 \circ G \circ F$. We want to prove that $b_3^w QP - QPb_1^w$ is bounded on L^2. It follows easily from

$$b_3^w QP - QPb_1^w = \{b_3^w Q - Qb_2^w\} P + Q \{b_2^w P - Pb_1^w\} ,$$

where the curled brackets are bounded on L^2, and actually are Fourier integral operators. The case of iterated twisted commutators follows by induction.

The proof of part (iii) is a consequence of the following result, which says that Fourier integral operators are concentrated on the graph of F. Its proof uses the geodesic temperance and the functions p_{YZ} introduced in the proof of theorem 3.6.

Proposition 4.6 *Let* $P \in \mathrm{FIO}(F, g_1, g_2)$.
(i) For (α_Y) *and* (β_Z) *respectively* g_1- *and* g_2-*uniformly confined in balls of radius* r, *one has for some* $C > 1$ *and for any* N

$$\|\beta_Z^w P \alpha_Y^w\|_{\mathcal{L}(L^2)} \le C_N \Delta_{2,Cr}(F(Y), Z)^{-N} .$$

(ii) If $Q \in \mathrm{FIO}(F^{-1}, g_2, g_1)$ *and* $\gamma^w = Q\beta^w P$, *one has*

$$\|\gamma\|_{k\,;\mathrm{Conf}(g_1, Y, Cr)} \le C' \, \|\beta\|_{l\,;\mathrm{Conf}(g_2, F(Y), r)} , \quad C' = C'(k) , \; l = l(k) .$$

Remark 4.7 The reader may be surprised to see that Fourier integral operators can be defined without any assumption about the symplectic character of F, but we have not proved that the class $\mathrm{FIO}(F, g_1, g_2)$ contains many operators. As we shall see in the theorem 4.8 below, when λ is not bounded, almost invertible Fourier integral operators can exist only for almost symplectic diffeomorphisms F. On the other hand, for symplectic metrics ($g = g^\sigma$), we shall see in section 7 that there exist non trivial Fourier integral operators associated to functions F very far from being symplectic.

This may be well understood considering the Lie algebra $S^+(1, g)/S(1, g)$. From proposition 3.2, we know that, for a and b belonging to this space, then $a\#b - b\#a$ is a well defined element of $S(\lambda, g)/S(1, g) \subset S^+(1, g)/S(1, g)$ and that it is equal to the class of $-i\{a, b\}$. The condition (18) below says precisely that F should preserve this structure of Lie algebra.

When g is symplectic, $S^+(1, g)/S(1, g)$ is not trivial as a space (in contrast with the space of principal symbols $S(\lambda, g)/S(1, g)$), but it is trivial as a Lie algebra : the bracket is identically 0.

Theorem 4.8 *Assume that there exist* $P \in \mathrm{FIO}(F, g_1, g_2)$ *and* $Q \in \mathrm{FIO}(F^{-1}, g_2, g_1)$ *such that* $QP = I + R$ *with* $R \in \Psi(\lambda^{-1}, g_1)$*. Then, for* $b_j \in S^+(1, g_2)$*, one has*

$$\{b_1 \circ F, b_2 \circ F\} \equiv \{b_1, b_2\} \circ F \pmod{S(1, g)} \tag{18}$$

$$|[dF(Y) \cdot T_1, dF(Y) \cdot T_2] - [T_1, T_2]| \leq C g_{1,Y}(T_1)^{1/2} g_{1,Y}(T_2)^{1/2} \tag{19}$$

The formula (19) is a consequence of the preceding one, for it is possible to find $b_j \in S^+(1, g)$ which coincide with $[X - Y, T_j]$ near Y. If the left hand side of this formula were zero, it would mean that $dF(Y)$ is symplectic. The actual condition says that this left hand side is λ times smaller than it could be. Indeed one has only $|[T_1, T_2]| \leq \lambda_1(Y) g_{1,Y}(T_1)^{1/2} g_{1,Y}(T_2)^{1/2}$ and it is the best possible estimate.

In order to prove (18), set $\widetilde{b_j} = b_j \circ F$ and let $B_j, \widetilde{B}_j$ denote the corresponding operators. One has

$$B_2 B_1 P = B_2 \left(P \widetilde{B}_1 + P_1\right) = P \widetilde{B}_2 \widetilde{B}_1 + P_2 \widetilde{B}_1 + P_1 \widetilde{B}_2 + P'$$

where $P_j = K_{b_j} P$ and P' belong to $\mathrm{FIO}(F, g_1, g_2)$. Taking the difference with the corresponding formula for $B_1 B_2 P$ and taking into account the fact that $\left[\widetilde{B}_1, \widetilde{B}_2\right] \in \Psi(\lambda_1, g_1)$, we get

$$\left[\widetilde{B}_1, \widetilde{B}_2\right] \equiv QP\left[\widetilde{B}_1, \widetilde{B}_2\right] \equiv Q[B_1, B_2]P \pmod{\Psi(1, g_1)}\ .$$

On the other hand, for $C = [B_1, B_2] \in S(\lambda, g_2)$, one has $CP = P\widetilde{C} + P''$ with $P'' \in \mathrm{FIO}(F, g_1, g_2)$ and thus

$$\widetilde{C} \equiv QP\widetilde{C} \equiv Q[B_1, B_2]P \pmod{\Psi(1, g_1)}\ .$$

One has thus $\left[\widetilde{B}_1, \widetilde{B}_2\right] \equiv \widetilde{C}$ which proves the theorem.

5 Relations with metaplectic operators

We shall denote by $\mathrm{ASp}(n)$ the affine symplectic group, i.e. the group of affine bijective maps in $\mathbf{R}^{2n}$ which preserve the symplectic form. Let us recall the theorem of Segal (see [Hö2] theorem 18.5.9) which says that, given $\chi \in \mathrm{ASp}(n)$, there exist a unitary operator U, which is also an automorphism of $\mathcal{S}$ and $\mathcal{S}'$, such that

$$\forall a \in \mathcal{S}'(\mathbf{R}^n)\ , \quad U^\star a^w U = (a \circ \chi)^w\ .$$

If U_1 and U_2 are such operators associated to χ, then $U_2 = e^{i\theta} U_1$ with $\theta \in \mathbf{R}$.

We shall denote by $\mathrm{AMp}(n)$ the group of these operators, which are called *metaplectic*. The map $U \mapsto \chi$ will be denoted by π. It is a homomorphism from $\mathrm{AMp}(n)$ onto $\mathrm{ASp}(n)$, whose kernel is the group $U(1)$ of complex numbers of modulus 1.

5.1 *The fiber bundle $\widetilde{\Omega}$.* Let Ω be the submanifold of $\mathbf{R}^{2n}\times\mathbf{R}^{2n}\times\mathrm{ASp}(n)$ made of points (Y,Z,χ) such that $Z=\chi(Y)$ and let $\widehat{\Omega}$ the corresponding submanifold of $\mathbf{R}^{2n}\times\mathbf{R}^{2n}\times\mathrm{AMp}(n)$ equipped with its natural projection still denoted by $\pi:\widehat{\Omega}\to\Omega$. It is a fiber bundle over Ω, with fiber $U(1)$. We shall denote by $\widetilde{\Omega}=\widehat{\Omega}\times_{U(1)}\mathbf{C}$ the fiber product over $U(1)$ of $\widehat{\Omega}$ and $\mathbf{C}$, that is the line bundle whose fiber over (X,Y,χ) is made of expressions μU, $\mu\in\mathbf{C}$, $\pi(U)=\chi$ with the relations $(e^{-i\theta}\mu)(e^{i\theta}U)=\mu U$.

5.2 *Extended graph and weights.* For a symplectic diffeomorphism F of $\mathbf{R}^{2n}$, let us denote by Γ_F its graph and by $\overline{\Gamma}_F$ its extended graph : the submanifold of Ω made of points $(Y,F(Y),DF(Y))$ where $DF(Y)$ is the affine symplectic map tangent to F at Y. If g_1 is a Hörmander metric and if (16) and (17) are valid with $g_2=F_\star g_1$, a function m defined on Γ_F will be called a weight if m_1 : $Y\mapsto m(Y,F(Y))$ is a g_1-weight, or equivalently if m_2 : $Z\mapsto m(F^{-1}(Z),Z)$ is a g_2-weight. The function λ will be defined on Γ_F by $\lambda(Y,F(Y))=\lambda_1(Y)=\lambda_2(F(Y))$.

5.3 *Classes of sections.* Let us define classes $S_{\overline{\Gamma}_F}(m;\widetilde{\Omega})$ of sections of $\widetilde{\Omega}$ over $\overline{\Gamma}_F$. Consider first a section $(Y,F(Y),DF(Y))\mapsto U_Y$ whose values are metaplectic operators. Then $U_Y^\star\,\langle T_0,\partial_Y\rangle\,U_Y$ takes its values in the Lie algebra of $\mathrm{AMp}(n)$, which is a vector space isomorphic to the space of real polynomials of degree at most 2. We shall say that this section belongs to $S_{\overline{\Gamma}_F}(1;\widetilde{\Omega})$ if one has

$$\|\langle T_1,\partial_Y\rangle\dots\langle T_k,\partial_Y\rangle\,\{U_Y^\star\,\langle T_0,\partial_Y\rangle\,U_Y\}\|\le C_k\,,\quad k\ge 0\,,$$

for some norm in the Lie algebra and for $g_Y(T_j)\le 1$.

We shall say that a general section $(Y,F(Y),DF(Y))\mapsto s_Y$ belongs to $S_{\overline{\Gamma}_F}(m;\widetilde{\Omega})$ if it can be written $s_Y=\mu(Y)U_Y$ with U_Y as above and $\mu\in S(m_1,g_1)$.

5.4 *Fourier integral operators associated to an affine symplectic map.* Assume that $F\in\mathrm{ASp}(n)$ and that g_1 is a geodesically tempered Hörmander metric with $g_1^\sigma=\lambda_1^2 g_1$. Let $g_2=F_\star g_1$. It is clear that the geometric assumptions of the previous section are valid for the triple (F,g_1,g_2).

Theorem 5.5 *Let $P\in\mathcal{L}(L^2)$ and $U\in\pi^{-1}(F)$. The three following properties are equivalent*
(i) $P\in\mathrm{FIO}(F,g_1,g_2)$.
(ii) There exist $A_1\in\Psi(1,g_1)$ with $P=UA_1$.
(iii) There exist $A_2\in\Psi(1,g_2)$ with $P=A_2U$.

One shows easily that (ii) and (iii) are equivalent : one has $A_1=U^\star A_2U$ and if a_1 and a_2 are their symbols, $a_1=a_2\circ F$. It is clear also that $U\in\mathrm{FIO}(F,g_1,g_2)$ for twisted commutators are null. By theorem 4.5 one has that (iii) imply (i).

Let now $P\in\mathrm{FIO}(F,g_1,g_2)$, and set $A_2=PU^\star$. For $b\in S^+(1,g_2)$, one has

$$b^wA_2-A_2b^w=(b^wP-P(b\circ F)^w)U^\star+P((b\circ F)^wU^\star-U^\star b^w)=P'U^\star+0$$

where $P'=K_b\cdot P$ is a Fourier integral operator. By induction, this proves that the iterated commutators $\prod\mathrm{ad}\,b_j^w\cdot A_2$ are bounded on L^2 and thus $A_2\in\Psi(1,g_2)$.

5.6 *Symbols.* In this particular case, the definition will be tautological, but it explains what will happen in the general case. The first idea would be to look for a symbol $P \mapsto \sigma(P)$ defined on the graph of F, such that the symbol of the adjoint is the complex conjugate, and with a product law for the principal symbol of the composition of two operators.

For $P = Ua_1^w$, these properties can be true only if the module of the symbol of P is $|a_1(Y)|$ (modulo symbols of weight λ^{-1}). So the symbol of P looks well defined apart from a factor of modulus 1, while U, and so a_1, are defined apart from a factor of the same kind. In some sense, this indetermination will be resolved by deciding that the symbol of U is U itself.

In our case, $\overline{\Gamma}_F$ is just the set of triples $T_Y = (Y, F(Y), F)$ and we will define the full symbol of $P = Ua_1^w$ as the section $T_Y \mapsto \sigma_P(T_Y) = a_1(Y)U$. It belongs to $S_{\overline{\Gamma}_F}(1;\widetilde{\Omega})$ and the principal symbol will be its equivalence class modulo sections belonging to $S_{\overline{\Gamma}_F}(\lambda^{-1};\widetilde{\Omega})$. The formula for adjoints is clear :

$$\sigma(P^\star)(F(Y), Y, F^{-1}) = \overline{\sigma(P)(Y, F(Y), F)}\,, \quad \text{defining } \overline{\mu U} = \overline{\mu} U^\star \tag{20}$$

If $G \in \mathrm{ASp}(n)$ and if $P \in \mathrm{FIO}(G, g_2, g_3)$, setting $g_3 = G_\star g_2$, one has

$$\sigma(QP)(Y, G{\circ}F(Y), G{\circ}F) \equiv \sigma(Q)(F(Y), G{\circ}F(Y), G) \times \sigma(P)(Y, F(Y), F)$$
$$\text{defining } (\mu U)\times(\nu V) = \mu\nu\, U{\circ}V\,, \tag{21}$$

with equality modulo sections belonging to $S_{\overline{\Gamma}_{G\circ F}}(\lambda^{-1};\widetilde{\Omega})$.

6 Symbolic calculus

For the sake of simplicity, we shall consider triples (F, g_1, g_2), called *admissibles triples*, where F is a symplectic diffeomorphism of $\mathbf{R}^{2n}$, where g_1 and g_2 are geodesically tempered Hörmander metrics satisfying $g_i^\sigma = \lambda_i^2 g_i$, such that $g_2 = F_\star g_1$ and that the geometric assumptions (15) to (17) of section 4 are satisfied. Actually, the full construction works when F is asymptotically symplectic as defined by (19), with minor modifications for giving asymptotic definitions of local almost generating functions, extended graphs and principal symbols.

Up to now we have considered only Fourier integral operators of weight 1. Thanks to the following lemma, which is a consequence of proposition 4.6, it is easy to define operators of weight m where m is a weight on the graph of F.

Lemma 6.1 *Let $F \in \mathrm{FIO}(F, g_1, g_2)$ and $A_j \in \Psi(\mu_j, g_j)$, $j = 1, 2$, where the weights satisfy $\mu_1(Y)\mu_2(F(Y)) = 1$. Then $A_2PA_1 \in \mathrm{FIO}(F, g_1, g_2)$.*

Definition 6.2 *We shall say that an operator P belongs to $\mathrm{FIO}(F, m, g_1, g_2)$ if, for any choice of operators $A_j \in \Psi(\mu_j, g_j)$, $j = 1, 2$, such that the weights satisfy $\mu_1(Y)\mu_2(F(Y)) = m(Y, F(Y))^{-1}$, one has $A_2PA_1 \in \mathrm{FIO}(F, g_1, g_2)$. It is sufficient to have this property for one such couple (A_1, A_2) made of invertible operators.*

The notation $\mathrm{FIO}(F, 1, g_1, g_2)$ is thus equivalent to $\mathrm{FIO}(F, g_1, g_2)$. The main result of this section is the following.

Theorem 6.3 *For any admissible triple (F, g_1, g_2), there exists a linear application $P \mapsto \sigma(P)$ from $\mathrm{FIO}(F, m, g_1, g_2)$ onto $S_{\overline{\Gamma}_F}(m; \widetilde{\Omega})/S_{\overline{\Gamma}_F}(m\lambda^{-1}; \widetilde{\Omega})$ such that*
(i) The map σ induces an isomorphism $\mathrm{FIO}(F, m, g_1, g_2)/\,\mathrm{FIO}(F, m\lambda^{-1}, g_1, g_2) \xrightarrow{\sim} S_{\overline{\Gamma}_F}(m; \widetilde{\Omega})/S_{\overline{\Gamma}_F}(m\lambda^{-1}; \widetilde{\Omega})$.
(ii) If F is the identity and $P = a^w$ with $a \in S(m, g_1)$, then $\sigma(P)$ is the class of $(Y, Y, I) \mapsto a(Y)I$, the first (resp. second) I being the unity element of $\mathrm{ASp}(n)$ (resp. $\mathrm{AMp}(n)$).
(iii) The symbol of the adjoint $P^\star$ is

$$(F(Y), Y, DF(Y)^{-1}) \longmapsto \overline{\sigma(P)(Y, F(Y), DF(Y))}\ ,$$

where the conjugate of a section of $\widetilde{\Omega}$ is defined in (20).
(iv) If (G, g_2, g_3) is another admissible triple and $Q \in \mathrm{FIO}(G, g_2, g_3)$, then $QP \in \mathrm{FIO}(H, g_1, g_3)$ with $H = G \circ F$ and

$$\sigma(QP)(X, Z, DH(X)) = \sigma(Q)(Y, Z, DG(Y)) \times \sigma(P)(X, Y, DF(X))$$

with $Y = F(X)$, $Z = H(X)$, the product $\times$ being defined in (21).

Corollary 6.4 *For any admissible (F, g_1, g_2), there exist $P \in \mathrm{FIO}(F, g_1, g_2)$ and $Q \in \mathrm{FIO}(F^{-1}, g_2, g_1)$ such that $QP = I + R_1$ and $PQ = I + R_2$, with $R_j \in \Psi(\lambda_j^{-N}, g_j)$ for any N.*

One can find a global section $s \in S_{\overline{\Gamma}_F}(1; \widetilde{\Omega})$ whose values belong to $\mathrm{AMp}(n)$. Let $P \in \mathrm{FIO}(F, g_1, g_2)$ such that $\sigma(P) = s$. Then $PP^\star$ and $P^\star P$ are pseudo-differential operators whose principal symbol is 1. One can then find $S \in \Psi(1, g_1)$ such that $SP^\star P = I + R$ with $R \in \Psi(\lambda_1^{-\infty}, g_1)$. It is easy to see that P and $Q = SP^\star$ satisfy the above conditions.

The rest of this section will be devoted to a sketch of the proof of theorem 6.3.

6.5 *Localization.* Given $\tau = (\tau_1, \tau_2)$ in the graph of F, one can give two almost equivalent definitions of a Fourier integral operator Q_τ localized at τ (or rather of a family depending on τ which is uniformly localized). The first one is to require that

$$Q = \beta^w P \alpha^w, \quad \alpha \in \mathrm{Conf}(g_1, \tau_1, r),\ \beta \in \mathrm{Conf}(g_2, \tau_2, Cr),\ P \in \mathrm{FIO}(F, g_1, g_2) \tag{22}$$

with uniform boundedness when τ varies. Here, C sufficiently large and $r < r_0$ sufficiently small depend only on the slowness constants of g_1 and on bounds for the first derivatives of F.

The second definition, like confinement for symbols, will depend only on local data : the quadratic forms g_{1,τ_1} and g_{2,τ_2} and the portion of the graph of F close to τ. It says that

$$Q = \beta^w \widetilde{P} \alpha^w, \quad \alpha \in \mathrm{Conf}(g_1, \tau_1, r),\ \beta \in \mathrm{Conf}(g_2, \tau_2, Cr) \tag{23}$$

where $\widetilde{P}$ satisfies

$$\lambda_1(\tau_1)^N \prod_1^N K_{\theta_j} \cdot \widetilde{P} \in \mathcal{L}(L^2) \quad \text{for } \theta_j \in S(1, g_{2,\tau_2}),\ \mathrm{Supp}(\theta_j) \subset B_{2,\tau_2,C^3 r} \tag{24}$$

with of course uniform bounds with respect to τ.

It is clear that the first definition imply the second one, for $\lambda_2(\tau_2)\theta_j \in S^+(1, g_2)$ (with control of the semi-norms). Conversely, any operator satisfying the second definition can be written as the sum of a convergent series of operators satisfying the first one (with a radius $r' = 2r$). This comes from the fact that α can be written (see [B&C]) as $\alpha = \sum \alpha'_\nu \# \alpha''_\nu$ with bounds for the semi-norms of α', α'' in $\mathrm{Conf}(g_1, \tau_1, 2r)$ and the same is true for β. So, what remains to be proved is that $P' = \beta'' \widetilde{P} \alpha' \in \mathrm{FIO}(F, g_1, g_2)$ i.e. that $K_b P'$ and twisted commutators of higher order are bounded on L^2 for $b \in S^+(1, g_2)$. After replacing b by $b(\cdot) - b(\tau_2)$, which does not change the commutators, write $b = b' + b''$ with b' supported in $B_{2,\tau_2,C^3 r}$ and b'' vanishing in $B_{2,\tau_2,C^2 r}$. The function $b'/\lambda_2(\tau_2)$ can be taken as a function θ_j in (24), while $b'' \# \beta''$ is confined and the proof follows.

Now, any $P \in \mathrm{FIO}(F, m, g_1, g_2)$ can be written as

$$P = \int m(\tau) Q_\tau \, |g_{1,\tau_1}|^{1/2} \, d\tau_1 + P' , \quad P' \in \mathrm{FIO}(F, m\lambda^{-\infty}, g_1, g_2)$$

where the Q_τ are uniformly localized in the sense of (22). We can first, using partitions of unity, write P as a double integral of $\beta'_Z{}^w P \alpha'_Y{}^w$. Then we can set $\tau = (Y, F(Y))$ and

$$\alpha_\tau = \alpha'_Y \ , \ \beta_\tau = \int_{g_{2,\tau_2}(Z-\tau_2) \le C r^2} \beta'_Z \, |g_{2,Z}|^{1/2} \, dZ \ , \ Q_\tau = \beta_\tau P \alpha_\tau \ .$$

The remaining part of the double integral, which corresponds to points far from Γ_F, will be called P'. The fact that $P' \in \mathrm{FIO}(F, m\lambda^{-\infty}, g_1, g_2)$ is a consequence of proposition 4.6.

The preceding considerations show that, in order to prove theorem 6.3, it is sufficient to define a symbol $\sigma(Q)$ in $S_{\widetilde{\Gamma}_F}(1; \widetilde{\Omega})$ for operators Q which are localized in a ball in the sense of (23) and (24), provided that the following conditions be satisfied : $\sigma(Q)$ is supported in this ball, any symbol supported in a smaller ball is in the image of σ (modulo symbols of order λ^{-1}), the algebraic relations for adjoint and products are satisfied; all this, of course, with uniform bounds.

6.6 *Reduction to a euclidean ball.* Assume now Q localized in a ball centered at $\widetilde{\tau}$. We shall write λ instead of $\lambda(\widetilde{\tau})$ for short. Let g_0 be the metric $(dx^2 + d\xi^2)/\lambda$ and B_r its ball of radius r centered at 0. For any τ close to $\widetilde{\tau}$, choose $\chi_\tau \in \mathrm{ASp}(n)$ mapping B_r onto the $g_{1,\widetilde{\tau}_1}$-ball of radius r centered at τ_1, set $\chi'_\tau = (DF(\tau_1) \circ \chi_\tau)$ and $\widetilde{F} = \chi'^{-1}_\tau \circ F \circ \chi_\tau$. Choose U_τ and V_τ in $\mathrm{AMp}(n)$ with $\pi(U_\tau) = \chi_\tau$ and $\pi(V_\tau) = \chi'_\tau$ and set $\widetilde{Q} = V^\star_\tau Q U_\tau$. These choices can be done with uniform bounds and depending smoothly (i.e. in a $S(1, g_1)$ manner) on τ.

Thus $\widetilde{F}$ is a symplectic diffeomorphism, tangent to the identity at 0, and $\widetilde{Q}$ is a localized Fourier integral operator associated to $(\widetilde{F}, g_0, g_0)$. We shall define below a principal symbol of $\widetilde{Q}$ at 0, which will be a scalar c (or more precisely cI). Then $\sigma(Q)(\tau)$ will be defined as $cV_\tau U_\tau^\star$. Our definition will be coherent if conjugating $\widetilde{Q}$ by a metaplectic operator associated to a linear symplectic application leaving invariant the euclidean ball do not change its symbol at 0 (modulo $O(\lambda^{-1})$).

Let us introduce the generating function S of $\widetilde{F}$ vanishing at 0 and defined in B_{C^3r}, which satisfy

$$\left(\tfrac{\partial S}{\partial \eta}(x,\eta),\eta\right) = \widetilde{F}\left(x, \tfrac{\partial S}{\partial x}(x,\eta)\right).$$

One has $S(x,\eta) = x\cdot\eta + O(|x,\eta|^3)$.

6.7 *Stationary phase.* This part of the argument owes much to an unpublished paper of N. Lerner [Le] where, in our situation, he considers operators

$$Lu(x) = \int e^{iS(x,\eta)-iz\cdot\eta} t(x,\eta)u(z)\,dz\,d\eta/(2\pi)^n \tag{25}$$

with $t \in S(1,g_0)$ supported in B_{Cr} and where he studies $L^\star a^w L$ for $a \in S(m,g_0)$.

First, we want to show that $\widetilde{L} = \beta^w L \alpha^w$ is a localized Fourier integral operator with respect to $(\widetilde{F}, g_0, g_0)$, when α and β are confined in B_{Cr}. We have to consider twisted commutators $K_{\lambda\theta}\cdot L = \lambda(\theta^w L - L(\theta\circ\widetilde{F})^w)$ with $\theta \in S(1,g_0)$ supported in B_{C^3r}. Introducing change of variables $x = \lambda^{1/2}x', \eta = \lambda^{1/2}\eta'$, we get integral expressions in a fixed ball with λ as a large parameter. A rather standard application of the method of stationary phase shows that $K_{\lambda\theta}\cdot L = L_1 + L_2 B$, with L_j are as in (25) and $B \in \Psi(1,g_0)$. Thus it is bounded on L^2 and the case of iterated twisted commutators follows by induction.

For the same reason, $\widetilde{L}^\star$ is a localized Fourier integral operator associated to $\widetilde{F}^{-1}$. Thus $\widetilde{L}\widetilde{L}^\star$ and $\widetilde{L}^\star\widetilde{Q}$ belong to $\Psi(1,g_0)$. Moreover, if we choose t, α and β equal to 1 near B_r, the principal symbol of $\widetilde{L}^\star\widetilde{L}$ does not vanish on the support of the principal symbol of $\widetilde{L}^\star\widetilde{Q}$. Using pseudo-differential symbolic calculus, one gets $\widetilde{Q} = B_1\widetilde{L}B_2 + R$ where B_j are pseudo-differential operators of order 0 and R is a localized Fourier integral operator of weight λ^{-1}.

Using the same argument of stationary phase, we obtain

$$\widetilde{Q}u(x) = \int e^{iS(x,\eta)-iz\cdot\eta} q(x,\eta)u(z)\,dz\,d\eta/(2\pi)^n + \widetilde{R}$$

where $q \in S(1,g_0)$ is supported in B_r and $\widetilde{R}$ is a localized Fourier integral operator of weight λ^{-1}. We shall define a symbol of $\widetilde{Q}$ at 0 as $q(0,0)I$.

As explained above, this gives the general definition of the principal symbol, but one should prove that this definition (modulo $O(\lambda^{-1})$) does not depend on our arbitrary choices, and that the rule for product is satisfied. Thanks to the fact that this rule is already known for operators correspondig to affine symplectic maps, one is reduced to the composition of operators corresponding to symplectic diffeomorphisms tangent to the identity at 0, which is still a matter of stationary phase.

7 Case of symplectic metrics

The results of the previous section are asymptotic in λ and give no information for symplectic metrics. However, there exists in this case another characterization of pseudo-differential operators and Fourier integral operators. We shall just state the results, referring to [Bo1] and [Bo2] for proofs.

The first result characterizes pseudo-differential operators by the fact that they are localized near the diagonal of $\mathbf{R}^{2n}$.

Theorem 7.1 *Let g be a symplectic Hörmander metric. Let $a \in \mathcal{S}'(\mathbf{R}^{2n})$ and set $A = a^w$. The three following properties are equivalent.*
(i) $a \in S(1, g)$
(ii) There exist (a_Y) and (b_Y) uniformly confined, and operators T_{YZ} such that

$$A = \iint a_Z^w \, T_{YZ} \, b_Y^w \, dY \, dZ \quad \text{with} \quad \forall N, \ \|T_{YZ}\|_{\mathcal{L}(L^2)} \le C_N \Delta(Y,Z)^{-N} .$$

(iii) For a convenient choice of uniformly confined families (ψ_Y) and (θ_Y), one has

$$\forall N, \quad \left\| \theta_Z^w \, A \, \psi_Y^w \right\|_{\mathcal{L}(L^2)} \le C_N \Delta(Y,Z)^{-N} .$$

For Fourier integral operators, we shall keep the assumptions of section 4 on g_1, g_2 and F, adding the symplectic character of g_1 and g_2. However (see [Bo2]), the whole theory works fine under weaker assumptions on F.

Theorem 7.2 *Let $P \in \mathcal{L}(L^2)$. The four following properties are equivalent.*
(i) $P \in \mathrm{FIO}(F, g_1, g_2)$.
(ii) Let U_Y be metaplectic operators associated to affine symplectic transformations mapping the g_1-unit ball centered at Y onto the g_2-unit ball centered at $F(Y)$. Then one has

$$P = \int U_Y \, a_Y^w \, dY ,$$

where (a_Y) is g_1-uniformly confined.
(iii) There exist (a_Y) and (b_Y) uniformly confined for g_2 and g_1 respectively, and operators T_{YZ} such that

$$A = \iint a_Z^w \, T_{YZ} \, b_Y^w \, dY \, dZ \quad \text{with } \|T_{YZ}\|_{\mathcal{L}(L^2)} \le C_N \Delta_2(F(Y), Z)^{-N} .$$

(iv) For convenient choices of uniformly confined families (ψ_Y) and (θ_Y) for g_1 and g_2 respectively, one has

$$\forall N, \quad \|\theta_Z^w \, P \, \psi_Y^w\|_{\mathcal{L}(L^2)} \le C_N \Delta_2(F(Y), Z)^{-N} .$$

Example 7.3 Let us consider, in dimension 2 for instance, the application F defined by

$$F(x_1, x_2, \xi_1, \xi_2) = (x_1, x_2, \xi_1, -\xi_2) ,$$

and let $g_1 = g_2$ be the metric of type $(0, 0)$, i.e. the euclidean metric on $\mathbf{R}^4$. It is clear that the geometric assumptions (15) to (17) are satisfied, and that F does not respect the symplectic form. Nethertheless, one can find operators in $\mathrm{FIO}(F, g_1, g_2)$ which are quite non trivial.

Let $\varphi \in C^\infty(\mathbf{R})$, symmetric and supported in $[-1/2, 1/2]$ and let Φ be its periodization : $\Phi(t) = \sum_{k\in\mathbf{Z}} \varphi(x-k)$. Let P be the following operator

$$Pu(x) = \sum_{\nu\in\mathbf{Z}} \varphi(x_2 - \nu)u(x_1, 2\nu - x_2) .$$

Using the characterization (ii) in the previous theorem, it is not difficult to see that $P \in \mathrm{FIO}(F, g, g)$. On the other hand, if we compute P^2 which should belong to $\Psi(1, g)$, we get the operator $u \mapsto \Phi^2(x_2)u$, which is far from being negligible (the sum of two such operators can be invertible).

References

[Be] R. Beals, *Characterization of pseudodifferential operators*, Duke Math. J. **44** (1977), 45–57.

[Bo1] J.-M. Bony, *Opérateurs intégraux de Fourier et calcul de Weyl-Hörmander (cas d'une métrique symplectique)*, Actes Journées E.D.P. St. Jean de Monts, 1994, exposé n° 9.

[Bo2] J.-M. Bony, *Weyl quantization and Fourier integral operators*, Partial differential equations and mathematical physics (L. Hörmander and A. Melin, eds.), Birkhaüser, 1995, pp. 45–57.

[B&C] J.-M. Bony and J.-Y. Chemin, *Espaces fonctionnels associés au calcul de Weyl-Hörmander*, Bull. Soc. Math. France **122** (1994), 77–118.

[B&L] J.-M. Bony and N. Lerner, *Quantification asymptotique et microlocalisations d'ordre supérieur I*, Ann. scient. Ec. Norm. Sup. 4e série **22** (1989), 377–433.

[Br] F. Bruyant, *Estimations pour la composition d'un grand nombre d'opérateurs pseudo-différentiels et applications, Thèse Univ. Reims*, 1979.

[H&S] B. Helffer and J. Sjöstrand, *Equation de Harper*, Lecture Notes in Physics **345**, Springer, 1989, pp. 118–197.

[Hö1] L. Hörmander, *The Weyl calculus of pseudo-differential operators*, Comm. Pure Appl. Math. **32** (1979), 359–443.

[Hö2] L. Hörmander, The analysis of linear partial differential operators, Springer-Verlag, 1985.

[Le] N. Lerner, private communication, 1994.

[Un] A. Unterberger, *Les opérateurs métadifférentiels*, Lecture Notes in Physics **126**, Springer, 1980, pp. 205–241.

The Wick calculus of pseudo-differential operators and energy estimates

Nicolas Lerner

Irmar, Campus de Beaulieu, Université de Rennes 1, 35 042 Rennes Cedex, France (e-mail: lerner@univ-rennes1.fr)

Acknowledgement. This is my pleasure to congratulate Professor Komatsu on his sixtieth birthday and to thank him for his invitation and the very warm welcome he gave us during our stay in Japan. I wish also to express my thanks to the other organizers of this meeting, Professors Bony and Tose.

1 Introduction

The main goal of this lecture is to explore how the Wick quantization can be used to prove energy estimates. Instead of giving right now our definition of this quantization, let us just summarize its main features. Let $a(x,\xi)$ be a Hamiltonian (e.g. a smooth compactly supported function on the phase space $\mathbb{R}^n_x \times \mathbb{R}^n_\xi$). The Wick quantization associates an operator a^{Wick} on $L^2(\mathbb{R}^n)$ to this function in such a way that the two following properties are satisfied :

(1.1) $a(x,\xi) \geq 0$ implies $a^{\text{Wick}} \geq 0$,

(1.2) $a^{\text{Wick}} - a(x,D_x)$ is $L^2(\mathbb{R}^n)$ bounded if a is a symbol of order 1.

Property (1.1) means that non-negative functions are quantified into non-negative operators, which is not true for the ordinary quantization $a \mapsto a(x,D_x)$ or the Weyl quantization $a \mapsto a^w$, where these operators are given by

$$(1.3)\quad a(x,D_x)u(x) = \iint e^{2i\pi x\xi} a(x,\xi)\hat{u}(\xi)d\xi, \qquad \hat{u}(\xi) = \int e^{-2i\pi x\xi} u(y)dy,$$

$$(1.4)\qquad a^w u(x) = \iint e^{2i\pi(x-y)\xi} a(\frac{x+y}{2},\xi)u(y)dyd\xi.$$

Property (1.2) is of course devised to keep track of basic rules for standard operators. For a to be a symbol of order 1 could mean

$$(1.5)\qquad |(D_x^\alpha D_\xi^\beta a)(x,\xi)| \le C_{\alpha\beta}(1+|\xi|)^{1-|\beta|},$$

but we shall use it in a more general context. When a is in $C_0^\infty(\mathbb{R}^{2n})$, the best constants $C_{\alpha\beta}$ in (1.5) are of course finite, and the meaning of (1.2) is then that there exists N and C, depending only on the dimension n, such that

$$\|a^{\mathrm{Wick}} - a^w\|_{\mathcal{L}(L^2(\mathbb{R}^n))} \le C \max_{|\alpha|+|\beta|\le N} C_{\alpha\beta}.$$

Gårding's inequality with gain of one derivative is a direct consequence of (1.1–2) : let a be a non-negative symbol of order 1 ; then, there exists γ such that, for any $u \in \mathcal{S}(\mathbb{R}^n)$,

$$\mathrm{Re}\langle\, a(x,D_x)u\ ,\ u\,\rangle_{L^2(\mathbb{R}^n)} + \gamma\|u\|^2{}_{L^2(\mathbb{R}^n)} \ge 0.$$

This is theorem 18.1.14 in [H1], and it is remarkable that this method provides a proof for systems as well (remark 2 in section 18.1 of [H1]). This wave-packets approach to lower bounds for pseudo-differential operators has a long history in soviet mathematics, going back to Berezin [Be], and was used extensively by B.Lascar in connection with partial differential equations in infinite dimension [La]. This method was revived by several papers of the late seventies ([CF][Un]) and was used by Bony ([Bo]) to handle a Gårding inequality for paradifferential operators. One should say, however, that deeper results on lower bounds for pseudo-differential operators, like Fefferman-Phong or Melin's inequality, could not be obtained through the use of this tool.

One could say that a "stationary" use of Wick quantization leads to Gårding's inequality. In fact, Wick quantization is given through a formula of the following type :

$$(1.6)\qquad a^{\mathrm{Wick}} = W^* a^\mu W,$$

where W is an isometric mapping from $L^2(\mathbb{R}^n)$ to $L^2(\mathbb{R}^{2n})$, W^* its adjoint, a^μ the multiplication by $a(x,\xi)$ on $L^2(\mathbb{R}^{2n})$. Precise formulas will be given below. Roughly speaking, this quantization amounts to replace the action of a differential operator on a function u by the multiplication of its *wave-packet* $(Wu)(x,\xi)$ by the symbol of the operator. We want to show here that a dynamical use of this tool can be made and that propagation phenomena are good material to be investigated by such a method. Typically, we shall be given an evolution equation

$$(1.7)\qquad D_t + iQ(t,x,D_x),$$

where t is one real variable, $x \in \mathbb{R}^n$ and $Q(t, x, \xi)$ a first-order real-valued symbol. We shall use the wave-packets in the space variables x, and study the ordinary differential equation $D_t + iQ(t, x, \xi)$. A priori estimates are easily proved for this ODE, and we show that, in various non-trivial cases, this is enough to get estimates for the operator (1.7). To illustrate these points, we give a complete proof of a brief notice (cf. remark 1.1 in [L2]) in one of our paper on local solvability for pseudo-differential operators. In fact, the theorem announced in this remark has been improved after an exchange of letters with Hörmander, whom I thank for useful discussions.

Let M be a smooth manifold and p a complex-valued smooth homogeneous function of degree 1 on a conic open subset of $T^*(M)\backslash 0$. We assume that p is of principal-type on this set : we have there

$$H_p \wedge L \neq 0, \tag{1.8}$$

where H_p is the hamiltonian vector field of p and L the Liouville vector field. Moreover, we assume that p satisfies Nirenberg-Treves' condition (ψ) :

$$\begin{cases} \text{Im } p \text{ does not change sign from } - \text{ to } + \text{ along the} \\ \text{oriented integral curves of } H_{\text{Re} p}. \end{cases} \tag{1.9}$$

It is known after the works of Nirenberg and Treves [NT], Beals and Fefferman [BF] that condition (ψ) is sufficient for local solvability of differential operators. The paper [NT] used an analyticity assumption, removed in [BF]. Hörmander proved a semi-global existence theorem in that case (th. 26.11.3 in [H1]). Later, the author proved the sufficiency of condition (ψ) for local solvability in two dimensions [L1] and for the oblique-derivative problem [L2]. The necessity of (ψ) for local solvability was proved in two dimensions by Moyer and in the general case by Hörmander (th. 26.4.7 in [H1]). The sufficiency of this condition for local solvability is an open problem.

However, one should note that in all the previously quoted sufficiency results, condition (ψ) implies "optimal solvability": the equation $Pu = f$ has a local solution u in H^{s+m-1} for a right-hand-side in H^s , where m is the order of P. This is a sharp contrast with the situation in dimension ≥ 3, where (ψ) does not imply optimal solvability, as proved in [L3]: there exists a first-order principal-type operator, satisfying (ψ) and $f \in L^2$ such that $Pu = f$ has no local solution in L^2.

2 Wick quantization

We describe in this section the main features of the Wick calculus of pseudo-differential operators. Most of the properties listed below are classical, with the possible exception of the composition formula (2.21). Let n be an integer, $\mathbb{R}^n_x \times \mathbb{R}^n_\xi$ the $2n$-dimensional phase space, σ its symplectic form given by

$$\sigma\Big((x,\xi),(y,\eta)\Big) = \langle \xi, y\rangle - \langle \eta, x\rangle. \tag{2.1}$$

Let $Y = (y,\eta) \in \mathbb{R}^{2n}$. We define the operator Σ_Y as

$$\Sigma_Y = \Big[2^n e^{-2\pi|\bullet - Y|^2}\Big]^w, \tag{2.2}$$

that is the operator whose Weyl symbol $a(x,\xi)$ is equal to $2^n e^{-2\pi[|x-y|^2+|\xi-\eta|^2]}$ (Weyl quantization formula is given in (1.4)). The operator Σ_Y is a rank-one orthogonal projection:

$$\Sigma_Y u = (Wu)(Y)\tau_Y\varphi \quad \text{with} \quad (Wu)(Y) = \langle\, u\;,\; \tau_Y\varphi\,\rangle_{L^2(\mathbb{R}^n)}, \tag{2.3}$$

$$(\tau_{y,\eta}\varphi)(x) = \varphi(x-y)e^{2i\pi\langle x-\frac{y}{2},\eta\rangle} \quad \text{and} \quad \varphi(x) = 2^{n/4}e^{-\pi|x|^2}. \tag{2.4}$$

It is easy to see that

$$\mathrm{Id} = \int_{\mathbb{R}^{2n}} \Sigma_Y dY. \tag{2.5}$$

We can now define, say for $a \in L^\infty(\mathbb{R}^{2n})$,

$$a^{\mathrm{Wick}} = \int_{\mathbb{R}^{2n}} a(Y)\Sigma_Y dY = W^* a^\mu W, \tag{2.6}$$

where W is the isometric mapping from $L^2(\mathbb{R}^n)$ to $L^2(\mathbb{R}^{2n})$ given in (2.3), and a^μ the operator of multiplication by a. It is easy to see that the operator W is not onto (compute the real part of the dot product of ix_ju with D_ju) and WW^* is the orthogonal projection on a closed proper subspace of $L^2(\mathbb{R}^{2n})$. Formula (2.6) also makes sense for $a \in \mathcal{S}'(\mathbb{R}^{2n})$, since for $u, v \in \mathcal{S}(\mathbb{R}^n)$, $Y \mapsto (\Sigma_Y u, v)_{L^2(\mathbb{R}^n)}$ is in $\mathcal{S}(\mathbb{R}^{2n})$. We see at once that the non-negativity condition (1.1) is satisfied, since the operators Σ_Y are non-negative. Formula (2.6) appears as a Gaussian regularization of Weyl quantization. In fact, formula (1.4) can be written as

$$a^w = 2^n \int_{\mathbb{R}^{2n}} a(X)\, \sigma_X dX, \tag{2.7}$$

where σ_X is the phase symmetry, a unitary and selfadjoint operator on $L^2(\mathbb{R}^n)$ given by

$$(\sigma_{x,\xi}u)(y) = u(2x-y)e^{-4i\pi(x-y)\xi}. \tag{2.8}$$

In contrast with Weyl quantization, we have from (2.5–6) and $\Sigma_Y \geq 0$,

$$\|a^{\mathrm{Wick}}\|_{\mathcal{L}(L^2(\mathbb{R}^n))} \leq \|a\|_{L^\infty(\mathbb{R}^{2n})}. \tag{2.9}$$

For future reference, it is useful to notice that, for $Y, Z \in \mathbb{R}^{2n}$ the Weyl symbol of $\Sigma_Y \Sigma_Z$ is, as a function of the variable $X \in \mathbb{R}^{2n}$,

$$e^{-\frac{\pi}{2}|Y-Z|^2} e^{-2i\pi[X-Y,X-Z]} 2^n e^{-2\pi|X-\frac{Y+Z}{2}|^2}. \tag{2.10}$$

Since for the Weyl quantization, (2.7) gives

$$\|a^w\|_{\mathcal{L}(L^2(\mathbb{R}^n))} \leq 2^n \|a\|_{L^1(\mathbb{R}^{2n})}, \tag{2.11}$$

we get

$$\|\Sigma_Y \Sigma_Z\|_{\mathcal{L}(L^2(\mathbb{R}^n))} \leq 2^n e^{-\frac{\pi}{2}|Y-Z|^2}. \tag{2.12}$$

We define now classes of symbols with a large parameter. For $m \in \mathbb{R}$, $\mathcal{T}^m$ is the set of smooth functions $a_\Lambda(X)$, defined for $\Lambda \geq 1$ and $X \in \mathbb{R}^{2n}$, such that, for any integer k, there exists $\gamma_k(a)$ so that, for all $\Lambda \geq 1$ and all $X \in \mathbb{R}^{2n}$,

$$|a_\Lambda^{(k)}(X)| \leq \gamma_k(a) \Lambda^{m-\frac{k}{2}}, \tag{2.13}$$

where the left-hand-side stands for the norm of the k-th derivative with respect to the canonical Euclidean norm of $\mathbb{R}^{2n}$. The set $\mathcal{T}^m$ is endowed with a Fréchet space structure, where the best constants in (2.13) are the semi-norms.

Proposition 2.1. *Let $m \in \mathbb{R}$ and a be a symbol in $\mathcal{T}^m$. Then*

$$a^{\mathrm{Wick}} = a^w + r(a)^w, \tag{2.14}$$

with $r(a) \in \mathcal{T}^{m-1}$ so that the mapping $a \mapsto r(a)$ is continuous from $\mathcal{T}^m$ to $\mathcal{T}^{m-1}$.

Proof. From (2.2) and (2.6), one has $a^{\mathrm{Wick}} = b^w$, with

$$\begin{aligned} b(X) &= \int_{\mathbb{R}^{2n}} a(X+Y)\, e^{-2\pi|Y|^2} 2^n dY \\ &= a(X) + \underbrace{\int_0^1 \int_{\mathbb{R}^{2n}} (1-\theta) a''(X+\theta Y) Y^2 e^{-2\pi|Y|^2} 2^n dY d\theta.}_{r(X)} \end{aligned}$$

thus we get from (2.13),

$$|r^{(k)}(X)| \le C_{k+2}\Lambda^{m-\frac{k+2}{2}} \int_{\mathbb{R}^{2n}} |Y|^2 e^{-2\pi|Y|^2} 2^{n-1} dY,$$

which implies $r \in \mathcal{T}^{m-1}$.□

Remark 2.2. This procedure can be applied to classical symbols $S^1_{1,0}$, after a suitable rescaling, and gives a remainder r in $S^0_{1,0}$, not in $S^0_{1,1/2}$ as in [CF]. One writes first a Littlewood-Paley partition of unity, with non-negative functions $\varphi_\nu \in C_0^\infty(\mathbb{R}^n)$ such that

$$1 = \sum_{\nu \ge 0} \varphi_\nu(\xi) ,$$

and for $\nu \ge 1$,

$$\operatorname{supp}\varphi_\nu \subset \{2^{\nu-1} \le |\xi| \le 2^{\nu+1}\}, \ |\varphi_\nu^{(k)}| \le C_{k,n} 2^{-\nu k}.$$

Proposition 2.1 implies that the following "Berezin -Wick" quantization, satisfies (1.1–2):

$$a^{\mathrm{BW}} = \sum_{\nu \ge 0} \iint \varphi_\nu(\eta) a(y,\eta) 2^n \left[e^{-2\pi(2^\nu|x-y|^2 + 2^{-\nu}|\xi-\eta|^2)} \right]^w dy d\eta. \tag{2.15}$$

As a matter of fact, using the Taylor expansion at any order $2N$ in the proof of proposition 1.2, we get that the Weyl symbol of the general term in the series (2.15) is $\varphi_\nu a + r_\nu$ where r_ν satisfies the estimates of $S^0_{1,0}$ on the support of φ_ν and is bounded above by $2^{-\nu(N-1)}$ outside of the support of φ_ν. Since the derivatives of r_ν satisfy similar estimates, this is enough for the summability. To get the $S^0_{1,0}$ estimates for the derivatives of $\sum r_\nu$ requires a more careful analysis. In particular, to estimate $\sum_\nu |\partial_\xi^\alpha r_\nu||\xi|^\alpha$, we must check,

$$\sum_\nu 2^{\nu(1-N)} (|\xi| 2^{-\nu})^{|\alpha|} \iiint_{\theta=0}^{1} (1 + 2^{-\nu}|\xi + \theta\eta 2^{\nu/2}|)^{-N_1} \ e^{-\pi(|y|^2+|\eta|^2)} dy d\eta d\theta,$$

which is bounded for $N \ge 2, N_1 \ge |\alpha|$. We obtain then

$$a^{\mathrm{BW}} = a^w + r^w, \qquad \text{with} \qquad r \in S^0_{1,0}. \tag{2.16}$$

Since (1.1) follows from (2.15) and (2.3), and (1.2) is (2.16), formula (2.15) actually provides a nonnegative quantization for classical symbols.

We want now to investigate the composition formula $a^{\mathrm{Wick}} b^{\mathrm{Wick}}$ when one of the symbols is only bounded measurable. Let $a \in L^\infty(\mathbb{R}^{2n}), b \in \mathcal{T}^1(2.13)$.

We have from (2.6), (2.5) and (2.2),

(2.17)

$$
\begin{aligned}
a^{\text{Wick}}\, b^{\text{Wick}} &= \iint a(Y)\, b(Z)\, \Sigma_Y \Sigma_Z\, dY dZ \\
&= \iint \Big[b(Y) + b'(Y)\cdot(Z-Y) + \int_0^1 (1-\theta) b''(Y+\theta(Z-Y)) d\theta (Z-Y)^2 \Big] \\
&\qquad\qquad\qquad a(Y)\Sigma_Y\Sigma_Z\, dYdZ \\
&= \int a(Y)b(Y)\Sigma_Y dY + \iint a(Y)\, b'(Y)\cdot(Z-Y)\, \Sigma_Y\Sigma_Z\, dYdZ + R,
\end{aligned}
$$

with

$$R = \iint \alpha(Y,Z)(Z-Y)^2\, \Sigma_Y\Sigma_Z\, dYdZ, \tag{2.18}$$

where the norm of the quadratic form $\alpha(Y,Z)$ is less than $\|a\|_{L^\infty}\gamma_2(b)$; here $\gamma_2(b)$ is a semi-norm of the symbol b in $\mathcal{T}^1$(see (2.13)). From (2.12) and Cotlar's lemma (see e.g. lemme 4.2.3 in [BL]), using $\Sigma_Y\Sigma_Z\Sigma_{Y'}\Sigma_{Z'} = (\Sigma_Y\Sigma_Z)(\Sigma_Z\Sigma_{Y'})(\Sigma_{Y'}\Sigma_{Z'})$, one gets that

$$\|R\|_{\mathcal{L}(L^2(\mathbb{R}^n))} \le C(n)\|a\|_{L^\infty}\gamma_2(b), \tag{2.19}$$

where $C(n)$ depends only on the dimension. We check now the second term in (2.17), using (2.2),

$$
\begin{aligned}
\int b'(Y)\cdot(Z-Y)\,\Sigma_Z\, dZ &= b'(Y)\cdot\Big[\int (\overbrace{Z-X}^{\textit{will give } 0} + X - Y) 2^n e^{-2\pi|X-Z|^2}\, dZ\Big]^w \\
&= b'(Y)\cdot L_Y^w,
\end{aligned} \tag{2.20}
$$

where L_Y is the (vector-valued) linear form $X-Y$. Note that, from the fact that $r(a)$ in (2.14) depends only of the second derivative of a, $L^w = L^{\text{Wick}}$ for a linear form. We state now, using the notation $2\text{Re}A = A + A^*$,

Proposition 2.3. *Let $a \in L^\infty(\mathbb{R}^{2n}), b \in \mathcal{T}^1$, be real-valued functions. Then*

$$\text{Re}\, a^{\text{Wick}}\, b^{\text{Wick}} = \Big[ab - \frac{1}{4\pi}\, a'(Y)\cdot b'(Y)\Big]^{\text{Wick}} + S, \tag{2.21}$$

where $\|S\|_{\mathcal{L}(L^2(\mathbb{R}^n))} \le c_n \|a\|_{L^\infty}\gamma_2(b)$. Here $\gamma_2(b)$ is a semi-norm of b in $\mathcal{T}^1$ defined in (2.13), and c_n depends only on the dimension.

Proof. From (2.17), (2.20), we get

$$\text{Re}\, a^{\text{Wick}}\, b^{\text{Wick}} = (ab)^{\text{Wick}} + \int a(Y)b'(Y)\cdot \text{Re}(L_Y^w \Sigma_Y) dY + \text{Re}R. \tag{2.22}$$

Now, since L_Y is a real linear form, we have

$$\operatorname{Re}(L_Y^w \Sigma_Y) = \left[(X-Y)2^n e^{-2\pi|X-Y|^2}\right]^w = \frac{1}{4\pi}\frac{\partial}{\partial Y}(\Sigma_Y). \tag{2.23}$$

An integration by parts, in the distribution sense, gives what we expect in (2.21), except possibly for

$$-\frac{1}{4\pi}\int a(Y) \text{ Trace } b''(Y)\Sigma_Y dY + \operatorname{Re}R. \tag{2.24}$$

The estimate (2.19), $a \in L^\infty$, $b \in \mathcal{T}^1$ and (2.9) applied to the integral in (2.24) prove the last statement. The proof of proposition 2.3 is complete.□

3 Energy estimates

Let $q_\Lambda(t,X)$ be a smooth function on $\mathbb{R}_t \times \mathbb{R}_X^{2n}$, defined for $\Lambda \geq 1$, supported in

$$\mathbf{B} = \{|t| \leq 1\} \times \{|X| \leq \Lambda^{1/2}\},$$

so that, for each k,

$$\sup_{t\in\mathbf{R}, X\in\mathbf{R}^{2n}, \Lambda\geq 1} |D_X^k q_\Lambda(t,X)|\Lambda^{-1+\frac{k}{2}} < \infty. \tag{3.1}$$

This means in particular that $q_\Lambda(t,X)$ is uniformly (with respect to t) in $\mathcal{T}^1$ (see (2.13)). We omit below the subscript Λ on q_Λ. We assume that $\tau - iq$ satisfies Nirenberg-Treves condition (ψ) :

$$q(t,X) > 0 \quad \text{and} \quad s > t \quad \Longrightarrow \quad q(s,X) \geq 0. \tag{3.2}$$

We define

$$\theta(X) = \begin{cases} \inf\ \{t \in (-1,+1), q(t,X) > 0\}, & \text{if this set is not empty,} \\ +1, & \text{otherwise.} \end{cases} \tag{3.3}$$

The function θ is bounded measurable (actually, θ is upper semi-continuous). We set

$$s(t,X) = \begin{cases} +1, & \text{if } t > \theta(X), \\ 0, & \text{if } t = \theta(X), \\ -1, & \text{if } t < \theta(X). \end{cases} \tag{3.4}$$

The function s is bounded measurable and (3.2) implies that

$$s(t,X)q(t,X) = |q(t,X)|. \tag{3.5}$$

Moreover, the distribution derivative $\frac{\partial s}{\partial t}$ is a positive measure satisfying

$$\langle \frac{\partial s}{\partial t}, \Psi(t,X)\rangle_{\mathcal{S}'(\mathbb{R}^{2n+1}),\mathcal{S}(\mathbb{R}^{2n+1})} = 2\int_{\mathbb{R}^{2n}} \Psi(\theta(X),X)dX. \tag{3.6}$$

We need to examine a couple of properties for the distribution

$$T = \sum_{1\le j\le 2n} \frac{\partial q}{\partial X_j}\frac{\partial s}{\partial X_j} = \frac{\partial q}{\partial X}\cdot\frac{\partial s}{\partial X} \quad . \tag{3.7}$$

First of all, we claim that

$$\operatorname{supp} T \subset \{(t,X)\in \mathbf{B}, \quad q(t,X)=0\}. \tag{3.8}$$

In fact, from (3.5), the restriction of s to the open set $\{q(t,X)>0\}$ (resp.$\{q(t,X$ $0\}$) is 1 (resp.-1). Thus the support of $\partial s/\partial X_j$ is included in $\{q(t,X)=0\}$. Since the restriction of q to the open set $\mathbf{B}^c$ is zero, (3.8) is proved. Let $\chi_0 : \mathbb{R}\to[0,1]$ be a smooth function, equal to 1 on $[-1,1]$, vanishing outside $(-2,2)$, and set

$$T_0 = \chi_0(|q'_X|^2)\,T \quad , \quad \omega = 1-\chi_0 \quad , \quad T_1 = \omega(|q'_X|^2)\,T. \tag{3.9}$$

We have from (3.8–9)

$$\operatorname{supp} T_1 \subset \{(t,X)\in \mathbf{B}, \quad q(t,X)=0 \text{ and } |q'_X(t,X)|\ge 1\} = \mathbf{K}. \tag{3.10}$$

We are able to give an explicit expression for T_1. The open set

$$\Omega = \{q'_X(t,X)\neq 0\}\cap\{|q(t,X)|<1\} \tag{3.11}$$

is a neighborhood of the compact $\mathbf{K}$. From (3.5) and the fact that the Lebesgue measure of $\Omega\cap\{q(t,X)=0\}$ is zero, the restriction $s_{|\Omega}$ of s to Ω is the L^∞ function $q/|q|$. This proves that

$$T_{1|\Omega} = \omega(|q'_X|^2)q'_X\cdot\frac{\partial}{\partial X}\left[\frac{q}{|q|}\right] = 2\delta(q)|q'_X|^2\omega(|q'_X|^2). \tag{3.12}$$

Since Ω is a neighborhood of the support of T_1, (3.12) determines completely T_1.

Lemma 3.1. *Let q and s be as above (3.1–4). We define, using (2.3–6),*

$$Q(t) = \int_{\mathbb{R}^{2n}} q(t,X)\Sigma_X dX = q(t,\cdot)^{\text{Wick}}, \tag{3.13}$$

$$J(t) = \int_{\mathbb{R}^{2n}} s(t,X)\Sigma_X dX = s(t,\cdot)^{\text{Wick}}.$$

Let $u(t,x) \in \mathcal{S}(\mathbb{R}\times\mathbb{R}^n)$, *set* $u(t)(x) = u(t,x)$, *and for* $(t,X) \in \mathbb{R}\times\mathbb{R}^{2n}$

$$\Phi(t,X) = [Wu(t)](X) = \langle u(t), \tau_X\varphi\rangle_{L^2(\mathbb{R}^n)}. \tag{3.14}$$

The function Φ *belongs to* $\mathcal{S}(\mathbb{R}\times\mathbb{R}^{2n})$ *and, with* $D_t = (2i\pi)^{-1}\partial/\partial t$, ω *defined in* (3.9), Ω *in* (3.11), $\Psi \in C_0^\infty(\Omega,[0,1])$, $\Psi \equiv 1$ *on a neighborhood of* $\mathbf{K}$ *(see* (3.10)*), we have*

$$\mathrm{Re}\langle\, D_t u \,,\, iJ(t)u(t)\,\rangle_{L^2(\mathbb{R}^{n+1})} = \frac{1}{2\pi}\int_{\mathbb{R}^{2n}} |\Phi(\theta(X),X)|^2 dX, \tag{3.15}$$

$$\mathrm{Re}\langle\, Q(t)u(t) \,,\, J(t)u(t)\,\rangle_{L^2(\mathbb{R}^{n+1})} \ge \iint_{\mathbb{R}_t\times\mathbb{R}_X^{2n}} |q(t,X)||\Phi(t,X)|^2 dtdX \tag{3.16}$$
$$-\frac{1}{2\pi}\langle\, \delta(q)|q'_X|^2\omega(|q'_X|^2) \,,\, \Psi(t,X)|\Phi(t,X)|^2\,\rangle_{\mathcal{D}'(\Omega),\mathcal{D}(\Omega)} - C\|u\|^2_{L^2(\mathbb{R}^{n+1})},$$

where C *is a constant depending only on the dimension and the semi-norms of* q.

Proof. Let us first notice that from (2.6) and (3.6) the left-hand-side of (3.15) is

$$-\frac{1}{4\pi}\iint \frac{\partial}{\partial t}\Big[\langle \Sigma_X u(t), u(t)\rangle_{L^2(\mathbb{R}^n)}\Big]\, s(t,X)\, dtdX = \frac{1}{2\pi}\int_{\mathbb{R}^{2n}} |\Phi(\theta(X),X)|^2 dX.$$

We use (3.13) and proposition 2.3 to write, with $L^2(\mathbb{R}^{n+1}) = L^2(\mathbb{R}_t, L^2(\mathbb{R}^n))$ dot products,

$$\mathrm{Re}\langle\, Q(t)u(t) \,,\, J(t)u(t)\,\rangle = \langle\, \mathrm{Re}[J(t)Q(t)]\, u(t) \,,\, u(t)\,\rangle$$
$$= \langle\, \Big[|q(t,\cdot)| - \frac{1}{4\pi}\frac{\partial q}{\partial X}(t,\cdot)\cdot\frac{\partial s}{\partial X}(t,\cdot)\Big]^{\mathrm{Wick}} u(t) \,,\, u(t)\,\rangle + \langle\, S(t)\, u(t) \,,\, u(t)\,\rangle,$$

where $\|S(t)\|_{\mathcal{L}(L^2(\mathbb{R}^n))} \le c_n\gamma_2(q)$. We get then the following inequality, using (3.7), (3.9), (3.12) and (3.14), with Ψ as in lemma 3.1,

$$\mathrm{Re}\langle\, Q(t)u(t) \,,\, J(t)u(t)\,\rangle \ge \iint_{\mathbb{R}_t\times\mathbb{R}_X^{2n}} |q(t,X)||\Phi(t,X)|^2 dX$$
$$-\frac{1}{2\pi}\langle\, \delta(q)|q'_X|^2\omega(|q'_X|^2) \,,\, \Psi(t,X)|\Phi(t,X)|^2\,\rangle_{\mathcal{D}'(\Omega),\mathcal{D}(\Omega)}$$
$$-\frac{1}{4\pi}\,\langle\, \chi_0(|q'_X|^2)\frac{\partial q}{\partial X}(t,X)\cdot\frac{\partial s}{\partial X}(t,X) \,,\, |\Phi(t,X)|^2\,\rangle_{\mathcal{S}'(\mathbb{R}^{2n+1}),\mathcal{S}(\mathbb{R}^{2n+1})}$$
$$-c_n\gamma_2(q)\, \|u\|^2_{L^2(\mathbb{R}^{n+1})}.$$

To obtain (3.16), we need only to check the duality bracket with χ_0. This term is

$$\frac{1}{4\pi}\iint s(t,X)\,\frac{\partial}{\partial X}\cdot\Big[\chi_0(|q'_X|^2)\frac{\partial q}{\partial X}(t,X)|\Phi(t,X)|^2\Big]dtdX$$
$$= \frac{1}{4\pi}\iint s(t,X)\,\frac{\partial}{\partial X}\cdot\Big[\chi_0(|q'_X|^2)\frac{\partial q}{\partial X}(t,X)\Big]|\Phi(t,X)|^2 dtdX \tag{3.17}$$
$$+\frac{1}{4\pi}\iint s(t,X)\,\chi_0(|q'_X|^2)\frac{\partial q}{\partial X}(t,X)\cdot\frac{\partial}{\partial X}\big[\langle \Sigma_X u(t), u(t)\rangle_{L^2(\mathbb{R}^n)}\big]\, dtdX.$$

We calculate
(3.18)
$$\frac{\partial}{\partial X}\cdot\left[\chi_0(|q'_X|^2)\frac{\partial q}{\partial X}(t,X)\right]=\chi'_0(|q'_X|^2)2q''_{XX}(q'_X,q'_X)+\chi_0(|q'_X|^2)\operatorname{Tr}q''_{XX}.$$

From (3.1) and the fact that the support of χ_0 is bounded by 2, we get that (3.18) is bounded by a semi-norm of q. This proves that the absolute value of the first term in the right-hand-side of (3.17) is bounded above by the product of a semi-norm of q with $\|u\|^2_{L^2(\mathbb{R}^{n+1})}$. We claim that, from Cotlar's lemma and (2.10),

$$\|\int_{\mathbb{R}^{2n}}\alpha(Y)\frac{\partial}{\partial Y_j}(\Sigma_Y)\,dY\|_{\mathcal{L}(L^2(\mathbb{R}^n))}\le\|\alpha\|_{L^\infty(\mathbb{R}^{2n})}d_n,\tag{3.19}$$

where d_n depends only on the dimension : in fact, from (2.10), the Weyl symbol of $\Sigma_Y\Sigma_Z$ is

$$p_{Y\,Z}(X)=e^{-\pi|X-Y|^2}e^{-\pi|X-Z|^2}e^{-2i\pi[X-Y,X-Z]}2^n.$$

This implies that the Weyl symbol of $\frac{\partial}{\partial Y_j}(\Sigma_Y)\frac{\partial}{\partial Z_j}(\Sigma_Z)=\frac{\partial^2}{\partial Y_j\partial Z_j}\Sigma_Y\Sigma_Z$ is

$$q_{Y\,Z}(X)=p_{Y\,Z}(X)L_j(Y-X,Z-X),$$

where L_j is a polynomial of degree 2. Now, we have

$$|q_{Y\,Z}(X)|\le 16\pi 2^{n/2}\sqrt{|p_{Y\,Z}(X)|}\le 16\pi 2^n e^{-\frac{\pi}{4}|Y-Z|^2}e^{-\pi|X-\frac{Y+Z}{2}|^2},\tag{3.20}$$

so that the $\mathcal{L}(L^2(\mathbb{R}^n))$ norm of $\frac{\partial}{\partial Y_j}(\Sigma_Y)\frac{\partial}{\partial Z_j}(\Sigma_Z)$ is bounded above by the $L^1(\mathbb{R}^{2n})$ norm of its symbol $q_{Y\,Z}$, which is estimated by $16\pi 2^n\ e^{-\frac{\pi}{4}|Y-Z|^2}$ from (3.20). Cotlar's lemma implies then (3.19). We note that

$$s(t,X)\chi_0(|q'_X|^2)\frac{\partial q}{\partial X}(t,X)$$

is bounded by 2, so that (3.19) implies that the absolute value of the second term in the right-hand-side of (3.17) is bounded above by $\pi^{-1}nd_n\|u\|^2_{L^2(\mathbb{R}^{n+1})}$. This concludes the proof of lemma 3.1.□

Theorem 3.2. *Let q,Q,J,u be as in lemma 3.1. We assume that*
(3.21)
$$q(t,X)=0\quad\text{and}\quad|q'_X(t,X)|^2\ge 1\quad\Longrightarrow\quad|q'_X(t,X)|^2\le q'_t(t,X).$$

Then, assuming supp $u\subset\{|t|\le T\}$ for a positive T, the following estimate holds (with $L^2(\mathbb{R}^{n+1})$ dot products and norms)

$$\operatorname{Re}\langle\, D_t u+iQ(t)u\ ,\ iJ(t)u+i\frac{t}{T}u\,\rangle\ge(4\pi T)^{-1}\|u\|^2-C\|u\|^2,\tag{3.22}$$

where C is the constant given in (3.16). Thus, for $0 < T \le \frac{1}{36\pi C}$,

$$\|D_t u + iQ(t)u\| \ge (9\pi T)^{-1}\|u\|. \tag{3.23}$$

Proof. Let (t_0, X_0) be a point in $\mathbf{K}$ (3.10). From (3.21), $q'_t(t_0, X_0) > 0$, so that the implicit function theorem and (3.3) give that, in an open neighborhood of (t_0, X_0),

$$q(t,X) = e(t,X)(t-\theta(X)) \quad \text{with} \quad e>0 \quad \text{and} \quad e,\theta \in C^\infty.$$

This implies that, on this neighborhood,

$$\delta(t-\theta(X)) = \delta(q)q'_t(t,X). \tag{3.24}$$

Eventually, (3.24) makes sense and is satisfied in an open neighborhood $\tilde{\Omega}$ of $\mathbf{K}$. Thus, setting $\Omega_0 = \Omega \cap \tilde{\Omega}$, where Ω is defined in (3.11), we obtain, with ω defined in (3.9) and $\Psi \in C_0^\infty(\Omega_0, [0,1])$, $\Psi \equiv 1$ in a neighborhood of $\mathbf{K}$, Φ given by (3.14),

(3.25)
$$\langle\ \delta(q)q'_t(t,X)\ ,\ \Psi(t,X)|\Phi(t,X)|^2\ \rangle_{\mathcal{D}'(\Omega_0),\mathcal{D}(\Omega_0)} \le \int_{\mathbf{R}^{2n}} |\Phi(\theta(X),X)|^2 dX.$$

Moreover, from the assumption (3.21) and (3.25), we have

(3.26)
$$\begin{aligned}
&\frac{1}{2\pi}\langle\ \delta(q)|q'_X|^2\omega(|q'_X|^2)\ ,\ \Psi(t,X)|\Phi(t,X)|^2\ \rangle_{\mathcal{D}'(\Omega_0),\mathcal{D}(\Omega_0)}\\
&\le \frac{1}{2\pi}\langle\ \delta(q)q'_t\ ,\ \omega(|q'_X|^2)\Psi(t,X)|\Phi(t,X)|^2\ \rangle_{\mathcal{D}'(\Omega_0),\mathcal{D}(\Omega_0)}\\
&\le \frac{1}{2\pi}\int_{\mathbf{R}^{2n}} |\Phi(\theta(X),X)|^2 dX.
\end{aligned}$$

Inequalities (3.25–26) and lemma 3.1 imply that

(3.27)
$$\begin{aligned}
&\mathrm{Re}\langle\ D_t u + iQ(t)u\ ,\ iJ(t)u + i\frac{t}{T}u\ \rangle \ge\\
&((4\pi T)^{-1} - C)\|u\|^2 + \iint_{\mathbf{R}_t\times\mathbf{R}^{2n}_X} \Big[|q(t,X)| + \frac{t}{T}q(t,X)\Big]|\Phi(t,X)|^2 dX dt.
\end{aligned}$$

Since u is supported in $|t| \le T$, so is Φ. The inequality (3.27) implies (3.22). The estimate (3.23) follows from (3.22), $|t| \le T, \|J(t)\| \le 1$. The proof of theorem 3.2 is complete.□

4 Comments

Going back to the homogeneous case, theorem 3.2 yields local solvability in L^2 for the operators

$$P = D_t + ia(t, x, D_x) + R \tag{4.1}$$

where $a(t, x, \xi)$ is real-valued homogeneous of degree one with respect to ξ, smooth outside of the zero section, such that $\tau + ia(t, x, \xi)$ satisfies condition (ψ) and so that, for some $\gamma < 1$,

(4.2)

$$a(t, x, \xi) = 0 \quad \Longrightarrow \quad |a'_x(t, x, \xi)|^2 |\xi|^{-1} + |a'_\xi(t, x, \xi)|^2 |\xi| \leq \gamma |a'_t(t, x, \xi)|.$$

In (4.1), R is any L^2 bounded operator. This condition bears only on the non-singular zero set

$$D = \{a = 0 \ , \ d_{x,\xi} a \neq 0\}, \tag{4.3}$$

and in particular, if D is empty, we get that condition (ψ) implies solvability. Condition (4.2) is also weaker than the classical condition $|a'_x(t, x, \xi)|^2 |\xi|^{-1} + |a'_\xi(t, x, \xi)|^2 |\xi| \leq C|a(t, x, \xi)|$, since the latter implies $D = \emptyset$. Note that condition (ψ) implies $a'_t \leq 0$ at $a = 0$. Moreover, both sides in the inequality of (4.2) are homogeneous of degree 1, so that it suffices to check it on the sphere. This condition is of course devised to secure assumption (3.21) in theorem 3.2, dealing with pseudo-differential operators with a large parameter. As a matter of fact, starting from an homogeneous symbol $a(t, x, \xi)$, vanishing for $|t| \geq 1$ and for $|x| \geq 1$, choosing a number $\rho_0 > 1$, using a modification of the quantization in remark 2.2 (replacing powers of 2 by powers of ρ_0, except for the unchanged 2^n and 2π in (2.15)), we get the symbol with large parameter Λ

$$q(t, x, \xi) = \varphi_\nu(\xi) a(t, x, \xi) \quad , \quad \Lambda = \rho_0^\nu. \tag{4.4}$$

We denote

$$\Gamma = \rho_0^\nu |dx|^2 + \rho_0^{-\nu} |d\xi|^2 \tag{4.5}$$

the square of the norm we use in (2.2) which becomes

$$\Sigma_Y = \left[2^n e^{-2\pi\Gamma(\,\bullet\, - Y)}\right]^w. \tag{4.6}$$

After these modifications, (3.21) reads at points $\varphi_\nu(\xi) a(t, x, \xi) = 0$,

$$\rho_0^\nu |\varphi'_\nu a + \varphi_\nu a'_\xi|^2 + \rho_0^{-\nu} |\varphi_\nu a'_x|^2 \leq |\varphi_\nu a'_t|. \tag{4.7}$$

If $\varphi_\nu(\xi)=0$, since this function is nonnegative, φ'_ν should vanish too and (4.7) is satisfied. We can assume $\varphi_\nu(\xi)>0$ and $a(t,x,\xi)=0$. Then the left-hand-side of (4.7) is

(4.8)

$$\varphi_\nu(\xi)^2\left[\rho_0^{-\nu}|\xi||a'_x(t,x,\xi)|^2|\xi|^{-1}+\rho_0^{\nu}|\xi|^{-1}|a'_\xi(t,x,\xi)|^2|\xi|\right]\le\gamma\rho_0\varphi_\nu(\xi)|a'_t(t,x,\xi)|,$$

from (4.2), the fact that $\rho_0^{\nu-1}\le|\xi|\le\rho_0^{\nu+1}$ on the support of φ_ν and φ_ν is valued in [0,1]. Inequality (4.8) gives (4.7) if one chooses

$$\rho_0=\frac{1}{\gamma} \tag{4.9}$$

In our remark 1.1 of [L2], we stated that the existence of a Lipschitz continuous function, homogeneous of degree 0, $\theta(x,\xi)$, so that

$$(t-\theta(x,\xi))\ a(t,x,\xi)\le 0 \tag{4.10}$$

would imply solvability of the operator (4.1). The hint given after this remark used as a multiplier the Wick quantization of the function $(t-\theta(x,\xi))$. This operator is somewhat simpler to handle than the rougher sign of the same quantity, used in the present paper. Let us show now that theorem 3.2 above implies solvability in this case. Differentiating the identity $a(\theta(x,\xi),x,\xi)\equiv 0$, we get

(4.11)

$$|a'_x(\theta(x,\xi),x,\xi)|^2|\xi|^{-1}+|a'_\xi(\theta(x,\xi),x,\xi)|^2|\xi|\le 2\|\theta\|^2_{Lip}a'_t(\theta(x,\xi),x,\xi)^2\ |\xi|^{-1},$$

which means that, at points where $a=0$ (in fact, we check here only the points $t=\theta(x,\xi)$ since on the open set $\{t<\theta\}\cup\{t>\theta\}$, a is of constant sign and thus vanishes at singular points),

$$|a'_x(t,x,\xi)|^2|\xi|^{-1}+|a'_\xi(t,x,\xi)|^2|\xi|\le Ca'_t(t,x,\xi)^2\ |\xi|^{-1}. \tag{4.12}$$

Since the case $a'_t(0,x_0,\xi_0)<0$ corresponds to the elementary 1/2 subelliptic estimate for D_t-ia, we can assume $a'_t(0,x_0,\xi_0)=0$ and find a conic neighborhood such that $C|a'_t||\xi|^{-1}\le 1/2$ so that (4.12) implies (4.2).

Moreover, condition (4.2) is unaffected by a time change of scale $t\mapsto\mu t$ which transforms P in (4.1) into $\mu^{-1}[D_t+i\mu q(\mu t,x,D_x)]$. This means that the limitation on the constant γ in (4.2) should be taken seriously. Actually, it is possible to remove the constraint $\gamma<1$ and to assume (4.2) with some arbitrary constant, but the proof is much more involved and requires a very precise microlocal analysis. A detailed article on this topic will be published somewhere.

References

[BF] R.Beals, C.Fefferman, *On local solvability of linear partial differential equations*, Ann. of Math. **97** (1973), 482-498.

[Be] F.A. Berezin, *Quantization*, Math. USSR, Izvest. **8** (1974), 1109-1165.

[Bo] J.M.Bony, *Calcul symbolique et propagation des singularités pour les équations aux dérivées partielles non linéaires*, Ann.Sci.Ec.N.Sup. **14** (1981), 209-256.

[BL] J.M.Bony, N.Lerner, *Quantification asymptotique et microlocalisations d'ordre supérieur*, Ann.Sci.Ec.N.Sup. **4, t.22** (1989), 377- 433.

[CF] A.Cordoba, C.Fefferman, *Wave packets and Fourier integral operators*, Comm. PDE **3(11)** (1978), 979-1005.

[FP] C.Fefferman, D.H.Phong, *On positivity of pseudo-differential operators*, Proc.Nat.Ac.Sc. (1978), 4673-4674.

[Fo] G.B.Folland, *Harmonic analysis in phase space*, Annals of Math. Studies **122**, Princeton University press, 1989.

[H1] L.Hörmander, *The analysis of linear partial differential operators*, Springer-Verlag, 1985.

[H2] L.Hörmander, *On the solvability of pseudodifferential equations*, 1995, (preprint).

[La] B.Lascar, *Condition nécessaire et suffisante d'ellipticité en dimension infinie*, Comm. PDE **2(1)** (1977), 31-67.

[L1] N.Lerner, *Sufficiency of condition* (ψ) *for local solvability in two dimensions,*, Ann. of Math. **128** (1988), 243-258.

[L2] ———, *An iff solvability condition for the oblique derivative problem*, exposé 18, 1990-91, Séminaire EDP, Ecole Polytechnique.

[L3] ———, *Nonsolvability in* L^2 *for a first order operator satisfying condition* (ψ), Ann. of Math. **139** (1994), 363-393.

[L4] ———, *Coherent states and evolution equations*, preprint 95-34,Université de Rennes (1995).

[NT] L.Nirenberg, F. Treves, *On local solvability of linear partial differential equations*, Comm. Pure Appl. Math. **23** (1970), 1-38, 459-509; **24** (1971), 279-288.

[Sh] M. Shubin, *Pseudo-differential operators and spectral theory*, Springer-Verlag, 1985.

[Un] A.Unterberger, *Oscillateur harmonique et opérateurs pseudo-différentiels*, Ann. Inst. Fourier **29** (1979), 201-221.

Eigen functions of the Laplacian of exponential type

Dedicated to Professor H. Komatsu on his 60th birthday

Mitsuo Morimoto[1] *and Keiko Fujita*[2]

[1] Department of Mathematics, Sophia University, Chiyoda-ku, Tokyo 102, Japan
[2] Faculty of Education, Saga University, Saga City, Saga 840, Japan

Introduction

Let $\tilde{\mathrm{E}} = \mathrm{C}^{n+1}$, $L(z)$ the Lie norm on $\tilde{\mathrm{E}}$ and $L^*(z)$ the dual Lie norm on $\tilde{\mathrm{E}}$. We denote by $\mathcal{O}(\tilde{\mathrm{E}})$ the space of entire functions on $\tilde{\mathrm{E}}$ and by $\Delta_z = \partial^2/\partial z_1^2 + \partial^2/\partial z_2^2 + \cdots + \partial^2/\partial z_{n+1}^2$ the complex Laplacian on $\tilde{\mathrm{E}}$. Let $r > 0$. For $F \in \mathcal{O}(\tilde{\mathrm{E}})$ we put

$$\|F\|_r = \sup\{|F(z)| \exp(-rL^*(z)); z \in \tilde{\mathrm{E}}\}.$$

In this paper we consider the following spaces of entire functions of exponential type:

$$\mathrm{Exp}(\tilde{\mathrm{E}}; (r)) = \{F \in \mathcal{O}(\tilde{\mathrm{E}}); \|F\|_{r'} < \infty \text{ for any } r' > r\}$$

and

$$\mathrm{Exp}(\tilde{\mathrm{E}}; [r]) = \{F \in \mathcal{O}(\tilde{\mathrm{E}}); \|F\|_{r'} < \infty \text{ for some } r' < r\}.$$

Let λ be a complex number. We put

$$\mathcal{O}_{\Delta+\lambda^2}(\tilde{\mathrm{E}}) = \{F \in \mathcal{O}(\tilde{\mathrm{E}}); (\Delta_z + \lambda^2)F(z) = 0\},$$
$$\mathrm{Exp}_{\Delta+\lambda^2}(\tilde{\mathrm{E}}; (r)) = \mathrm{Exp}(\tilde{\mathrm{E}}; (r)) \cap \mathcal{O}_{\Delta+\lambda^2}(\tilde{\mathrm{E}})$$

and

$$\mathrm{Exp}_{\Delta+\lambda^2}(\tilde{\mathrm{E}}; [r]) = \mathrm{Exp}(\tilde{\mathrm{E}}; [r]) \cap \mathcal{O}_{\Delta+\lambda^2}(\tilde{\mathrm{E}}).$$

These spaces of eigen functions of the complex Laplacian of exponential type appear as the Fourier-Borel images of spaces of analytic functionals on the complex sphere.

Let $\tilde{S}_\lambda = \{\zeta \in \tilde{\mathrm{E}}; \zeta^2 = \zeta_1^2 + \zeta_2^2 + \cdots + \zeta_{n+1}^2 = \lambda^2\}$ be the complex sphere of radius λ and $S_1 = \{\zeta \in \tilde{S}_1; \zeta = \overline{\zeta}\}$ the real unit sphere. We denote by $\tilde{B}(r)$ and $\tilde{B}[r]$ the open and the closed Lie balls of radius $r > 0$. The Lie sphere $\Sigma_r = \{re^{i\theta}\omega; \theta \in \mathrm{R},\ \omega \in \mathrm{S}_1\}$ is the Shilov boundary of $\tilde{B}[r]$. For $r > |\lambda|$ we put $\tilde{S}_\lambda(r) = \tilde{S}_\lambda \cap \tilde{B}(r)$ and $\tilde{S}_\lambda[r] = \tilde{S}_\lambda \cap \tilde{B}[r]$. Let $\mathcal{O}(\tilde{S}_\lambda(r))$ be the space of holomorphic functions on $\tilde{S}_\lambda(r)$ and $\mathcal{O}(\tilde{S}_\lambda[r])$ the space of germs of holomorphic functions on $\tilde{S}_\lambda[r]$. We denote by $\mathcal{O}'(\tilde{S}_\lambda(r))$ and $\mathcal{O}'(\tilde{S}_\lambda[r])$ their dual spaces.

For $T \in \mathcal{O}'(\tilde{S}_\lambda(r))$ we put

$$\mathcal{F}_\lambda T(z) = \langle T_\zeta, \exp(iz\cdot\zeta)\rangle, \quad z \in \tilde{\mathrm{E}}$$

and call it the Fourier-Borel transform of T. $\mathcal{F}_\lambda T$ belongs to $\mathrm{Exp}_{\Delta+\lambda^2}(\tilde{\mathrm{E}};[r])$. We call the mapping

$$\mathcal{F}_\lambda \ : \ \mathcal{O}'(\tilde{S}_\lambda(r)) \to \mathrm{Exp}_{\Delta+\lambda^2}(\tilde{\mathrm{E}};[r]) \tag{1}$$

the Fourier-Borel transformation. It is known that (1) is a topological linear isomorphism. (See [12].)

We denote by

$$\mathrm{Exp}'_{\Delta+\lambda^2}(\tilde{\mathrm{E}};(r)) \quad (\text{resp. } \mathrm{Exp}'_{\Delta+\lambda^2}(\tilde{\mathrm{E}};[r]))$$

the dual space of

$$\mathrm{Exp}_{\Delta+\lambda^2}(\tilde{\mathrm{E}};(r)) \quad (\text{resp. } \mathrm{Exp}_{\Delta+\lambda^2}(\tilde{\mathrm{E}};[r])).$$

Let $\Lambda \in \mathrm{Exp}'_{\Delta+\lambda^2}(\tilde{\mathrm{E}};(r))$. Then there is $r' > r$ such that

$$|\langle \Lambda, F\rangle| \le C\|F\|_{r'} \text{ for } F \in \mathrm{Exp}_{\Delta+\lambda^2}(\tilde{\mathrm{E}};(r)).$$

Take r'' with $r < r'' < r'$. For $\zeta \in \tilde{S}_\lambda[r'']$ the function $z \mapsto \exp(-iz\cdot\zeta)$ belongs to $\mathcal{O}_{\Delta+\lambda^2}(\tilde{\mathrm{E}})$ and satisfies $\|\exp(-iz\cdot\zeta)\|_{r'} < \infty$. Therefore, the spherical Fourier-Borel transform of Λ

$$\mathcal{F}^S_\lambda \Lambda(\zeta) = \langle \Lambda_z, \exp(-iz\cdot\zeta)\rangle$$

can be defined for $\zeta \in \tilde{S}_\lambda[r'']$. That is, $\mathcal{F}^S_\lambda\Lambda$ is a holomorphic function on a neighborhood of $\tilde{S}_\lambda[r]$. We call the mapping

$$\mathcal{F}^S_\lambda \ : \ \mathrm{Exp}'_{\Delta+\lambda^2}(\tilde{\mathrm{E}};(r)) \to \mathcal{O}(\tilde{S}_\lambda[r]) \tag{2}$$

the spherical Fourier-Borel transformation.

It was proved in [9] that the spaces $\mathcal{O}(\tilde{S}_\lambda(r))$ and $\mathcal{O}(\tilde{S}_\lambda[r])$ are dual to each other with respect to the symbolic integral form

$$\langle f, g\rangle_{\tilde{S}_{\lambda,r}} = \int_{\tilde{S}_{\lambda,r}} f(\zeta)g(\overline{\zeta})d\tilde{S}_{\lambda,r}(\zeta),$$

where $d\tilde{S}_{\lambda,r}(\zeta)$ is the normalized invariant measure on $\tilde{S}_{\lambda,r} = \partial\tilde{S}_\lambda[r]$. We recall these results in §2 (Theorem 13).

In §3 we construct a duality bilinear form $\langle\langle F, G\rangle\rangle^{\lambda,r}$ on

$$\mathrm{Exp}_{\Delta+\lambda^2}(\tilde{\mathrm{E}};(r)) \times \mathrm{Exp}_{\Delta+\lambda^2}(\tilde{\mathrm{E}};[r])$$

and prove that the Fourier-Borel transformation (1) and the spherical Fourier-Borel transformation (2) are inverse to each other via the bilinear forms $\langle f, g\rangle_{\tilde{S}_{\lambda,r}}$

and $\langle\langle F, G\rangle\rangle^{\lambda,r}$ (Theorem 22). This proves, among others, that (2) is a topological linear isomorphism. (See [1] for a different proof.)

In [9] we considered the spherical Fourier-Borel transformation

$$\mathcal{F}_\lambda^S \; : \; \mathcal{O}'_{\Delta+\lambda^2}(\tilde{B}(r)) \to \mathrm{Exp}(\tilde{S}_\lambda; [r])$$

and the Fourier-Borel transformation

$$\mathcal{F}_\lambda \; : \; \mathrm{Exp}'(\tilde{S}_\lambda; (r)) \to \mathcal{O}_{\Delta+\lambda^2}(\tilde{B}[r])$$

and proved $\mathcal{F}_\lambda^S$ and $\mathcal{F}_\lambda$ are inverse to each other via the bilinear form $\langle F, G\rangle_r^\lambda$ on

$$\mathcal{O}_{\Delta+\lambda^2}(\tilde{B}(r)) \times \mathcal{O}_{\Delta+\lambda^2}(\tilde{B}[r])$$

and the bilinear form $\langle\langle f, g\rangle\rangle_{\lambda,r}$ on $\mathrm{Exp}(\tilde{S}_\lambda; (r)) \times \mathrm{Exp}(\tilde{S}_\lambda; [r])$.

The case where $\lambda = 0$ was treated in [7] and [8].

1 Holomorphic functions on $\tilde{B}(r)$

Let $\|x\|$ be the Euclidean norm on $\mathrm{E} = \mathrm{R}^{n+1}$, $n \geq 2$. The cross norm $L(z)$ on $\tilde{\mathrm{E}} = \mathrm{C}^{n+1}$ corresponding to $\|x\|$ is the Lie norm defined by

$$L(z) = L(x+iy) = \left\{\|x\|^2 + \|y\|^2 + 2\sqrt{\|x\|^2\|y\|^2 - (x \cdot y)^2}\right\}^{1/2},$$

where $z = x + iy$, $x, y \in \mathrm{E}$ and $x \cdot y = x_1y_1 + x_2y_2 + \cdots + x_{n+1}y_{n+1}$. We denote by $L^*(z)$ the dual Lie norm

$$\begin{aligned} L^*(z) &= \sup\{|z \cdot \zeta|; L(\zeta) \leq 1\} \\ &= \frac{1}{\sqrt{2}}\left\{\|x\|^2 + \|y\|^2 + \sqrt{(\|x\|^2 - \|y\|^2)^2 + 4(x \cdot y)^2}\right\}^{1/2}. \end{aligned}$$

Note that $L^*(z) \leq L(z) \leq 2L^*(z)$ for all $z \in \tilde{\mathrm{E}}$. The open and the closed Lie balls of radius $r > 0$ with center at 0 are defined by

$$\tilde{B}(r) = \{z \in \tilde{\mathrm{E}}; L(z) < r\}, \quad \tilde{B}[r] = \{z \in \tilde{\mathrm{E}}; L(z) \leq r\}.$$

We denote by $\mathcal{O}(\tilde{B}(r))$ the space of holomorphic functions on $\tilde{B}(r)$ with the topology of uniform convergence on compact sets. It is an FS (Fréchet-Schwartz) space.

Let γ be a complex number. We call

$$\mathcal{O}_{\Delta+\gamma^2}(\tilde{B}(r)) = \{F \in \mathcal{O}(\tilde{B}(r)); (\Delta_z + \gamma^2)F(z) = 0\}$$

the space of γ-harmonic functions on $\tilde{B}(r)$, where Δ_z is the complex Laplacian on $\tilde{\mathrm{E}}$. Because the space $\mathcal{O}_{\Delta+\gamma^2}(\tilde{B}(r))$ is a closed subspace of $\mathcal{O}(\tilde{B}(r))$, it is an FS space.

For $r > 0$ we put

$$\mathcal{O}(\tilde{B}[r]) = \text{ind}\ \lim\{\mathcal{O}(\tilde{B}(r')); r' > r\},$$
$$\mathcal{O}_{\Delta+\gamma^2}(\tilde{B}[r]) = \text{ind}\ \lim\{\mathcal{O}_{\Delta+\gamma^2}(\tilde{B}(r')); r' > r\}.$$

$\mathcal{O}(\tilde{B}[r])$ and $\mathcal{O}_{\Delta+\gamma^2}(\tilde{B}[r])$ are DFS (dual Fréchet Schwartz) spaces.

A continuous linear functional on $\mathcal{O}_{\Delta+\gamma^2}(\tilde{B}(r))$ (resp. $\mathcal{O}_{\Delta+\gamma^2}(\tilde{B}[r])$) is called a γ-harmonic functional on $\tilde{B}(r)$ (resp. $\tilde{B}[r]$). We denote by $\mathcal{O}'_{\Delta+\gamma^2}(\tilde{B}(r))$ (resp. $\mathcal{O}'_{\Delta+\gamma^2}(\tilde{B}[r])$) the dual space of $\mathcal{O}_{\Delta+\gamma^2}(\tilde{B}(r))$ (resp. $\mathcal{O}_{\Delta+\gamma^2}(\tilde{B}[r])$).

We denote by $\mathcal{P}^k(\tilde{\mathrm{E}})$ the space of k-homogeneous polynomials of $n+1$ variables with complex coefficients.

Let $S_1 = \{x \in \mathrm{E}; \|\mathrm{x}\| = 1\}$ be the real unit sphere. For a complex number λ we call $S_\lambda = \lambda S_1 = \{\lambda\omega; \omega \in S_1\}$ the real sphere of complex radius λ. The integration over S_λ will be performed with respect to the normalized invariant measure on S_λ:

$$\int_{S_\lambda} f(x) dS_\lambda(x) = \int_{S_1} f(\lambda\omega) dS_1(\omega), \quad \int_{S_\lambda} dS_\lambda(x) = 1.$$

Let $r > 0$. The Shilov boundary of $\tilde{B}[r]$ is the Lie sphere

$$\Sigma_r = \{re^{i\theta}\omega; \theta \in \mathrm{R}, \omega \in \mathrm{S}_1\}.$$

We consider the Lie norm because of the following lemma.

Lemma 1 (Lemma 5.5 in [4]). *For $F \in \mathcal{P}^k(\tilde{\mathrm{E}})$ we have*

$$\sup\{|F(z)|; z \in \tilde{B}[r]\} = \sup\{|F(z)|; z \in \Sigma_r\} = \sup\{|F(z)|; z \in S_r\}.$$

Note that $e^{i(\theta+\pi)}(-\omega) = e^{i\theta}\omega$ and $\Sigma_1 \cong ([0, 2\pi) \times S_1)/\sim$, where $(\theta, \omega) \sim (\theta + \pi, -\omega)$.

Let $F(z)$ be a holomorphic function on $\tilde{B}(r)$. We denote by F_k the k-homogeneous component of F:

$$F_k(z) = \frac{1}{2\pi} \oint_{|t|=s} \frac{F(tz)}{t^{k+1}} dt.$$

The integral is defined for sufficiently small $s > 0$ and is independent of s. Then $F_k(z)$ is a k-homogeneous polynomial and we have

$$F(z) = \sum_{k=0}^{\infty} F_k(z), \tag{3}$$

where the convergence is uniform on compact sets in $\tilde{B}(r)$.

For polynomials f, g on $\tilde{\mathrm{E}}$ we denote $\langle f, g\rangle_{S_\lambda} = \int_{S_\lambda} f(x) g(\overline{x}) dS_\lambda(x)$ and $\|f\|_{S_\lambda} = \sqrt{\langle f, \overline{f}\rangle_{S_\lambda}}$. Note that, if f, g are k-homogeneous, then $\langle f, g\rangle_{S_\lambda} = |\lambda|^{2k} \langle f, g\rangle_{S_1}$ and $\|f\|_{S_\lambda} = |\lambda|^k \|f\|_{S_1}$.

Proposition 2 (Lemma 4.1 in [4]). *Let* (3) *be the expansion of* F *in homogeneous polynomials.*

1) $F \in \mathcal{O}(\tilde{B}(r))$ *if and only if* $\limsup_{k\to\infty} \sqrt[k]{\|F_k(z)\|_{S_r}} \leq 1.$

2) $F \in \mathcal{O}(\tilde{B}[r])$ *if and only if* $\limsup_{k\to\infty} \sqrt[k]{\|F_k(z)\|_{S_r}} < 1.$

If $F,\ G \in \mathcal{O}(\tilde{B}[r])$, then we have

$$\int_{\Sigma_r} F(z)G(\overline{z})d\Sigma_r(z) = \sum_{k=0}^{\infty}\langle F_k, G_k\rangle_{S_r},$$

where $d\Sigma_r$ is the normalized invariant measure on the Lie sphere Σ_r. Because of Proposition 2, the right-hand side converges for $F \in \mathcal{O}(\tilde{B}(r))$ and $G \in \mathcal{O}(\tilde{B}[r])$. We define $\langle F, G\rangle_{\Sigma_r}$ by

$$\langle F, G\rangle_{\Sigma_r} = \sum_{k=0}^{\infty}\langle F_k, G_k\rangle_{S_r} = \sum_{k=0}^{\infty} r^{2k}\langle F_k, G_k\rangle_{S_1} \tag{4}$$

and call it the symbolic integral form on Σ_r. We sometimes write

$$\langle F, G\rangle_{\Sigma_r} = \int_{\Sigma_r} F(z)G(\overline{z})d\Sigma_r(z).$$

The Szëgo kernel $S_r(z,w)$ for $\tilde{B}[r]$ is defined by $S_r(z,w) = S_1(z/r, w/r)$, where

$$S_1(z,w) = \frac{1}{(1+z^2w^2-2z\cdot w)^{(n+1)/2}}.$$

For fixed $z \in \tilde{B}(r)$ the function $w \mapsto S_r(z,w)$ is holomorphic in a neighborhood of $\tilde{B}[r]$. We have the following reproducing formula:

Theorem 3 (Theorem 5.2 in [3]). *Let* $F \in \mathcal{O}(\tilde{B}(r))$. *Then we have*

$$F(z) = \int_{\Sigma_r} F(w)S_r(z,\overline{w})d\Sigma_r(w), \quad z \in \tilde{B}(r),$$

where the right-hand side is the symbolic integral form on Σ_r.

We denote by $\mathcal{P}^k_\Delta(\tilde{\mathrm{E}}) = \{F \in \mathcal{P}^k(\tilde{\mathrm{E}}); \Delta_z F(z) = 0\}$ the space of complex harmonic polynomials of degree k. The dimension of $\mathcal{P}^k_\Delta(\tilde{\mathrm{E}})$ is given by

$$N(k) = \frac{(2k+n-1)(k+n-2)!}{k!(n-1)!} = O(k^{n-1}).$$

Let $P_k(t)$ be the Legendre polynomial of degree k and of dimension $n+1$. The highest coefficient $\gamma(k)$ of $P_k(t)$ is given by

$$\gamma(k) \equiv \frac{\Gamma(k+(n+1)/2)2^k}{N(k)\Gamma((n+1)/2)k!}.$$

We define the extended Legendre polynomial of degree k by

$$\tilde{P}_k(z,w) = (\sqrt{z^2})^k(\sqrt{w^2})^k P_k\left(\frac{z}{\sqrt{z^2}} \cdot \frac{w}{\sqrt{w^2}}\right).$$

Then $\tilde{P}_k(z,w)$ is a symmetric homogeneous polynomial of degree k in z and in w, and satisfies $\Delta_z \tilde{P}_k(z,w) = \Delta_w \tilde{P}_k(z,w) = 0$ and $|\tilde{P}_k(z,w)| \le L(z)^k L(w)^k$. If $z^2 = 0$ or $w^2 = 0$, then we have $\tilde{P}_k(z,w) = \gamma(k)(z \cdot w)^k$.

Lemma 4. *Let $F_k \in \mathcal{P}^k_\Delta(\tilde{\mathrm{E}})$ and $F_j \in \mathcal{P}^j_\Delta(\tilde{\mathrm{E}})$. Then we have*

$$\int_{\Sigma_r} F_k(z) F_j(\overline{z}) d\Sigma_r(z) = \int_{S_r} F_k(x) F_j(x) dS_r(x) = 0, \quad k \ne j$$

and

$$F_k(z) = \frac{N(k)}{r^{2k}} \int_{\Sigma_r} F_k(w) \tilde{P}_k(z,\overline{w}) d\Sigma_r(w) = \frac{N(k)}{r^{2k}} \int_{S_r} F_k(x) \tilde{P}_k(z,x) dS_r(x).$$

Lemma 5. *Suppose $F \in \mathcal{O}_\Delta(\tilde{B}[r])$. The k-homogeneous component F_k of F is harmonic and coincides with the k-harmonic component of F:*

$$F_k(z) = \frac{N(k)}{r^{2k}} \int_{\Sigma_r} F(w) \tilde{P}_k(z,\overline{w}) d\Sigma_r(w).$$

The Poisson kernel $K_r(z,w)$ is defined by $K_r(z,w) = K_1(z/r, w/r)$, where

$$K_1(z,w) = \frac{1 - z^2 w^2}{(1 + z^2 w^2 - 2z \cdot w)^{(n+1)/2}} = \sum_{k=0}^{\infty} N(k) \tilde{P}_k(z,w).$$

Then $K_r(z,w)$ is a symmetric holomorphic function on

$$\{(z,w) \in \tilde{\mathrm{E}} \times \tilde{\mathrm{E}}; L(z)L(w) < r^2\}$$

and satisfies $\Delta_z K_r(z,w) = \Delta_w K_r(z,w) = 0$. We have the following Poisson integral formula.

Theorem 6. *Let $F \in \mathcal{O}_\Delta(\tilde{B}(r))$. Then we have*

$$F(z) = \int_{\Sigma_r} F(w) K_r(z,\overline{w}) d\Sigma_r(w) = \int_{S_r} F(x) K_r(z,x) dS_r(x), \quad z \in \tilde{B}(r),$$

where the right-hand side is the symbolic integral forms on Σ_r and on S_r.

Let μ be a complex number and $\tilde{J}_\mu(t)$ the entire Bessel function defined by

$$\tilde{J}_\mu(t) = \sum_{k=0}^{\infty} \frac{(-1)^k \Gamma(\mu+1)}{\Gamma(\mu+k+1)k!} \left(\frac{t}{2}\right)^{2k} = \Gamma(\mu+1)\left(\frac{t}{2}\right)^{-\mu} J_\mu(t).$$

Note that $\tilde{J}_\mu(0) = 1$, $\tilde{J}_\mu(t) = \tilde{J}_\mu(-t)$ and $|\tilde{J}_\mu(t)| \le e^{|t|}$. We put further

$$\tilde{j}_k(t) = \tilde{J}_{k+(n-1)/2}(t).$$

Let $\gamma \in \mathbf{C}$ and $F \in \mathcal{O}_{\Delta+\gamma^2}(\tilde{B}(r))$. Then we can expand F as follows:

$$F(z) = \sum_{k=0}^{\infty} \tilde{j}_k(\gamma\sqrt{z^2}) F_{k,k}(z), \tag{5}$$

where $F_{k,k}$ is the k-harmonic component of the k-homogeneous component F_k of F. $F_{k,k}(z)$ is a k-homogeneous harmonic polynomial and will be called the (k,k)-component of F. The series (5) converges uniformly on compact sets of $\tilde{B}(r)$.

Proposition 7 (Theorem 2.1 in [12]). *Let* (5) *be the expansion of* F *in* (k,k)-*components.*

1) $F \in \mathcal{O}_{\Delta+\gamma^2}(\tilde{B}(r))$ *if and only if* $\limsup_{k\to\infty} \sqrt[k]{\|F_{k,k}\|_{S_r}} \le 1$.

2) $F \in \mathcal{O}_{\Delta+\gamma^2}(\tilde{B}[r])$ *if and only if* $\limsup_{k\to\infty} \sqrt[k]{\|F_{k,k}\|_{S_r}} < 1$.

Suppose F, $G \in \mathcal{O}_{\Delta+\gamma^2}(\tilde{B}[r])$ and denote by $F_{k,k}$ and $G_{k,k}$ their (k,k)-components. Then we have

$$\langle F, G\rangle_{\Sigma_r} = \sum_{k=0}^{\infty} \tilde{j}_k^2(\gamma r)\langle F_{k,k}, G_{k,k}\rangle_{S_r}, \tag{6}$$

where we denote

$$\begin{aligned}\tilde{j}_k^2(\gamma r) &= \frac{1}{2\pi}\int_0^{2\pi} \tilde{j}_k(\gamma r e^{i\theta})\tilde{j}_k(\gamma r e^{-i\theta})d\theta \\ &= \sum_{\ell=0}^{\infty}\left(\frac{\Gamma(k+(n+1)/2)}{\Gamma(k+\ell+(n+1)/2)\ell!}\right)^2 \left(\frac{\gamma r}{2}\right)^{4\ell}.\end{aligned}$$

Because we have $|\tilde{j}_k^2(\gamma r)| \le e^{2|\gamma| r}$, the right-hand side of (6) converges for $F \in \mathcal{O}_{\Delta+\gamma^2}(\tilde{B}(r))$ and $G \in \mathcal{O}_{\Delta+\gamma^2}(\tilde{B}[r])$. If $\gamma = 0$, then $\tilde{j}_k^2(\gamma r) = 1$, $F_{k,k}(z) = F_k(z)$, $G_{k,k}(z) = G_k(z)$ and (6) reduces to (4).

The symbolic integral form

$$\langle F, G\rangle_{\Sigma_r} = \int_{\Sigma_r} F(w)G(\overline{w})d\Sigma_r(w) = \sum_{k=0}^{\infty} \tilde{j}_k^2(\gamma r)\langle F_{k,k}, G_{k,k}\rangle_{S_r}$$

is a separately continuous bilinear form on $\mathcal{O}_{\Delta+\gamma^2}(\tilde{B}(r)) \times \mathcal{O}_{\Delta+\gamma^2}(\tilde{B}[r])$. In particular, if $G \in \mathcal{O}_{\Delta+\gamma^2}(\tilde{B}[r])$ is given, then

$$\Lambda_G : F \mapsto \langle F, G\rangle_{\Sigma_r}, \quad F \in \mathcal{O}_{\Delta+\gamma^2}(\tilde{B}(r))$$

is a γ-harmonic functional on $\tilde{B}(r)$.

Let us define the γ-Poisson kernel

$$K_r^\gamma(z,w) = \sum_{k=0}^{\infty} \frac{N(k)\tilde{j}_k(\gamma\sqrt{z^2})\tilde{j}_k(\gamma\sqrt{w^2})}{r^{2k}\tilde{j}_k^2(\gamma r)} \tilde{P}_k(z,w).$$

Then it is symmetric and holomorphic on $\{(z,w) \in \tilde{\mathrm{E}} \times \tilde{\mathrm{E}}; L(z)L(w) < r^2\}$ and satisfies

$$(\Delta_z + \gamma^2)K_r^\gamma(z,w) = (\Delta_w + \gamma^2)K_r^\gamma(z,w) = 0.$$

Note that $K_r^0(z,w) = K_r(z,w)$.

The Poisson integral formula (Theorem 6) can be generalized as follows:

Theorem 8. *Let $F \in \mathcal{O}_{\Delta+\gamma^2}(\tilde{B}(r))$. Then we have the following reproducing formula:*

$$F(z) = \int_{\Sigma_r} F(w)K_r^\gamma(z,\overline{w})d\Sigma_r(w), \quad z \in \tilde{B}(r),$$

where the right-hand side is the symbolic integral form.

Let $\Lambda \in \mathcal{O}'_{\Delta+\gamma^2}(\tilde{B}(r))$. If $w \in \tilde{B}[r]$, then the function $z \mapsto K_r^\gamma(z,w)$ is a γ-harmonic function on $\tilde{B}(r)$. We define the γ-Poisson transform $\mathcal{P}^\gamma\Lambda$ of Λ by

$$\mathcal{P}^\gamma\Lambda(w) = \langle \Lambda_z, K_r^\gamma(z,w)\rangle, \quad w \in \tilde{B}[r].$$

We call the mapping $\mathcal{P}^\gamma : \Lambda \mapsto \mathcal{P}^\gamma\Lambda$ the γ-Poisson transformation.

Theorem 9. *The γ-Poisson transformation establishes the following topological linear isomorphisms:*

$$1) \quad \mathcal{P}^\gamma : \mathcal{O}'_{\Delta+\gamma^2}(\tilde{B}(r)) \xrightarrow{\sim} \mathcal{O}_{\Delta+\gamma^2}(\tilde{B}[r]),$$
$$2) \quad \mathcal{P}^\gamma : \mathcal{O}'_{\Delta+\gamma^2}(\tilde{B}[r]) \xrightarrow{\sim} \mathcal{O}_{\Delta+\gamma^2}(\tilde{B}(r)).$$

We have

$$\langle \Lambda, F\rangle = \langle F, \mathcal{P}^\gamma\Lambda\rangle_{\Sigma_r}$$

for $\Lambda \in \mathcal{O}'_{\Delta+\gamma^2}(\tilde{B}(r))$ and $F \in \mathcal{O}_{\Delta+\gamma^2}(\tilde{B}(r))$, or for $\Lambda \in \mathcal{O}'_{\Delta+\gamma^2}(\tilde{B}[r])$ and $F \in \mathcal{O}_{\Delta+\gamma^2}(\tilde{B}[r])$.

For the proof Theorems 8 and 9 we refer the reader to [9].

2 Holomorphic functions on $\tilde{S}_\lambda(r)$

Let λ be a complex number. We denote by $\tilde{S}_\lambda = \{z \in \tilde{E}; z^2 = \lambda^2\}$ the complex sphere of radius λ. For $r > |\lambda|$ we put $\tilde{S}_\lambda(r) = \tilde{S}_\lambda \cap \tilde{B}(r)$ and $\tilde{S}_\lambda[r] = \tilde{S}_\lambda \cap \tilde{B}[r]$. The boundary $\tilde{S}_{\lambda,r} = \partial\tilde{S}_{\lambda,r}$ of $\tilde{S}_\lambda[r]$ is isomorphic to the cotangential sphere bundle over S_1 and the orthogonal group $\mathrm{O}(n+1)$ acts transitively on it. Hence there is a unique normalized invariant measure on $\tilde{S}_{\lambda,r}$. Note that $\tilde{S}_\lambda \cap \tilde{B}[|\lambda|] = S_\lambda$ is isomorphic to the real sphere S_1 if $\lambda \neq 0$. Note also that $\tilde{S}_\lambda \cap \tilde{B}(|\lambda|) = \emptyset$.

We denote by $\mathcal{O}(\tilde{S}_\lambda(r))$ the space of holomorphic functions on $\tilde{S}_\lambda(r)$ equipped with the topology of uniform convergence on compact sets. It is an FS space. We put

$$\mathcal{O}(\tilde{S}_\lambda[r]) = \text{ind}\lim\{\mathcal{O}(\tilde{S}_\lambda(r')); r' > r\}.$$

It is a DFS space. A continuous linear functional on $\mathcal{O}(\tilde{S}_\lambda(r))$ (resp. $\mathcal{O}(\tilde{S}_\lambda[r])$) is called an analytic functional on $\tilde{S}_\lambda(r)$ (resp. $\tilde{S}_\lambda[r]$). We denote by $\mathcal{O}'(\tilde{S}_\lambda(r))$ (resp. $\mathcal{O}'(\tilde{S}_\lambda[r])$) the dual space of $\mathcal{O}(\tilde{S}_\lambda(r))$ (resp. $\mathcal{O}(\tilde{S}_\lambda[r])$).

If $\lambda \neq 0$, then $\tilde{S}_\lambda(r)$ is a complex neighborhood of the real sphere S_λ. The spaces $\mathcal{O}(\tilde{S}_\lambda(r))$ and $\mathcal{O}(\tilde{S}_\lambda[r])$ are subspaces of the space $\mathcal{A}(S_\lambda)$ of real analytic functions on S_λ. Hence we can expand $f \in \mathcal{O}(\tilde{S}_\lambda(r))$ into spherical harmonics.

We denote by $\mathcal{P}^k(\tilde{S}_\lambda) = \{P|_{\tilde{S}_\lambda}; P \in \mathcal{P}^k_\Delta(\tilde{E})\}$ the space of k-spherical harmonics on $\tilde{S}_\lambda$. It is known that $\dim\mathcal{P}^k(\tilde{S}_\lambda) = \dim\mathcal{P}^k_\Delta(\tilde{E}) = N(k)$.

Lemma 10. *Let $0 < |\lambda| < r$. Let $f_k \in \mathcal{P}^k(\tilde{S}_\lambda)$ and $f_j \in \mathcal{P}^j(\tilde{S}_\lambda)$. Then we have*

$$\int_{S_\lambda} f_k(x)f_j(\overline{x})dS_\lambda(x) = \int_{\tilde{S}_{\lambda,r}} f_k(z)f_j(\overline{z})d\tilde{S}_{\lambda,r}(z) = 0, \quad k \neq j$$

and

$$\begin{aligned} f_k(z) &= \frac{N(k)}{|\lambda|^{2k}}\int_{S_\lambda} f_k(x)\tilde{P}_k(z,\overline{x})dS_\lambda(x) \\ &= \frac{N(k)}{|\lambda|^{2k}P_{k,\lambda,r}}\int_{\tilde{S}_{\lambda,r}} f_k(w)\tilde{P}_k(z,\overline{w})d\tilde{S}_{\lambda,r}(w), \end{aligned}$$

where

$$P_{k,\lambda,r} = P_k((r^2/|\lambda|^2 + |\lambda|^2/r^2)/2). \tag{7}$$

Let $f \in \mathcal{O}(\tilde{S}_\lambda(r))$. We denote by f_k the k-spherical harmonic component of f. Then for $z \in \tilde{S}_\lambda$ we have

$$f_k(z) = \frac{N(k)}{|\lambda|^{2k}}\int_{S_\lambda} f(x)\tilde{P}_k(z,\overline{x})dS_\lambda(x) = \frac{N(k)}{|\lambda|^{2k}P_{k,\lambda,s}}\int_{\tilde{S}_{\lambda,s}} f(w)\tilde{P}_k(z,\overline{w})d\tilde{S}_{\lambda,r}(w), \tag{8}$$

where the right-hand side is independent of s with $|\lambda| < s < r$. Then $f_k \in \mathcal{P}^k(\tilde{S}_\lambda)$ and

$$f(z) = \sum_{k=0}^{\infty} f_k(z), \quad z \in \tilde{S}_\lambda(r) \tag{9}$$

where the convergence is uniform on compact sets in $\tilde{S}_\lambda(r)$.

Note that $P_{k,\lambda,r}|_{|\lambda|=r} = 1$ and $|\lambda|^{2k} P_{k,\lambda,r}|_{\lambda=0} = r^{2k}2^{-k}\gamma(k)$. We have

$$|\lambda|^{2k} P_{k,\lambda,r} \sim r^{2k}2^{-k}\gamma(k) \sim r^{2k} \text{ as } k \to \infty.$$

If we interprete $|\lambda|^{2k} P_{k,\lambda,r} = r^{2k}2^{-k}\gamma(k)$ when $\lambda = 0$, the formulas for $\tilde{S}_{\lambda,r}$ in Lemma 10 and in (8) are still valid for $\lambda = 0$.

For $f \in \mathcal{P}^k(\tilde{\mathrm{E}})$ we put $\langle f, g\rangle_{\tilde{S}_{\lambda,r}} = \int_{\tilde{S}_{\lambda,r}} f(z)g(\overline{z})d\tilde{S}_{\lambda,r}(z)$ and $\|f\|_{\tilde{S}_{\lambda,r}} = \sqrt{\langle f, \overline{f}\rangle_{\tilde{S}_{\lambda,r}}}$. Note that, if $f, g \in \mathcal{P}^k(\tilde{\mathrm{E}})$, we have

$$\|f\|_{\tilde{S}_{\lambda,r}} = r^k\|f\|_{S_1} = (r/|\lambda|)^k\|f\|_{S_\lambda}.$$

Proposition 11. *Let $|\lambda| < r$ and (9) be the expansion of f in spherical harmonics.*

1) $f \in \mathcal{O}(\tilde{S}_\lambda(r))$ *if and only if* $\limsup_{k\to\infty} \sqrt[k]{\|f_k\|_{S_{\lambda,r}}} \le 1$.

2) $f \in \mathcal{O}(\tilde{S}_\lambda[r])$ *if and only if* $\limsup_{k\to\infty} \sqrt[k]{\|f_k\|_{S_{\lambda,r}}} < 1$.

Suppose $f,\ g \in \mathcal{O}(\tilde{S}_\lambda[r])$. Then Lemma 10 implies

$$\langle f, g\rangle_{\tilde{S}_{\lambda,r}} = \int_{\tilde{S}_{\lambda,r}} f(z)g(\overline{z})d\tilde{S}_{\lambda,r}(z) = \sum_{k=0}^{\infty}\langle f_k, g_k\rangle_{\tilde{S}_{\lambda,r}} = \sum_{k=0}^{\infty} P_{k,\lambda,r}\langle f_k, g_k\rangle_{S_\lambda}. \tag{10}$$

The third and the fourth terms are meaningful for $f \in \mathcal{O}(\tilde{S}_\lambda(r))$ and $g \in \mathcal{O}(\tilde{S}_\lambda[r])$ so we put

$$\langle f, g\rangle_{\tilde{S}_{\lambda,r}} = \sum_{k=0}^{\infty}\langle f_k, g_k\rangle_{\tilde{S}_{\lambda,r}} = \sum_{k=0}^{\infty} P_{k,\lambda,r}\langle f_k, g_k\rangle_{S_\lambda}.$$

We call it the symbolic integral form on $\tilde{S}_{\lambda,r}$ and sometimes write

$$\langle f, g\rangle_{\tilde{S}_{\lambda,r}} = \int_{\tilde{S}_{\lambda,r}} f(z)g(\overline{z})d\tilde{S}_{\lambda,r}(z).$$

The bilinear form $\langle f, g\rangle_{\tilde{S}_{\lambda,r}}$ is separately continuous on $\mathcal{O}(\tilde{S}_\lambda(r)) \times \mathcal{O}(\tilde{S}_\lambda[r])$. This implies that, for a given $g \in \mathcal{O}(\tilde{S}_\lambda[r])$,

$$T_g\ :\ f \mapsto \langle f, g\rangle_{\tilde{S}_{\lambda,r}},\quad f \in \mathcal{O}(\tilde{S}_\lambda(r))$$

is an analytic functional on $\tilde{S}_\lambda(r)$.

For $|\lambda| < r$ we define the Cauchy-Poisson kernel for $\tilde{S}_\lambda(r)$ by

$$K_{\lambda,r}(z, w) = \sum_{k=0}^{\infty} \frac{N(k)}{|\lambda|^{2k}P_{k,\lambda,r}}\tilde{P}_k(z, w). \tag{11}$$

The function $K_{\lambda,r}(z,w)$ is symmetric and holomorphic on

$$\{(z,w) \in \tilde{E} \times \tilde{E}; L(z)L(w) < r^2\}$$

and satisfies $\Delta_z K_{\lambda,r}(z,w) = \Delta_w K_{\lambda,r}(z,w) = 0$. The function $K_{\lambda,r}(z,w)$ is the Cauchy kernel for $\mathcal{O}(\tilde{S}_\lambda(r))$ and we have the following Cauchy integral formula. (See [5].)

Theorem 12. *Let $|\lambda| < r$ and $f \in \mathcal{O}(\tilde{S}_\lambda(r))$. Then we have*

$$f(z) = \int_{\tilde{S}_{\lambda,r}} f(w)K_{\lambda,r}(z,\overline{w})d\tilde{S}_{\lambda,r}(w), \quad z \in \tilde{S}_\lambda(r),$$

where the right-hand side is the symbolic integral form.

Let $T \in \mathcal{O}'(\tilde{S}_\lambda(r))$. If $z \in \tilde{S}_\lambda[r]$, then the function $K_{\lambda,r}(\,\cdot\,,z)$ belongs to $\mathcal{O}(\tilde{S}_\lambda(r))$. We define the Cauchy transform $\mathcal{C}T$ of T by

$$\mathcal{C}T(z) = \langle T_w, K_{\lambda,r}(w,z)\rangle, \quad z \in \tilde{S}_\lambda[r].$$

We call the mapping $\mathcal{C} : T \mapsto \mathcal{C}T$ the Cauchy transformation.

Theorem 13. *Let $|\lambda| < r$. The Cauchy transformation $\mathcal{C}$ establishes the following topological linear isomorphisms:*

$$\mathcal{C} : \mathcal{O}'(\tilde{S}_\lambda(r)) \xrightarrow{\sim} \mathcal{O}(\tilde{S}_\lambda[r]),$$
$$\mathcal{C} : \mathcal{O}'(\tilde{S}_\lambda[r]) \xrightarrow{\sim} \mathcal{O}(\tilde{S}_\lambda(r)).$$

We have

$$\langle T, f\rangle = \langle f, \mathcal{C}T\rangle_{\tilde{S}_{\lambda,r}}$$

for $T \in \mathcal{O}'(\tilde{S}_\lambda(r))$ and $f \in \mathcal{O}(\tilde{S}_\lambda(r))$, or for $T \in \mathcal{O}'(\tilde{S}_\lambda[r])$ and $f \in \mathcal{O}(\tilde{S}_\lambda[r])$.

For the proof we refer the reader to [9].

Consider the kernel $K_{\lambda,r}(z,w)$ as a function on

$$\{(z,w) \in \tilde{E} \times \tilde{S}_\lambda; L(z)L(w) < r^2\}.$$

Then for $T \in \mathcal{O}'(\tilde{S}_\lambda(r))$ the function $\mathcal{C}T(z) = \langle T_w, K_{\lambda,r}(z,w)\rangle$ belongs to $\mathcal{O}_\Delta(\tilde{B}[r])$ and $\mathcal{C}$ establishes the topological linear isomorphisms of $\mathcal{O}'(\tilde{S}_\lambda(r))$ onto $\mathcal{O}_\Delta(\tilde{B}[r])$ and of $\mathcal{O}'(\tilde{S}_\lambda[r])$ onto $\mathcal{O}_\Delta(\tilde{B}(r))$. An important special case is

$$\mathcal{C} \;:\; \mathcal{A}'(S_\lambda) \xrightarrow{\sim} \mathcal{O}_\Delta(\tilde{B}(|\lambda|)).$$

This means that the space of hyperfunctions on the real sphere S_λ is isomorphic to the space of (complex) harmonic functions on the open Lie ball $\tilde{B}(|\lambda|)$.

If $\lambda = 0$, then

$$K_{0,r}(z,w) = \sum_{k=0}^{\infty} \frac{2^k N(k)}{r^{2k}}(z\cdot w)^k = K_0(z/r, w/r),$$

where

$$K_0(z,w) = \frac{1+2z\cdot w}{(1-2z\cdot w)^n}, \quad z \in \tilde{\mathrm{E}},\ w \in \tilde{S}_0$$

is the Cauchy kernel for the complex light cone. (See [6].)

It is known that the restriction mapping $F \mapsto f = F|_{\tilde{S}_\lambda(r)}$ is a topological linear isomorphism of $\mathcal{O}_\Delta(\tilde{B}(r))$ onto $\mathcal{O}(\tilde{S}_\lambda(r))$. The following Cauchy-Poisson integral formula (12) gives the inverse of this restriction mapping. The function F is called the harmonic extension of f.

Theorem 14. *Let $F \in \mathcal{O}_\Delta(\tilde{B}(r))$ and $|\lambda| < r$. Then we have*

$$F(z) = \int_{\tilde{S}_{\lambda,r}} F(w)K_{\lambda,r}(z,\overline{w})d\tilde{S}_{\lambda,r}(w), \quad z \in \tilde{B}(r), \tag{12}$$

where the right-hand side is the symbolic integral form.

More generally, we know the following theorem: (See [11].)

Theorem 15. *Let $|\lambda| < r$. Then the restriction mapping*

$$\mathcal{O}_{\Delta+\gamma^2}(\tilde{B}[r]) \xrightarrow{\sim} \mathcal{O}(\tilde{S}_\lambda[r])$$

is a topological linear isomorphism provided

$$\tilde{j}_k(\gamma\lambda) \neq 0, \quad k = 0,1,2,\cdots. \tag{13}$$

Note that the zeros of Bessel "small" function $\tilde{j}_\mu$, $\mu > -1$, are real. (Watson: Theory of Bessel Functions, 1966).

Proof. The proof relies on Propositions 7, 11 and the following property of "entire" Bessel function:

$$\lim_{\nu\to\infty} |\tilde{J}_\nu(t)| = 1, \qquad t \in \mathrm{C},$$

$$|\tilde{J}_\nu(t)| \leq \exp(|t|), \qquad t \in \mathrm{C}.$$

Suppose (13) is satisfied. Define the function $K^\gamma_{\lambda,r}(z,w)$ as follows:

$$K^\gamma_{\lambda,r}(z,w) = \sum_{k=0}^\infty \frac{N(k)}{|\lambda|^{2k}P_{k,\lambda,r}} \frac{\tilde{j}_k(\gamma\sqrt{z^2})}{\tilde{j}_k(\gamma\lambda)} \frac{\tilde{j}_k(\gamma\sqrt{w^2})}{\tilde{j}_k(\gamma\lambda)} \tilde{P}_k(z,w).$$

$K^0_{\lambda,r}(z,w) = K_{\lambda,r}(z,w)$ is the Cauchy-Poisson kernel. It is a holomorphic function on $\{(z,w) \in \tilde{\mathrm{E}} \times \tilde{\mathrm{E}}; L(z)L(w) < r^2\}$ and satisfies

$$(\Delta_z + \gamma^2)K^\gamma_{\lambda,r}(z,w) = (\Delta_w + \gamma^2)K^\gamma_{\lambda,r}(z,w) = 0.$$

For $f \in \mathcal{O}(\tilde{S}_\lambda(r))$ we put

$$F(z) = \int_{\tilde{S}_{\lambda,r}} f(w)K^\gamma_{\lambda,r}(z,\overline{w})d\tilde{S}_{\lambda,r}(w).$$

Then F is a holomorphic function on $\tilde{B}(r)$ and satisfies $F|_{\tilde{S}_\lambda} = f$ and the differential equation $(\Delta_z + \gamma^2)F(z) = 0$.

3 λ-harmonic entire functionals

For $0 < r \leq \infty$ we denote by

$$\begin{aligned}\mathrm{Exp}(\tilde{\mathrm{E}};[r]) = \{F \in \mathcal{O}(\tilde{\mathrm{E}});\ &\text{there is } r' < r \text{ such that} \\ &\sup\{|F(z)|\exp(-r'L^*(z)); z \in \tilde{\mathrm{E}}\} < \infty\}\end{aligned}$$

the space of entire functions on $\tilde{\mathrm{E}}$ of exponential type $[r]$ with respect to the dual Lie norm $L^*(z)$. Put $\mathrm{Exp}_{\Delta+\lambda^2}(\tilde{\mathrm{E}};[r]) = \mathrm{Exp}(\tilde{\mathrm{E}};[r]) \cap \mathcal{O}_{\Delta+\lambda^2}(\tilde{\mathrm{E}})$. Furthermore, for $0 \leq r < \infty$ we put

$$\begin{aligned}\mathrm{Exp}(\tilde{\mathrm{E}};(r)) = \{F \in \mathcal{O}(\tilde{\mathrm{E}});\ &\text{for all } r' > r \text{ we have} \\ &\sup\{|F(z)|\exp(-r'L^*(z)); z \in \tilde{\mathrm{E}}\} < \infty\}\end{aligned}$$

and $\mathrm{Exp}_{\Delta+\lambda^2}(\tilde{\mathrm{E}};(r)) = \mathrm{Exp}(\tilde{\mathrm{E}};(r)) \cap \mathcal{O}_{\Delta+\lambda^2}(\tilde{\mathrm{E}})$.

A continuous linear functional on $\mathrm{Exp}_{\Delta+\lambda^2}(\tilde{\mathrm{E}};[r])$ or $\mathrm{Exp}_{\Delta+\lambda^2}(\tilde{\mathrm{E}};(r))$ will be called a λ-harmonic entire functional on $\tilde{\mathrm{E}}$. We denote by

$$\mathrm{Exp}'_{\Delta+\lambda^2}(\tilde{\mathrm{E}};[r]) \quad (\text{resp.}\mathrm{Exp}_{\Delta+\lambda^2}(\tilde{\mathrm{E}};(r)))$$

the dual space of $\mathrm{Exp}_{\Delta+\lambda^2}(\tilde{\mathrm{E}};[r])$ (resp. $\mathrm{Exp}_{\Delta+\lambda^2}(\tilde{\mathrm{E}};(r))$).

Proposition 16. *Let* $|\lambda| < r$ *and* (5) *be the expansion of* F *in* (k,k)*-components.*

1) $F \in \mathrm{Exp}_{\Delta+\lambda^2}(\tilde{\mathrm{E}};[r])$ *if and only if* $\limsup_{k\to\infty} \sqrt[k]{k!\|F_{k,k}\|_{S_1}} < r/2$.

2) $F \in \mathrm{Exp}_{\Delta+\lambda^2}(\tilde{\mathrm{E}};(r))$ *if and only if* $\limsup_{k\to\infty} \sqrt[k]{k!\|F_{k,k}\|_{S_1}} \leq r/2$.

Define $C^{\lambda,r}(k)$, $k = 0, 1, 2, \cdots$ by

$$C^{\lambda,r}(k) = \frac{N(k)^2\gamma(k)^2(k!)^2}{|\lambda|^{2k}P_{k,\lambda,r}} = \frac{4^k\Gamma(k+(n+1)/2)^2}{|\lambda|^{2k}P_{k,\lambda,r}\Gamma((n+1)/2)^2}, \tag{14}$$

where $P_{k,\lambda,r}$ is given in (7).

Let

$$F(z) = \sum_{k=0}^{\infty} \tilde{j}_k(\lambda\sqrt{z^2})F_{k,k}(z) \in \mathrm{Exp}_{\Delta+\lambda^2}(\tilde{\mathrm{E}};[r]) \tag{15}$$

and

$$G(z) = \sum_{k=0}^{\infty} \tilde{j}_k(\lambda\sqrt{z^2})G_{k,k}(z) \in \mathrm{Exp}_{\Delta+\lambda^2}(\tilde{\mathrm{E}};(r)), \tag{16}$$

where $F_{k,k}$ and $G_{k,k}$ are (k,k)-components of F and G. We define

$$\langle\!\langle F, G\rangle\!\rangle^{\lambda,r} = \langle\!\langle F(z), G(z)\rangle\!\rangle_z^{\lambda,r} = \sum_{k=0}^{\infty} C^{\lambda,r}(k)\langle F_{k,k}, G_{k,k}\rangle_{S_1}. \tag{17}$$

Because $C^{\lambda,r}(k) \sim (2/r)^{2k}(k!)^2$, the left-hand side of (17) converges and defines a separately continuous bilinear form on $\mathrm{Exp}_{\Delta+\lambda^2}(\tilde{\mathrm{E}};[r]) \times \mathrm{Exp}_{\Delta+\lambda^2}(\tilde{\mathrm{E}};(r))$.

Remark. Suppose $\rho^{\lambda,r}(s)$ is a function on $(0,\infty)$ such that

$$\int_0^\infty \tilde{j}_k^2(\lambda s)s^{2k+n-1}\rho^{\lambda,r}(s)ds = C^{\lambda,r}(k) \tag{18}$$

for $k = 0, 1, 2, \cdots$.

If $\lambda = 0$, then (18) reduced to the following formula. (See [7].) Namely,

$$\int_0^\infty s^{2k+n-1}\rho^{0,r}(s)ds = C^{0,r}(k) \equiv \frac{N(k)\gamma(k)(k!)^2 2^k}{r^{2k}}$$
$$= \frac{N(k)k!\Gamma(k+(n+1)/2)2^k}{r^{2k}\Gamma((n+1)/2)}$$

One such function $\rho^{0,r}(s)$ on $(0,\infty)$ is the Ii-Wada functions given by

$$\rho^{0,r}(s) = \begin{cases} \displaystyle\sum_{\ell=0}^{(n-1)/2} a_\ell r^{\ell+n+1}s^{\ell+1}K_\ell(rs), & n \text{ is odd,} \\ \displaystyle\sum_{\ell=0}^{n/2} a_\ell r^{\ell+n+1/2}s^{\ell+1/2}K_\ell(rs), & n \text{ is even,} \end{cases}$$

where a_ℓ are constants and

$$K_\nu(s) = \int_0^\infty \exp(-s\cosh t)\cosh \nu t\, dt$$

is the modified Bessel function. (See [2] and [10].)

Construct a measure $d\mu^{\lambda,r}(z)$ on

$$\Sigma = \bigcup_{s>0} \Sigma_s = \{se^{i\theta}\omega; s > 0, \theta \in \mathrm{R}, \omega \in \mathrm{S}\}$$

as follows: Set $z = sz'$, $s > 0$, $z' \in \Sigma_1$.

$$\int_\Sigma f(z)d\mu^{\lambda,r}(z) = \int_0^\infty \left(\int_{\Sigma_1} f(sz')d\dot{\Sigma}_1(z')\right)\rho^{\lambda,r}(s)s^{n-1}ds.$$

If $F(z) \in \mathrm{Exp}_{\Delta+\lambda^2}(\tilde{\mathrm{E}};[r])$ and $G(z) \in \mathrm{Exp}_{\Delta+\lambda^2}(\tilde{\mathrm{E}};(r))$ are given by (15) and (16), then we have

$$\langle\langle F, G\rangle\rangle^{\lambda,r} = \int_\Sigma F(z)G(\overline{z})d\mu^{\lambda,r}(z). \tag{19}$$

In fact, by (6) we have

$$\int_{\Sigma_s} F(z)G(\overline{z})d\Sigma_s(z) = \int_0^{2\pi}\int_{S_1} F(se^{i\theta}\omega)G(se^{-i\theta}\omega)dS_1(\omega)\frac{d\theta}{2\pi}$$
$$= \sum_{k=0}^\infty \tilde{j}_k^2(\lambda s)s^{2k}\langle F_{k,k}, G_{k,k}\rangle_{S_1}.$$

Therefore, (18) implies (19). We shall call $d\mu^{\lambda,r}$ the Plancherel measure on Σ.

Lemma 17. *Let* $r > |\lambda|$. *We have*

$$K_{\lambda,r}(\xi,\eta) = \langle\langle \exp(iz\cdot\xi), \exp(-iz\cdot\eta)\rangle\rangle_z^{\lambda,r}$$

for $\xi, \eta \in \tilde{S}_\lambda$ *with* $L(\xi)L(\eta) < r^2$.

Proof. The following formulas are well-known: (See [4].)

$$\exp(iz\cdot\xi) = \sum_{k=0}^{\infty} \frac{i^k}{k!\gamma(k)} \tilde{j}_k(\sqrt{\xi^2}\sqrt{z^2})\tilde{P}_k(\xi, z) \tag{20}$$

and

$$\exp(-iz\cdot\eta) = \sum_{k=0}^{\infty} \frac{(-i)^k}{k!\gamma(k)} \tilde{j}_k(\sqrt{\eta^2}\sqrt{z^2})\tilde{P}_k(\eta, z). \tag{21}$$

If $\xi^2 = \eta^2 = \lambda^2$, then by (11) we have

$$\begin{aligned}\langle\langle \exp(iz\cdot\xi), \exp(-iz\cdot\eta)\rangle\rangle_z^{\lambda,r} &= \sum_{k=0}^{\infty} \frac{C^{\lambda,r}(k)}{(k!)^2\gamma(k)^2} \int_{S_1} \tilde{P}_k(\xi,\omega)\tilde{P}_k(\eta,\omega)dS_1(\omega) \\ &= \sum_{k=0}^{\infty} \frac{C^{\lambda,r}(k)}{(k!)^2\gamma(k)^2 N(k)} \tilde{P}_k(\xi,\eta) \\ &= \sum_{k=0}^{\infty} \frac{N(k)}{|\lambda|^{2k} P_{k,\lambda,r}} \tilde{P}_k(\xi,\eta) = K_{\lambda,r}(\xi,\eta). \square\end{aligned}$$

We define the F-Cauchy kernel $E^{\lambda,r}(z,w)$ by

$$E^{\lambda,r}(z,w) = \int_{\tilde{S}_{\lambda,r}} \exp(iz\cdot\xi)\exp(-iw\cdot\overline{\xi})d\tilde{S}_{\lambda,r}(\xi) \tag{22}$$

for z, $w \in \tilde{\mathrm{E}}$. By the definition, $E^{\lambda,r}(z,w)$ is a symmetric entire function on $\tilde{\mathrm{E}} \times \tilde{\mathrm{E}}$ and satisfies $(\Delta_z + \lambda^2)E^{\lambda,r}(z,w) = 0$ and

$$|E^{\lambda,r}(z,w)| \leq \exp(rL^*(z) + rL^*(w)), \quad (z,w) \in \tilde{\mathrm{E}} \times \tilde{\mathrm{E}}.$$

That is, for w fixed, the function $z \mapsto E^{\lambda,r}(z,w)$ belongs to $\mathrm{Exp}_{\Delta+\lambda^2}(\tilde{\mathrm{E}}; (r))$.

Lemma 18. *We have*

$$\begin{aligned}E^{\lambda,r}(z,w) &= \sum_{k=0}^{\infty} \frac{\tilde{j}_k(\lambda\sqrt{z^2})\tilde{j}_k(\lambda\sqrt{w^2})}{(k!)^2\gamma(k)^2 N(k)} |\lambda|^{2k} P_{k,\lambda,r}\tilde{P}_k(z,w) \\ &= \sum_{k=0}^{\infty} \frac{\tilde{j}_k(\lambda\sqrt{z^2})\tilde{j}_k(\lambda\sqrt{w^2})}{C^{\lambda,r}(k)} N(k)\tilde{P}_k(z,w).\end{aligned}$$

Proof. By (20) and (21) we have

$$E^{\lambda,r}(z,w)$$
$$= \int_{\tilde{S}_{\lambda,r}} \left(\sum_{k=0}^{\infty} \frac{\tilde{j}_k(\lambda\sqrt{z^2})}{k!\gamma(k)} i^k \tilde{P}_k(\xi,z) \right) \left(\sum_{\ell=0}^{\infty} \frac{\tilde{j}_\ell(\lambda\sqrt{w^2})}{\ell!\gamma(\ell)} (-i)^\ell \tilde{P}_\ell(\bar{\xi},w) \right) d\tilde{S}_{\lambda,r}(\xi)$$
$$= \sum_{k=0}^{\infty} \frac{\tilde{j}_k(\lambda\sqrt{z^2})\tilde{j}_k(\lambda\sqrt{w^2})}{(k!)^2\gamma(k)^2N(k)} |\lambda|^{2k} P_{k,\lambda,r} \tilde{P}_k(z,w). \square$$

Theorem 19. *Let $r > |\lambda|$ and $F \in \mathrm{Exp}_{\Delta+\lambda^2}(\tilde{\mathrm{E}};(r))$. Then we have*

$$F(z) = \langle\!\langle F(w), E^{\lambda,r}(z,w)\rangle\!\rangle_w^{\lambda,r}, \quad z \in \tilde{\mathrm{E}}.$$

Proof. We can prove the reproducing formula by a calculation similar to the previous one.

Now we define the F-Cauchy transform $\mathcal{E}\Lambda$ of $\Lambda \in \mathrm{Exp}'_{\Delta+\lambda^2}(\tilde{\mathrm{E}};(r))$, $r > |\lambda|$, by

$$\mathcal{E}\Lambda(w) = \langle \Lambda_z, E^{\lambda,r}(z,w)\rangle.$$

It is easy to check that $\mathcal{E}\Lambda$ is an entire function on $\tilde{\mathrm{E}}$ and satisfies $(\Delta_w + \lambda^2)\mathcal{E}\Lambda(w) = 0$.

Theorem 20. *Let $|\lambda| < r$. The F-Cauchy transformation $\Lambda \mapsto \mathcal{E}\Lambda$ establishes the following topological linear isomorphisms:*

$$1) \quad \mathcal{E} : \mathrm{Exp}'_{\Delta+\lambda^2}(\tilde{\mathrm{E}};(r)) \xrightarrow{\sim} \mathrm{Exp}_{\Delta+\lambda^2}(\tilde{\mathrm{E}};[r]),$$
$$2) \quad \mathcal{E} : \mathrm{Exp}'_{\Delta+\lambda^2}(\tilde{\mathrm{E}};[r]) \xrightarrow{\sim} \mathrm{Exp}_{\Delta+\lambda^2}(\tilde{\mathrm{E}};(r)).$$

We have

$$\langle \Lambda, F\rangle = \langle\!\langle F, \mathcal{E}\Lambda\rangle\!\rangle^{\lambda,r}$$

for $\Lambda \in \mathrm{Exp}'_{\Delta+\lambda^2}(\tilde{\mathrm{E}};(r))$ and $F \in \mathrm{Exp}_{\Delta+\lambda^2}(\tilde{\mathrm{E}};(r))$, or for $\Lambda \in \mathrm{Exp}'_{\Delta+\lambda^2}(\tilde{\mathrm{E}};[r])$ and $F \in \mathrm{Exp}_{\Delta+\lambda^2}(\tilde{\mathrm{E}};[r])$

Proof. We prove only 1). Let $\Lambda \in \mathrm{Exp}'_{\Delta+\lambda^2}(\tilde{\mathrm{E}};(r))$. By the continuity of Λ, there are $r' > r$ and $C \geq 0$ such that

$$|\langle \Lambda, F\rangle| \leq C \sup\{|F(z)| \exp(-r'L^*(z)); z \in \tilde{\mathrm{E}}\}$$

for any $F \in \mathrm{Exp}_{\Delta+\lambda^2}(\tilde{\mathrm{E}};(r))$. This implies $\mathcal{E}\Lambda \in \mathrm{Exp}_{\Delta+\lambda^2}(\tilde{\mathrm{E}};[r])$.

Let $F \in \mathrm{Exp}_{\Delta+\lambda^2}(\tilde{\mathrm{E}};(r))$. By Theorem 19 and the Fubini Theorem we have

$$\langle \Lambda, F\rangle = \langle \Lambda_z, \langle\!\langle F(w), E^{\lambda,r}(z,w)\rangle\!\rangle_w^{\lambda,r}\rangle$$
$$= \langle\!\langle F(w), \langle \Lambda_z, E^{\lambda,r}(z,w)\rangle \rangle\!\rangle_w^{\lambda,r} = \langle\!\langle F(w), \mathcal{E}\Lambda(w)\rangle\!\rangle_w^{\lambda,r}.$$

Thus, $\mathcal{E}$ is a continuous injection.

Conversely, let $G \in \mathrm{Exp}_{\Delta+\lambda^2}(\tilde{\mathrm{E}};[r])$. Define $\Lambda_G \in \mathrm{Exp}'_{\Delta+\lambda^2}(\tilde{\mathrm{E}};(r))$ by

$$\langle \Lambda_G, F\rangle = \langle\!\langle F, G\rangle\!\rangle^{\lambda,r}$$

for $F \in \mathrm{Exp}_{\Delta+\lambda^2}(\tilde{\mathrm{E}};(r))$. Then we have

$$\mathcal{E}\Lambda_G(w) = \langle\!\langle E^{\lambda,r}(z,w), G(z)\rangle\!\rangle_z^{\lambda,r} = G(w)$$

by Theorem 19. Thus, $\mathcal{E}$ is surjective.

The continuity of $\mathcal{E}^{-1} : G \mapsto \Lambda_G$ is clear.

4 Main theorem

The Fourier-Borel transform $\mathcal{F}_\lambda T$ of $T \in \mathcal{O}'(\tilde{S}_\lambda(r))$ is defined by

$$\mathcal{F}_\lambda T(z) = \langle T_\zeta, \exp(iz\cdot\zeta)\rangle, \quad z \in \tilde{\mathrm{E}}. \tag{23}$$

By the Hahn-Banach theorem, there is a Radon measure μ on $\tilde{S}_\lambda(r)$ with $\mathrm{supp}\mu \subset \tilde{S}_\lambda[r']$, $r' < r$, such that $\mathcal{F}_\lambda T(z) = \int_{\tilde{S}_\lambda(r)} \exp(iz\cdot\zeta)d\mu(\zeta)$. Therefore, $\mathcal{F}_\lambda T$ is an entire function on $\tilde{\mathrm{E}}$, satisfies $(\Delta_z + \lambda^2)(\mathcal{F}_\lambda T)(z) = 0$ and is of exponential type:

$$|\mathcal{F}_\lambda T(z)| \le \|\mu\| \sup\{|\exp(iz\cdot\zeta)|; \zeta \in \tilde{S}_\lambda[r']\} = \|\mu\| \exp(r'L^*(z)). \tag{24}$$

The Fourier-Borel transform $\mathcal{F}_\lambda T$ of $T \in \mathcal{O}'(\tilde{S}_\lambda[r])$ is also defined by (23). By the Hahn-Banach theorem, for any $r' > r$ there is a Radon measure with compact support $\mu_{r'}$ on $\tilde{S}_\lambda(r')$ such that $\mathcal{F}_\lambda T(z) = \int_{\tilde{S}_\lambda(r')} \exp(iz\cdot\zeta)d\mu_{r'}(\zeta)$. Therefore, $\mathcal{F}_\lambda T$ is an entire function on $\tilde{\mathrm{E}}$, satisfies $(\Delta_z + \lambda^2)(\mathcal{F}_\lambda T)(z) = 0$ and

$$|\mathcal{F}_\lambda T(z)| \le \|\mu_{r'}\| \exp(r'L^*(z)).$$

By the definition we have the following lemma:

Lemma 21. *Let $|\lambda| < r$. The Fourier-Borel transformation $\mathcal{F}_\lambda$ establishes the following continuous linear mappings*:

$$1)\quad \mathcal{F}_\lambda \ :\ \mathcal{O}'(\tilde{S}_\lambda(r)) \to \mathrm{Exp}_{\Delta+\lambda^2}(\tilde{\mathrm{E}};[r]),$$
$$2)\quad \mathcal{F}_\lambda \ :\ \mathcal{O}'(\tilde{S}_\lambda[r]) \to \mathrm{Exp}_{\Delta+\lambda^2}(\tilde{\mathrm{E}};(r)).$$

We shall prove that these mappings are topological linear isomorphisms, constructing explicitly their inverse mappings.

If $\zeta \in \tilde{S}_\lambda(r)$, then

$$\exp(-iz\cdot\zeta) \in \mathrm{Exp}_{\Delta+\lambda^2}(\tilde{\mathrm{E}};[r]).$$

In fact, $|\exp(-iz\cdot\zeta)| \le \exp(L(\zeta)L^*(z))$. Therefore, for $\Lambda \in \mathrm{Exp}'_{\Delta+\lambda^2}(\tilde{\mathrm{E}};[r])$ we can define the spherical Fourier-Borel transform

$$\mathcal{F}_\lambda^S \Lambda(\zeta) = \langle \Lambda_z, \exp(-iz\cdot\zeta)\rangle, \quad \zeta \in \tilde{S}_\lambda(r).$$

It is clear that the spherical Fourier-Borel transformation $\mathcal{F}_\lambda^S : \Lambda \mapsto \mathcal{F}_\lambda^S \Lambda$ is a continuous linear mapping

$$\mathcal{F}_\lambda^S \ : \ \mathrm{Exp}'_{\Delta+\lambda^2}(\tilde{\mathrm{E}};[r]) \to \mathcal{O}(\tilde{S}_\lambda(r)).$$

If $\zeta \in \tilde{S}_\lambda[r]$, then

$$\exp(-iz\cdot\zeta) \in \mathrm{Exp}_{\Delta+\lambda^2}(\tilde{\mathrm{E}};(r)).$$

Therefore, for $\Lambda \in \mathrm{Exp}'_{\Delta+\lambda^2}(\tilde{\mathrm{E}};(r))$,

$$\mathcal{F}_\lambda^S \Lambda(\zeta) = \langle \Lambda_\zeta, \exp(-iz\cdot\zeta)\rangle$$

is defined for $\zeta \in \tilde{S}_\lambda[r]$. We claim $\mathcal{F}_\lambda^S \Lambda \in \mathcal{O}(\tilde{S}_\lambda[r])$. In fact, by the continuity of Λ, there is $r' > r$ such that Λ is continuous on $\mathrm{Exp}_{\Delta+\lambda^2}(\tilde{\mathrm{E}};(r'))$. Therefore, $\mathcal{F}_\lambda^S \Lambda$ can be extended to a holomorphic function in a neighborhood of $\tilde{S}_\lambda[r]$. It is clear that

$$\mathcal{F}_\lambda^S \ : \ \mathrm{Exp}'_{\Delta+\lambda^2}(\tilde{\mathrm{E}};(r)) \to \mathcal{O}(\tilde{S}_\lambda[r])$$

is a continuous linear mapping.

Now we can state our main theorem.

Theorem 22. *Let $|\lambda| < r$. The following diagram is commutative (hence, we have explicit formulas for the inverse mappings of $\mathcal{F}_\lambda$ and of $\mathcal{F}_\lambda^S$).*

$$1)\quad \begin{array}{ccc} \mathcal{O}'(\tilde{S}_\lambda(r)) & \overset{\mathcal{C}^{-1}}{\longleftarrow} & \mathcal{O}(\tilde{S}_\lambda[r]) \\ \downarrow \mathcal{F}_\lambda & & \uparrow \mathcal{F}_\lambda^S \\ \mathrm{Exp}_{\Delta+\lambda^2}(\tilde{\mathrm{E}};[r]) & \overset{\mathcal{E}^{-1}}{\longrightarrow} & \mathrm{Exp}'_{\Delta+\lambda^2}(\tilde{\mathrm{E}};(r)), \end{array}$$

$$2)\quad \begin{array}{ccc} \mathcal{O}'(\tilde{S}_\lambda[r]) & \overset{\mathcal{C}^{-1}}{\longleftarrow} & \mathcal{O}(\tilde{S}_\lambda(r)) \\ \downarrow \mathcal{F}_\lambda & & \uparrow \mathcal{F}_\lambda^S \\ \mathrm{Exp}_{\Delta+\lambda^2}(\tilde{\mathrm{E}};(r)) & \overset{\mathcal{E}^{-1}}{\longrightarrow} & \mathrm{Exp}'_{\Delta+\lambda^2}(\tilde{\mathrm{E}};[r]). \end{array}$$

Proof. We prove only the first diagram. Since $\mathcal{C}$ and $\mathcal{E}$ are topological linear isomorphisms (Theorems 13 and 20), we have only to show that

$$1)\quad \mathcal{F}_\lambda \circ \mathcal{C}^{-1} \circ \mathcal{F}_\lambda^S \circ \mathcal{E}^{-1} = \mathrm{id}, \quad 2)\quad \mathcal{F}_\lambda^S \circ \mathcal{E}^{-1} \circ \mathcal{F}_\lambda \circ \mathcal{C}^{-1} = \mathrm{id}.$$

1) Let $F \in \mathrm{Exp}_{\Delta+\lambda^2}(\tilde{\mathrm{E}};[r])$. Then we have

$$f(\zeta) = \mathcal{F}_\lambda^S(\mathcal{E}^{-1}F)(\zeta) = \langle\langle F(w), \exp(-iw\cdot\zeta)\rangle\rangle_w^{\lambda,r} \in \mathcal{O}(\tilde{S}_\lambda[r]).$$

Therefore, by (22) and Theorem 19 we have

$$\begin{aligned} (\mathcal{F}_\lambda \circ \mathcal{C}^{-1} \circ \mathcal{F}_\lambda^S \circ \mathcal{E}^{-1})F(z) &= \int_{\tilde{S}_{\lambda,r}} f(\zeta)\exp(iz\cdot\overline{\zeta})d\tilde{S}_{\lambda,r}(\zeta) \\ &= \int_{\tilde{S}_{\lambda,r}} \langle\langle F(w), \exp(-iw\cdot\zeta)\rangle\rangle_w^{\lambda,r} \exp(iz\cdot\overline{\zeta})d\tilde{S}_{\lambda,r}(\zeta) \\ &= \langle\langle F(w), \int_{\tilde{S}_{\lambda,r}} \exp(-iw\cdot\zeta)\exp(iz\cdot\overline{\zeta})d\tilde{S}_{\lambda,r}(\zeta)\rangle\rangle_w^{\lambda,r} \\ &= \langle\langle F(w), E^{\lambda,r}(z,w)\rangle\rangle_w^{\lambda,r} = F(z), \quad z \in \tilde{\mathrm{E}}. \end{aligned}$$

2) Let $f \in \mathcal{O}(\tilde{S}_\lambda[r])$. Then we have

$$F(z) = \mathcal{F}_\lambda(\mathcal{C}^{-1}f)(z) = \int_{\tilde{S}_{\lambda,r}} f(\eta)\exp(iz\cdot\overline{\eta})d\tilde{S}_{\lambda,r}(\eta) \in \mathrm{Exp}_{\Delta+\lambda^2}(\tilde{\mathrm{E}};[r]).$$

By Lemma 17 and Theorem 12, we have

$$\begin{aligned}(\mathcal{F}_\lambda^S \circ \mathcal{E}^{-1} \circ \mathcal{F}_\lambda \circ \mathcal{C}^{-1})f(\xi) &= \langle\langle F(z), \exp(-iz\cdot\xi)\rangle\rangle_z^{\lambda,r} \\ &= \langle\langle \int_{\tilde{S}_{\lambda,r}} f(\eta)\exp(iz\cdot\overline{\eta})d\tilde{S}_{\lambda,r}(\eta), \exp(-iz\cdot\xi)\rangle\rangle_z^{\lambda,r} \\ &= \int_{\tilde{S}_{\lambda,r}} f(\eta)\langle\langle \exp(iz\cdot\overline{\eta}), \exp(-iz\cdot\xi)\rangle\rangle_z^{\lambda,r} d\tilde{S}_{\lambda,r}(\eta) \\ &= \int_{\tilde{S}_{\lambda,r}} f(\eta)K_{\lambda,r}(\overline{\eta},\xi)d\tilde{S}_{\lambda,r}(\eta) = f(\xi), \quad \xi \in \tilde{S}_\lambda(r).\square\end{aligned}$$

References

1. K.Fujita and M.Morimoto, *Integral representation for eigen functions of the Laplacian*, (in preparation).
2. K.Ii, *On the Bargmann-type transform and a Hilbert space of holomorphic functions*, Tôhoku Math. J., **38**(1986), 57–69.
3. M.Morimoto, *Analytic functionals on the Lie sphere*, Tokyo J. Math., **3**(1980), 1–35.
4. M.Morimoto, *Analytic functionals on the sphere and their Fourier-Borel transformations*, Complex Analysis, Banach Center Publications **11** PWN-Polish Scientific Publishers, Warsaw, 1983, pp. 223–250.
5. M.Morimoto, *Entire functions of exponential type on the complex sphere*, Trudy Matem. Inst. Steklova **203**(1994), 334–364. (Proc. Steklov Math. 1995, Issue **3**, 281–303.)
6. M.Morimoto and K.Fujita, *Analytic functionals and entire functionals on the complex light cone*, Hiroshima Math. J., **25**(1995), 493–512.
7. M.Morimoto and K.Fujita, *Conical Fourier-Borel transformation for harmonic functionals on the Lie ball*, to appear in *Generalizations of Complex Analysis and their Applications in Physics*, Banach Center Publication.
8. M.Morimoto and K.Fujita, *Analytic functionals and harmonic functionals*, to appear in *Complex Analysis, Harmonic Analysis and Applications*, Addison Wesley Longman, London, 1996.
9. M.Morimoto and K.Fujita, *Analytic functionals on the complex sphere and eigen functions of the Laplacian on the Lie ball*, Structure of Solution of Differential Equations (M.Morimoto and T.Kawi , eds.), World Scientific, 1996, pp. 287-305.
10. R.Wada, *On the Fourier-Borel transformations of analytic functionals on the complex sphere*, Tôhoku Math. J., **38**(1986), 417–432.
11. R.Wada, *Holomorphic functions on the complex sphere*, Tokyo J. Math., **11**(1988), 205–218.

12. R.Wada and M.Morimoto, *A uniqueness set for the differential operator* $\Delta_z + \lambda^2$, Tokyo J. Math., **10**(1987), 93–105.

Wavelet transforms and operators in various function spaces

Shinya Moritoh

Department of Mathematics, Nara Women's University, Kita-Uoya Nishi-machi, Nara 630, Japan (e-mail: `moritoh@cc.nara-wu.ac.jp`)

Introduction

We define a class of wavelet transforms as a continuous and micro-local version of the Littlewood-Paley decompositions. Hörmander's wave front sets (see [3]) as well as the Besov and Triebel-Lizorkin spaces (see [6] and [7]) may be characterized in terms of our wavelet transforms. By using the results obtained above (see [4]), we characterize the wave front sets in the sense of the Besov-Triebel-Lizorkin regularity in terms of our wavelet transforms. Finally, Päivärinta's results on the continuity of pseudodifferential operators in the Besov-Triebel-Lizorkin spaces (see [9]) may be microlocalized. In other words, we show the pseudo-microlocal properties in the sense of the Besov-Triebel-Lizorkin regularity. We remark that the components of our decompositions are not linearly independent but can be treated as if they were.

First, we define our wavelet transforms as follows:

Definition 1. Suppose that a function $\psi(x)$ (called a wavelet) has the following properties: $\psi(x) \in \mathcal{S}(\mathbb{R}^n)$, $\hat{\psi}(\xi) \in C_0^\infty(\mathbb{R}^n)$ and $\hat{\psi}(\xi) \geqq 0$. Let Ω=supp$\hat{\psi}(\xi)$ be in a neighbourhood of $(0,\cdots,0,1)$. When $n=1$, $\Omega \subset (0,\infty)$, while when $n \geqq 2$, Ω does not contain the origin 0 and $\psi(x) = \psi(rx)$ for any $r \in SO(n)$ satisfying $r(0,\cdots,0,1) = (0,\cdots,0,1)$. Let r_ξ be any rotation which sends $\xi/|\xi|$ to $(0,\cdots,0,1)$. Then our wavelet transform is defined as follows: for $f(t) \in \mathcal{S}'(\mathbb{R}^n)$, $(x,\xi) \in \mathbb{R}^{2n}$,

$$W_\psi f(x,\xi) = \begin{cases} \int_{\mathbb{R}} f(t)|\xi|^{1/2}\overline{\psi(\xi(t-x))}dt, & \text{if } n=1, \\ \int_{\mathbb{R}^n} f(t)|\xi|^{n/2}\overline{\psi(|\xi|r_\xi(t-x))}dt, & \text{if } n \geqq 2. \end{cases}$$

Here $\mathcal{S}(\mathbb{R}^n)$ stands for the Schwartz class and $C_0^\infty(\mathbb{R}^n)$ consists of functions which are smooth and compactly supported.

Remark 1. $W_\psi f(x,\xi)$ is rewritten as follows:

$$\int_{\mathbb{R}^n} \hat{f}(\tau)\cdot|\xi|^{-n/2}\hat{\psi}(|\xi|^{-1}r_\xi\tau)\cdot e^{i\tau x}d\tau.$$

From this, the meaning of our wavelet transforms is clear.
Remark 2. Our wavelet transforms in $\mathbb{R}^n$ are the reduced versions of those defined by Murenzi[8].
Remark 3. The domain of a wavelet transformation is usually the L_2-space(see [1]), but can be extended to $\mathcal{S}'(\mathbb{R}^n)$, that is, the dual space of $\mathcal{S}(\mathbb{R}^n)$.

Now, we define our wave front set $WF_\psi(f)(\subset \mathbf{R}^n_x\times\mathbf{R}^n_\xi)$ of $f\in\mathcal{S}'(\mathbb{R}^n)$ as follows:

Definition 2. We say $(x_0,\xi^0)\notin WF_\psi(f)$ if there exists a neighbourhood $U(x_0)$ of x_0 and a conic neighbourhood $\Gamma(\xi^0)$ of ξ^0 such that $W_\psi f(x,\xi)=O(|\xi|^{-N})$ as $|\xi|$ tends to ∞ for any $N\in\mathbf{N}$ in $U(x_0)\times\Gamma(\xi^0)$. Here $\mathbf{N}$ stands for the set of all positive integers.

Moreover, we define the refinement $WF^{(s)}_\psi(f)$ as follows:

Definition 3.

$$(x_0,\xi^0)\notin WF^{(s)}_\psi(f)\Leftrightarrow \iint_{U(x_0)\times\Gamma(\xi^0)}|W_\psi f(x,\xi)|^2(1+|\xi|^2)^s dxd\xi<\infty.$$

It is easy to prove that if $f\in L_2(\mathbb{R}^n)$, then $WF_\psi(f)$ is contained in the closure of $\bigcup_{s\geqq 0} WF^{(s)}_\psi(f)$.

We need the following definition to state our theorems.

Definition 4. For Ω=suppψ, let cone$\Omega=\{t\xi;\xi\in\Omega,t>0\}$. Let W be a subset of $\mathbb{R}^n_x\times\mathbb{R}^n_\xi$ conical in the ξ variables and denote by proj$_x W$ the projection of W onto the x-space. We say $(x_0,\xi^0)\notin\overline{W}^\psi$ if $x_0\notin$ proj$_x W$ and $\xi^0\in\mathbb{R}^n$, or $x_0\in$ proj$_x W$ and r(coneΩ) does not intersect $\{\xi\in\mathbb{R}^n;(x_0,\xi)\in W\}$ for any $r\in SO(n)$ with r(coneΩ) containing ξ^0.

That is to say, the set $\overline{W}^\psi$ is the expanded set of W only in the frequency space.

Theorem 1. *Let $f\in L_2(\mathbb{R}^n)$, and $s\geqq 0$. When $n=1$, $WF^{(s)}_\psi(f)=WF^{(s)}(f)$. When $n\geqq 2$, $WF^{(s)}_\psi(f)\subseteq\overline{WF^{(s)}(f)}^\psi$ and $WF^{(s)}(f)\subseteq\overline{WF^{(s)}_\psi(f)}^\psi$. We have the same inclusion relations between $WF_\psi(f)$ and $WF(f)$.*

Next, we recall the definitions of those spaces by Peetre[6] and Triebel[7].

Definition 5. Let $\phi(x)$ be a rapidly decreasing function whose Fourier transform is compactly supported in $1/2\leqq|\xi|\leqq 2$. Moreover, suppose that any half line starting from the origin intersects supp $\hat{\phi}(\xi)$. Let $\phi_r(x)$ be $r^n\phi(rx)$. Then $\hat{\phi_r}(\xi)$ is equal to $\hat{\phi}(\xi/r)$.

Definition. A function f is said to belong to the Besov space $\dot{B}^s_{p,q}(\mathbb{R}^n)$ ($s > 0$, $1 \leqq p,q \leqq \infty$) if

$$\left\| \|r^s(\phi_r * f(x))\|_{L_p(dx)} \right\|_{L_q(dr/r)} < \infty.$$

Definition. A function f is said to belong to the Triebel-Lizorkinspace $\dot{F}^s_{p,q}(\mathbb{R}^n)$
($s > 0$, $1 \leqq p < \infty$, $1 \leqq q \leqq \infty$) if

$$\left\| \|r^s(\phi_r * f(x))\|_{L_q(dr/r)} \right\|_{L_p(dx)} < \infty.$$

Theorem 2. *A function f belongs to $\dot{B}^s_{p,q}(\mathbb{R}^n)$ ($s > 0$, $1 \leqq p,q \leqq \infty$) if and only if the following condition holds:*

$$\left\| \left\| |\xi|^{s+n/2} W_\psi f(x,\xi) \right\|_{L_p(dx)} \right\|_{L_q(d\xi/|\xi|^n)} < \infty.$$

Theorem 3. *A function f belongs to $\dot{F}^s_{p,q}(\mathbb{R}^n)$ ($s > 0$, $1 \leqq p < \infty$, $1 \leqq q \leqq \infty$) if and only if the following condition holds:*

$$\left\| \left\| |\xi|^{s+n/2} W_\psi f(x,\xi) \right\|_{L_q(d\xi/|\xi|^n)} \right\|_{L_p(dx)} < \infty.$$

Moreover, for $f \in \mathcal{S}'(\mathbb{R}^n)$, we define our wave front sets $WF_\psi(\dot{B}^s_{p,q})(f)(\subset \mathbb{R}^n_x \times \mathbb{R}^n_\xi)$ and $WF_\psi(\dot{F}^s_{p,q})(f)(\subset \mathbb{R}^n_x \times \mathbb{R}^n_\xi)$ of as follows (we abbreviate the domain of integration as below):

Definition 6. We say $(x_0, \xi^0) \notin WF_\psi(\dot{B}^s_{p,q})(f)$ ($s > 0$, $1 \leqq p,q \leqq \infty$) if there exists a neighbourhood $U(x_0)$ of x_0 and a conic neighbourhood $\Gamma(\xi^0)$ of ξ^0 such that

$$\left\| \left\| |\xi|^{s+n/2} W_\psi f(x,\xi) \right\|_{L_p(dx;U(x_0))} \right\|_{L_q(d\xi/|\xi|^n;\Gamma(\xi^0))} < \infty.$$

Definition 7. We say $(x_0, \xi^0) \notin WF_\psi(\dot{F}^s_{p,q})(f)$ ($s > 0$, $1 \leqq p < \infty$, $1 \leqq q \leqq \infty$) if there exists a neighbourhood $U(x_0)$ of x_0 and a conic neighbourhood $\Gamma(\xi^0)$ of ξ^0 such that

$$\left\| \left\| |\xi|^{s+n/2} W_\psi f(x,\xi) \right\|_{L_q(d\xi/|\xi|^n;\Gamma(\xi^0))} \right\|_{L_p(dx;U(x_0))} < \infty.$$

Our results are as follows:

Theorem 4. *Let $f \in \mathcal{S}'(\mathbb{R}^n)$ and $s > 0$, $1 \leqq p, q \leqq \infty$. When $n = 1$, $WF_\psi(\dot{B}^s_{p,q})(f) = WF(\dot{B}^s_{p,q})(f)$. When $n \geqq 2$,*

$$WF_\psi(\dot{B}^s_{p,q})(f) \subseteq \overline{WF(\dot{B}^s_{p,q})(f)}^{\psi} \text{ and } WF(\dot{B}^s_{p,q})(f) \subseteq \overline{WF_\psi(\dot{B}^s_{p,q})(f)}^{\psi}.$$

Theorem 5. *Let $f \in \mathcal{S}'(\mathbb{R}^n)$ and $s > 0$, $1 \leqq p < \infty$, $1 \leqq q \leqq \infty$. When $n = 1$, $WF_\psi(\dot{F}^s_{p,q})(f) = WF(\dot{F}^s_{p,q})(f)$. When $n \geqq 2$, $WF_\psi(\dot{F}^s_{p,q})(f) \subseteq \overline{WF(\dot{F}^s_{p,q})(f)}^{\psi}$ and $WF(\dot{F}^s_{p,q})(f) \subseteq \overline{WF_\psi(\dot{F}^s_{p,q})(f)}^{\psi}$.*

Finally, we microlocalize the continuity of pseudodifferential operators in the Besov-Triebel-Lizorkin spaces. We define the directional Triebel-Lizorkin spaces as follows:

Definition 8. A function f is said to belong to the directional Triebel-Lizorkin space $\dot{F}^s_{p,q}(\mathbb{R}^n)(\Xi_{\epsilon,\psi})$ $(0 < s < \infty, 1 \leqq p < \infty, 1 \leqq q \leqq \infty, 0 \leqq \epsilon \leqq 2, \Xi \in \mathbb{R}^n, |\Xi| = 1$ and ψ is a wavelet which has the property that the quantity d defined by $\max\{|\xi_1 - \xi_2|; \xi_1, \xi_2 \in \text{cone}\Omega \cap S^{n-1}\}$, where S^{n-1} denotes the unit sphere in $\mathbb{R}^n$, is sufficiently small compared with ϵ) if the following condition holds:

$$\left\| \left\| |\xi|^{s+n/2} W_\psi f(x,\xi) \right\|_{L_q(d\xi/|\xi|^n; |\xi/|\xi| - \Xi| \leqq \epsilon)} \right\|_{L_p(dx)} < \infty,$$

where we abbreviate the domain of integration with respect to the ξ variables as above.

Let this quantity denote $||f||_{\dot{F}^s_{p,q}(\mathbb{R}^n)(\Xi_{\epsilon,\psi})}$. Let us call the quantity d defined above the size of a wavelet ψ.

Theorem 6. *Let a pseudodifferential operator $a(x,\xi) \in S^m_{1,\delta}$ $(m > 0, 0 \leqq \delta < 1)$ satisfy*

$$|D^\alpha_\xi D^\beta_x a(x,\xi)| \leqq C_{\alpha,\beta}(1 + |\xi|)^{-|\alpha| + \delta|\beta|}, \quad |\xi/|\xi| - \Xi| \leqq \epsilon, \quad \text{for some positive } \epsilon.$$

If $\epsilon_1 < \epsilon_0 < \epsilon$, $s' < s$, $1 \leqq p < \infty, 1 \leqq q \leqq \infty$ and the quantity d, the size of a wavelet ψ, is sufficiently small compared with $\epsilon - \epsilon_0, \epsilon_0 - \epsilon_1, \epsilon_1$, then for any $f \in \dot{F}^{s'}_{p,q}(\mathbb{R}^n) \cap \dot{F}^s_{p,q}(\Xi_{\epsilon_0,\psi})$ there exists a constant C such that

$$||af||_{\dot{F}^s_{p,q}(\mathbb{R}^n)(\Xi_{\epsilon_1,\psi})} + ||af||_{\dot{F}^{s'-m}_{p,q}} \leqq C(||f||_{\dot{F}^s_{p,q}(\mathbb{R}^n)(\Xi_{\epsilon_0,\psi})} + ||f||_{\dot{F}^{s'}_{p,q}}),$$

where $af(x)$ denotes $1/(2\pi)^n \int a(x,\xi) \cdot \hat{f}(\xi) \cdot e^{ix\cdot\xi} d\xi$.

As a consequence, we get the following:

Theorem 7. *For a pseudodifferential operator* $a(x,\xi) \in S^0_{1,\delta}$ $(0 \leqq \delta < 1)$, *we have the pseudo-microlocal property in the sense of the Triebel-Lizorkin regularity:* $WF(\dot{F}^s_{p,q})(af) \subset WF(\dot{F}^s_{p,q})(f)$.

Remark 4. We can define the directional Besov spaces and state the theorems corresponding to the directional Besov spaces in the same way.

Remark 5. For details concerning the contents of this paper, see [5].

Acknowledgements. The author would like to express his sincere gratitude to Professors Hikosaburou Komatsu, Kiyoomi Kataoka, Masao Yamazaki and Kenji Asada as well as Doctors Susumu Tanabe and Susumu Yamazaki for many valuable suggestions and encouragement. He would like to express his sincere gratitude also to the referee of Tôhoku Mathematical Journal for valuable advice.

Proof of Theorems 5, 6 and 7

We omit the proof of Theorems 1, 2, 3 and 4. Concerning the proof of Theorems 1, 2 and 3, see [4]. The proof of Theorem 4 is quite similar to that of Theorem 5. We can prove Theorem 5 by using Theorem 3, the continuous version of the multiplier theorem (see [2] and [7]) and the techniques of the proof of Theorem 1.

Proof of Theorem 5. Step 1. We may assume $n \geqq 2$. Suppose that $(0,\xi^0)$ does not belong to the set $\overline{WF(\dot{F}^s_{p,q})(f)}^{\psi}$. We suppose $0 \in \mathrm{proj}_x WF(\dot{F}^s_{p,q})(f)$. Let $\Gamma(\xi^0)$ be the union of $r(\mathrm{cone}\Omega)$ for all rotations r such that $\xi^0 \in r(\mathrm{cone}\Omega)$. Then there exist a function $\chi(x) \in C_0^\infty(\mathbb{R}^n)$ such that $\chi = 1$ near $x = 0$ and a neighbourhood $\Gamma'(\xi^0) \subset \Gamma(\xi^0)$ such that $\left\| \|r^s(\phi_r * (\chi f)(x))\|_{L_q(dr/r)} \right\|_{L_p(dx)} < \infty$, where $\hat{\phi}(\xi) = 1$ in a neighbourhood of $\Gamma'(\xi^0) \cap \{\xi \in \mathbb{R}^n; 1/2 \leqq |\xi| \leqq 2\}$. What we want to say is that there exist a conic neighbourhood $\tilde{\Gamma}(\xi^0)$ of ξ^0 and a neighbourhood $U(0)$ of 0, satisfying

$$\left\| \left\| |\xi|^{s+n/2} W_\psi f(x,\xi) \right\|_{L_q(d\xi/|\xi|^n;\tilde{\Gamma}(\xi^0))} \right\|_{L_p(dx;U(0))} < \infty.$$

Using the inversion formula (see [1] and [4]), we divide $W_\psi f(x,\xi)$ into two parts:

$$\begin{aligned} W_\psi f(x,\xi) = & |\xi|^{n/2} \int (\chi f)(t) \cdot \overline{\psi(|\xi| r_\xi (t-x))} dt \\ & + |\xi|^{n/2} \int ((1-\chi) f)(t) \cdot \overline{\psi(|\xi| r_\xi (t-x))} dt. \end{aligned}$$

If we take a set $U(0) \Subset \{\chi(x) \equiv 1\}$, then, the second term is rapidly decresing in $|\xi|$ uniformly in $x \in U(0)$. Therefore, $(0,\xi_0) \notin WF_\psi(\dot{F}^s_{p,q})((1-\chi)f)$. On the other hand, if we take $\tilde{\Gamma}(\xi^0)$ sufficiently small, then we obtain

$$\begin{aligned} &\left\| \left\| |\xi|^{s+n/2} W_\psi(\chi f)(x,\xi) \right\|_{L_q(d\xi/|\xi|^n;\tilde{\Gamma}(\xi^0))} \right\|_{L_p(dx;U(0))} \\ &\leqq \left\| \left\| \phi_{|\xi|} * |\xi|^{s+n/2} W_\psi(\chi f)(x,\xi) \right\|_{L_q(d\xi/|\xi|^n;\tilde{\Gamma}(\xi^0))} \right\|_{L_p(dx)} \\ &\leqq \left\| \left\| \phi_{|\xi|} * |\xi|^{s}(\chi f)(x) \right\|_{L_q(d\xi/|\xi|^n)} \right\|_{L_p(dx)} \\ &= \left\| \left\| \phi_r * r^s(\chi f)(x) \right\|_{L_q(dr/r)} \right\|_{L_p(dx)} \end{aligned}$$

The first inequality is due to the facts that $\hat{\phi}(\xi) = 1$ in a neighbourhood of $\Gamma'(\xi^0) \cap \{\xi \in \mathbb{R}^n; 1/2 \leqq |\xi| \leqq 2\}$ and that we took $\tilde{\Gamma}(\xi^0)$ sufficiently small. The second inequality is obtained by using the continuous version of the multiplier theorem. Recall that our wavelet transform is of convolution type. For simplicity, let

$$\psi_{|\xi|,r_\xi} * f(x) = |\xi|^{n/2} W_\psi f(x,\xi).$$

Then, we have

$$\begin{aligned} &\left\| \left\| \phi_{|\xi|} * |\xi|^{s+n/2} W_\psi(\chi f)(x,\xi) \right\|_{L_q(d\xi/|\xi|^n;\tilde{\Gamma}(\xi^0))} \right\|_{L_p(dx)} \\ &= \left\| \left\| \phi_{|\xi|} * |\xi|^{s} \cdot \psi_{|\xi|,r_\xi} * (\chi f)(x) \right\|_{L_q(d\xi/|\xi|^n;\tilde{\Gamma}(\xi^0))} \right\|_{L_p(dx)} \end{aligned}$$

The support of the Fourier transform of $\phi_{|\xi|} * |\xi|^s \cdot (\chi f)(x)$ with respect to x is contained in the ball $\{y \in \mathbb{R}^n; |y| \leqq C|\xi|\}$ for some constant C. On the other hand, the Fourier transform of $\psi_{|\xi|,r_\xi}(x)$ with respect to x is $\hat{\psi}(|\xi|^{-1} r_\xi \cdot \tau)$. Therefore, the quantity $\sup_\xi \left\| \hat{\psi}(|\xi|^{-1} r_\xi C|\xi| \cdot \tau) \right\|_{H^\kappa}$ is bounded, where H^κ denotes the Sobolev space and κ is some positive constant. From this fact and the multiplier theorem, we have the second inequality. Therefore, we have $(0,\xi^0) \notin WF_\psi(\dot{F}^s_{p,q})(\chi f)$.

Step 2. Suppose that $(0,\xi^0)$ does not belong to the set $\overline{WF_\psi(\dot{F}^s_{p,q})(f)}^{\psi}$. If we take a conic neighbourhood $\Gamma'(\xi^0)$ of ξ^0 as in Step 1, then there exists a neighbourhood $U(0)$ of $x = 0$ such that

$$\left\| \left\| |\xi|^{s+n/2} W_\psi(f)(x,\xi) \right\|_{L_q(d\xi/|\xi|^n;\Gamma'(\xi^0))} \right\|_{L_p(dx;U(0))} < \infty.$$

Now using the inversion formula, we divide f into two parts

$$f = f_{\Gamma'} + f_{\Gamma'^c},$$

where

$$f_{\Gamma'}(t) = C_\psi^{-1} \iint_{\Gamma'(\xi^0)\times\mathbb{R}^n_x} W_\psi f(x,\xi)\cdot|\xi|^{n/2}\psi(|\xi|r_\xi(t-x))dxd\xi,$$

$$f_{\Gamma'^c}(t) = C_\psi^{-1} \iint_{\Gamma'(\xi^0)^c\times\mathbb{R}^n_x} W_\psi f(x,\xi)\cdot|\xi|^{n/2}\psi(|\xi|r_\xi(t-x))dxd\xi.$$

Then

$$\widehat{f_{\Gamma'^c}}(\tau) = C_\psi^{-1} \int_{\Gamma'(\xi^0)^c}\int_{\mathbb{R}^n_x} W_\psi f(x,\xi)\cdot|\xi|^{-n/2}\hat{\psi}(|\xi|^{-1}r_\xi\tau)e^{-i\tau\cdot x}dxd\xi.$$

If we take a sufficiently small conic neighbourhood $\tilde{\Gamma}(\xi^0)$ of ξ^0, then we obtain

$$\hat{\psi}(|\xi|^{-1}r_\xi\tau) \equiv 0 \text{ for any } \tau\in\tilde{\Gamma}(\xi^0) \text{ and for any } \xi\in\Gamma'(\xi^0)^c.$$

Therefore, it follows that $(0,\xi^0)\notin WF(\dot{F}^s_{p,q})(f_{\Gamma'^c})$.

Next, we choose a function $\chi(x)\in C_0^\infty(\mathbb{R}^n)$ such that $\mathrm{supp}\chi(x)$ is compactly supported in $U(0)$ and that $\chi(x)\equiv 1$ in some neighbourhood $U_1(0)$ of 0. Then, we further divide $f_{\Gamma'}(t)$ into two parts $f_{\Gamma'} = f_{\Gamma',\chi} + f_{\Gamma',1-\chi}$, where

$$f_{\Gamma',\chi}(t) = C_\psi^{-1} \iint_{\Gamma'(\xi^0)\times\mathbb{R}^n_x} \chi(x)\cdot W_\psi f(x,\xi)\cdot|\xi|^{n/2}\psi(|\xi|r_\xi(t-x))dxd\xi,$$

$$f_{\Gamma',1-\chi}(t) = C_\psi^{-1} \iint_{\Gamma'(\xi^0)\times\mathbb{R}^n_x} (1-\chi(x))\cdot W_\psi f(x,\xi)\cdot|\xi|^{n/2}\psi(|\xi|r_\xi(t-x))dxd\xi.$$

Let $U_2(0)\Subset\{\chi(x)\equiv 1\}$. Then we can easily see that $f_{\Gamma',1-\chi}(t)\in C^\infty(U_2(0))$ by the exchange of order of differentiation and integration. Therefore, it follows that $(0,\xi^0)\notin WF(\dot{F}^s_{p,q})(f_{\Gamma',1-\chi})$. Lastly, we want to show that $(0,\xi^0)\notin WF(\dot{F}^s_{p,q})(f_{\Gamma',\chi})$. This is the heart of the matter in proving Theorem 5. In fact, we can show more strongly that $f_{\Gamma',\chi}$ belongs globally to the Triebel-Lizorkin space $\dot{F}^s_{p,q}(\mathbb{R}^n)$. If we put $g(x,\xi) = \chi(x)\cdot W_\psi f(x,\xi)\cdot \mathbf{1}_{\Gamma'(\xi^0)}(\xi)$, where $\mathbf{1}_A$ denotes a characteristic function on a set A, then it follows from the hypothesis and from the fact that $\mathrm{supp}\chi(x)$ is contained in $U(0)$ that

$$\left\| \left\| |\xi|^{s+n/2} g(x,\xi) \right\|_{L_q(d\xi/|\xi|^n)} \right\|_{L_p(dx)} < \infty.$$

Our problem now reduces to showing the following:

$$\iint g(x,\xi)\cdot|\xi|^{n/2}\psi(|\xi|r_\xi(t-x))dxd\xi \in \dot{F}^s_{p,q}(\mathbb{R}^n_t).$$

Let us define the inversion wavelet transform W_ψ^{-1} as follows: for $g(x,\xi) \in \mathcal{S}'(\mathbb{R}^{2n})$,

$$W_\psi^{-1}g(t) = C_\psi^{-1}\iint g(x,\xi)\cdot|\xi|^{n/2}\psi(|\xi|r_\xi(t-x))dxd\xi.$$

Then, we have the following:

$$W_\psi W_\psi^{-1}g(x,\xi) = \mathrm{proj}_H g(x,\xi),$$

where H denotes the space $W_\psi(\mathcal{S}'(\mathbb{R}^n))$ that is the image of $\mathcal{S}'(\mathbb{R}^n)$ under our wavelet transform, and proj_H the projection onto the space H. This fact and the characterization of the Triebel-Lizorkin spaces (See Theorem 3) yield that

$$\begin{aligned}\left\|W_\psi^{-1}g\right\|_{\dot{F}^s_{p,q}} &\leqq \left\|\left\||\xi|^{s+n/2}W_\psi W_\psi^{-1}g(x,\xi)\right\|_{L_q(d\xi/|\xi|^n)}\right\|_{L_p(dx)}\\ &= \left\|\left\||\xi|^{s+n/2}\mathrm{proj}_H g(x,\xi)\right\|_{L_q(d\xi/|\xi|^n)}\right\|_{L_p(dx)}\\ &\leqq \left\|\left\||\xi|^{s+n/2}g(x,\xi)\right\|_{L_q(d\xi/|\xi|^n)}\right\|_{L_p(dx)} < \infty.\end{aligned}$$

q.e.d

Proof of Theorems 6 and 7. Let ϕ_0 be a smooth function which is supported in $\{\xi \in \mathbb{R}^n; |\xi/|\xi| - \Xi| \leqq \epsilon\}$ and is equal to 1 in a conic neighbourhood of $\{\xi \in \mathbb{R}^n; |\xi/|\xi| - \Xi| \leqq \epsilon_0\} \cup K_0$, where K_0 denotes the conic set $\{\xi \in \mathbb{R}^n; \xi = r_\eta^{-1}\mathrm{cone}\Omega, |\eta/|\eta| - \Xi| = \epsilon_0\}$. Recall that r_η denotes any rotation which sends $\eta/|\eta|$ to $(0,\cdots,0,1)$ and that $\Omega=\mathrm{supp}\hat{\psi}$. Let ϕ_1 be a smooth function which is supported in $\{\xi \in \mathbb{R}^n; |\xi/|\xi| - \Xi| \leqq \epsilon_0\} \setminus K_0$ and is equal to 1 in a conic neighbourhood of $\{\xi \in \mathbb{R}^n; |\xi/|\xi| - \Xi| \leqq \epsilon_1\} \cup K_1$, where K_1 denotes the conic set $\{\xi \in \mathbb{R}^n; \xi = r_\eta^{-1}\mathrm{cone}\Omega, |\eta/|\eta| - \Xi| = \epsilon_1\}$. Let ϕf denote the inverse Fourier transform of $\phi\cdot\hat{f}$, the multiplication of ϕ and $\hat{f}$. Then there exist constants C_1, C_2 such that

$$\begin{aligned}(1)\qquad &\|af\|_{\dot{F}^s_{p,q}(\Xi_{\epsilon_1})}\\ &\leqq C_1\|\phi_1(a(\phi_0+1-\phi_0)f)\|_{\dot{F}^s_{p,q}(\mathbb{R}^n)}\\ &\leqq C_2(\|f\|_{\dot{F}^s_{p,q}(\Xi_{\epsilon_0})} + \|f\|_{\dot{F}^{s'}_{p,q}}).\end{aligned}$$

Before proving the inequalities (1), let us recall two inequalities obtained from Theorem 2.

$$(2)\qquad \left\| \left\| |\xi|^{s+n/2} W_\psi f(x,\xi) \right\|_{L_q(d\xi/|\xi|^n;\tilde{\Gamma}(\xi^0))} \right\|_{L_p(dx)} \leqq \left\| \left\| r^s \cdot \phi_r * f(x) \right\|_{L_q(dr/r)} \right\|_{L_p(dx)},$$

where $\tilde{\Gamma}(\xi^0)$ is some conic neighbourhood of ξ^0 and ϕ is a rapidly decreasing function whose Fourier transform is equal to 1 in a neighbourhood of $\Gamma'(\xi^0) \cap \{\xi \in \mathbb{R}^n; 1/2 \leqq |\xi| \leqq 2\}$. Let us recall that $\Gamma'(\xi^0)$ denotes some conic neighbourhood of ξ^0 which is contained in and slightly smaller than $\Gamma(\xi^0)$ and that $\Gamma(\xi^0)$ is the union of $r\mathrm{cone}\Omega$ for all rotations such that $\xi^0 \in r\mathrm{cone}\Omega$.

$$(3)\qquad \left\| \left\| r^s \cdot \phi_r * f(x) \right\|_{L_q(dr/r)} \right\|_{L_p(dx)} \leqq \left\| \left\| |\xi|^{s+n/2} W_\psi f(x,\xi) \right\|_{L_q(d\xi/|\xi|^n;\Gamma'(\xi^0))} \right\|_{L_p(dx)},$$

where ϕ is a rapidly decreasing function whose Fourier transform is equal to 1 in a neighbourhood of $\tilde{\Gamma}(\xi^0) \cap \{\xi \in \mathbb{R}^n; 1/2 \leqq |\xi| \leqq 2\}$ for some conic neighbourhood $\tilde{\Gamma}(\xi^0)$ of ξ^0. $\Gamma'(\xi^0)$ has the same meaning as in (2).

Hereafter we frequently use Päivärinta's results (see [9]) to prove the inequalities (1). The first inequality of (1) is obtained by the following argument:

We have to show that there exists a constant C_1 such that

$$\|af\|_{\dot{F}^s_{p,q}(\Xi_{\epsilon_1})} \leqq C_1 \, \|\phi_1 af\|_{\dot{F}^s_{p,q}(\mathbb{R}^n)} \, .$$

However, because of the suitable choice of the cut-off function ϕ_1, this inequality is obtained in the same way as the inequality (2) is obtained. The second inequality is obtained by the following argument:

First we show that there exists a constant C_3 such that

$$||\phi_1 a(1-\phi_0)||_{\dot{F}^s_{p,q}(\mathbb{R}^n)} \leqq C_3 ||f||_{F^{s'}_{p,q}} .$$

Since $\mathrm{supp}\,\phi_1$ is separated from $\mathrm{supp}(1-\phi_0)$, by [2, Theorem 2.7], the operator $\phi_1 a(1-\phi_0)$ is a smoothing operator. Therefore we get the above inequality. Secondly we show that there exists a constant C_4 such that $||\phi_1 a\phi_0 f||_{\dot{F}^s_{p,q}(\mathbb{R}^n)} \leqq C_4 \, ||f||_{\dot{F}^s_{p,q}(\Xi_{\epsilon_0})}$. Because of the suitable choice of the cut-off functions ϕ_0 and ϕ_1, we have $\phi_1 a \phi_0 f = a\phi_1 f$. Since $a(x,D)$ is an operator of order zero in the cone $\{\xi \in \mathbb{R}^n; |\xi/|\xi| - \Xi| \leqq \epsilon\}$ in which $\operatorname{supp}\phi_1$ is contained, there exists a constant C_5 such that $\|a\phi_1 f\|_{\dot{F}^s_{p,q}(\mathbb{R}^n)} \leqq C_5 \|\phi_1 f\|_{\dot{F}^s_{p,q}(\mathbb{R}^n)}$. Because of the suitable choice of the cut-off function ϕ_1 and the inequality (3), there exists a constant C_6 such that $\|\phi_1 f\|_{\dot{F}^s_{p,q}(\mathbb{R}^n)} \leqq C_6 \|f\|_{\dot{F}^s_{p,q}(\Xi_{\epsilon_0})}$. Consequently, we obtain the second inequality. Of course, the inequality $\|af\|_{\dot{F}^{s'-m}_{p,q}} \leqq C_2 \|f\|_{\dot{F}^{s'}_{p,q}}$ is nothing but Päivärinta's result. Combining these inequalities, we have Theorem 3. As a consequence of letting ϵ tend to 0 and of localizing the x-space in the same way, we have Theorem 4.

q.e.d.

Remark 6. For wavelets ψ_1 and ψ_2, we have the following relation because of the inequalities (2) and (3). Let the sizes of wavelets ψ_1 and ψ_2 be d_1 and d_2. Then we have $\dot{F}^s_{p,q}(\mathbb{R}^n)(\Xi_{\epsilon_1,\psi_1}) \subseteq \dot{F}^s_{p,q}(\mathbb{R}^n)(\Xi_{\epsilon_2,\psi_2})$, where $\epsilon_1 \geqq \epsilon_2 + d_1 + d_2$. Let us call a sequence of wavelets $\{\psi_j\}_{j=1}^{\infty}$ satisfying the condition $\epsilon_1 = \epsilon_2 + \sum_{j=1}^{\infty} d_j$, where d_j is the size of a wavelet ψ_j, an admissible family of wavelets. Then we have $\cup_j \dot{F}^s_{p,q}(\mathbb{R}^n)(\Xi_{\epsilon_1,\psi_j}) \subseteq \cap_j \dot{F}^s_{p,q}(\mathbb{R}^n)(\Xi_{\epsilon_2,\psi_j})$. Let Ψ be a set of all admissible families of wavelets. Then we can present two problems whether we have the relation $\cup_{\{\psi_j\}\in\Psi} \cup_j \dot{F}^s_{p,q}(\mathbb{R}^n)(\Xi_{\epsilon_1,\psi_j}) \subseteq \cap_{\{\psi_j\}\in\Psi} \cap_j \dot{F}^s_{p,q}(\mathbb{R}^n)(\Xi_{\epsilon_2,\psi_j})$ or not and what these spaces (independent of wavelets) are.

References

1. I.Daubechies, *Ten Lectures on Wavelets*, SIAM CBMS-NSF regional conference series in applied mathematics-61, Philadelphia, 1992.
2. C.Fefferman and E.M.Stein, *Some maximal inequalities*, Amer. J. Math. **93** (1971), 107–115.
3. L.Hörmander, *Fourier integral operators.I*, Acta.Math. **127** (1971), 79–183.
4. S. Moritoh, *Wavelet transforms in Euclidean spaces — their relation with wave front sets and Besov, Triebel-Lizorkin spaces —*, Tôhoku Mathematical Journal **47** (1995), 555–565.
5. S. Moritoh, *Wave front sets in the sense of Besov-Triebel-Lizorkin regularity and pseudo-microlocal property*, (preprint).
6. J.Peetre, *New Thoughts on Besov Spaces*, Duke Univ., Durham, 1976.
7. H.Triebel, *Theory of Function Spaces*, Birkhäuser, Basel, 1983.
8. R.Murenzi, *Wavelet transforms associated to the n-dimensional Euclidean group with dilations: signals in more than one dimension*, Wavelets (J.M. Combes, A.Grossmann and Ph. Tchamitchian, eds.), Springer-Verlag, Berlin, Heidelberg, New York, 1989.
9. L. Päivärinta, *Pseudodifferential operators in Hardy-Triebel spaces*, Zeitschrift für Analysis und ihre Anwendungen **2(3)** (1983), 235-242.

Characteristic Cauchy problems in the complex domain

Yasunori Okada [1] *and Hideshi Yamane* [2]

[1]Department of Mathematics and Informatics, Faculty of Science,
Chiba University,Yayoi-cho, Inage-ku, Chiba 263, Japan
(e-mail: okada@math.s.chiba-u.ac.jp)
[2]Mathematics, Chiba Institute of Technology, Shibazono, Narashino, Chiba 275, Japan (e-mail: yamane@cc.it-chiba.ac.jp)

0 Introduction

Gårding-Kotake-Leray showed that in a certain characteristic Cauchy problem

$$Pu = v \in \mathcal{O} \quad \text{(the sheaf of holomorphic functions)}$$

with zero Cauchy data on a hypersurface S, u can be ramified. Moreover, u is of the form

$$(*) \qquad u(x) = \sum_{l=0}^{q-1} u_l(x)[k(x)]^{l/q}$$

where q is a positive integer ≥ 2 and u is ramified around $K : k(x) = 0$. Here K is tangent to S at characteristic points of S. Let us denote by $\mathcal{N}^m_{q,K}$ the class of functions which have the form $(*)$ and whose first m traces on S vanish.

Dunau generalized the above result. He proved that

$$P\mathcal{N}^m_{q,K} \supset \mathcal{N}^0_{q,K} \underset{\neq}{\supset} \mathcal{O}$$

under a weaker hypothesis on P with ord $P = m$.

In this paper, we generalize Dunau's result: we introduce a suitable class of functions for the study of this kind of (unique) solvability results. In fact, P induces an isomorphism between $\mathcal{N}^m_{q,K}$ and this new class. We make full use of the method introduced by Dunau. Microdifferential operators also play a very important role.

On the other hand, Hamada treated a different kind of characteristic Cauchy problem, in which the solution can have an essential singularity. We try an approach to this phenomenon from the viewpoint of microdifferential calculus.

This paper is an extended version of [O-Y].

1 Statement of the results –the first kind of characteristic Cauchy problem

Let X be an open neighborhood of $0 \in \mathbb{C}^n$ and $\mathcal{O}$ be the sheaf of holomorphic functions on X. The coordinate system of $\mathbb{C}^n$ is denoted by $x = (x_1, x_2, \ldots, x_n) = (x_1, x_2, x')$. Let $K \ni 0$ be a nonsingular hypersurface of X defined by $k(x) = 0$. Here $k(x)$ is a holomorphic function with $dk \neq 0$ on K. For a positive integer $q \geq 2$, we define the $\mathcal{O}_0$-module $\mathcal{N}_{q,K}$ of germs of ramified functions of order q around K: if $u \in \mathcal{N}_{q,K}$, u is of the form

$$u(x) = \sum_{l=0}^{q-1} u_l(x)[k(x)]^{l/q}$$

where $u_l \in \mathcal{O}_0$, and $\mathcal{O}_0$ is the stalk of $\mathcal{O}$ at 0.

Consider another nonsingular hypersurface S of X and we consider the following condition on the relative position of K and S.

$$(H_0) \qquad \begin{cases} S \text{ and } K \text{ intersect each other along a submanifold } T \\ \text{of codimension 2 in } X, \text{ with a contact along } T \text{ of order} \\ q-1 \text{ (thus, } S \text{ and } K \text{ don't intersect each other outside } T). \end{cases}$$

Set

$$\mathcal{N}_{q,K}^l = \{f \in \mathcal{N}_{q,K}; f \text{ vanishes up to order } l \text{ on } S\}, \quad l \in \mathbb{N}.$$

We consider a differential operator P of order m with holomorphic coefficients. We denote by $p_m(x;\xi)$ its principal symbol and $p_{m-k}(x;\xi)$, $k = 1, \ldots, m$ its other symbols.

We consider the following two conditions on P :

(H_1)

$$\begin{cases} \text{There exists a positive integer } r,\ 1 \leq r \leq m-1 \text{ such that} \\ \text{for } k = 0, 1, \ldots, r-1,\ p_{m-k}(x;\xi) \text{ vanishes up to order} \\ r-k \text{ on } \overset{\circ}{T}{}^*_K X = T^*_K X \backslash K, \text{ where } T^*_K X \text{ is the conormal} \\ \text{bundle and } K \text{ is identified with its zero section (Levi} \\ \text{condition).} \end{cases}$$

(H_2)

$$\begin{cases} \text{The restriction of } p_m(x;\xi) \text{ to } \overset{\circ}{T}{}^*_T X \text{ does not vanish up to} \\ \text{order } r+1 \text{ at any point of } \overset{\circ}{T}{}^*_T X \cap \overset{\circ}{T}{}^*_K X. \end{cases}$$

Under these conditions we are going to prove:

Theorem A. *The operator P induces an isomorphism*

$$P : \mathcal{N}_{q,K}^m \xrightarrow{\sim} Q(x,D)\mathcal{N}_{q,K}^{m-r}.$$

Here Q is an arbitrary differential operator of order $m-r$ for which K is non-characteristic.

Theorem B. *We have*

(B-1) $$[k(x)]^{-(q-1)/q}\mathcal{N}_{q,K} \subset Q(x,D)\mathcal{N}^{m-r}_{q,K},$$

where equality holds if $r = m-1$. *Moreover, with* $Q(x,D)$ *as in Theorem A,*

(B-2) $$[k(x)]^{-l/q} \not\subset Q(x,D)\mathcal{N}^{m-r}_{q,K},\ l \geq q.$$

Remark. The introduction of a class of ramified functions is inevitable. In the following example, *a ramified solution appears even for a regular right hand side.*

$$\begin{cases} D_1D_2u(x) = 1 \\ u|_S = 0, \quad D_1u|_S = 0, \quad S = \{x_1 - x_2^2 = 0\}. \end{cases}$$

In fact, we have

$$u(x) = x_2(x_1 - x_2^2) - \frac{2}{3}(x_1^{\frac{3}{2}} - x_2^3).$$

2 A system of coordinates

Dunau has proved that under the condition (H_0) , *we may assume, by a suitable change of coordinates preserving the origin,*

1. $K = \{x_1 = 0\}$
2. $T = \{x_1 = x_2 = 0\}$
3. $S = \{x_1 - x_2^q = 0\}$.

Moreover he has shown

Proposition 1. (1) *The condition* (H_1) *implies that*

$$P = P_1(x, x_1D_1, D_2, D')D_1^{m-r} + P_2(x,D),\ D' = (D_3, \ldots, D_n)$$

where $P_1(x, D_t, D_2, D')$ *is a differential operator of order* r *and* $P_2(x,D)$ *is of order at most* m *and of degree less than* $m-r$ *in* D_1.
(2) *The condition* (H_2) *implies that* $\sigma(P_1)(x;\xi) \neq 0$ *at* $x = 0, \xi = (0,1,0,\ldots,0)$, *where* $\sigma(P_1)$ *is the principal symbol of* P_1.

We will need a slightly different version of the above proposition.

Proposition 1'. *The operator* P *can be written in the form*

$$P = D_1^{m-r}P_1'(x, x_1D_1, D_2, D') + P_2'(x,D)$$

where P_1' *(resp.* P_2'*) satisfies the same condition as* P_1 *(resp.* P_2*).*

Proof. Put $\theta = x_1D_1$. Then we have $\theta D_1^j = D_1^j(\theta - j)$, as is easily seen by induction. Moreover we have $\theta^k D_1^j = D_1^j(\theta - j)^k$, which is proved by induction on k. □

3 Dunau's symbol class $\mathcal{E}_K(W_{\varepsilon_1\varepsilon_2})$

Let ε_1 and ε_2 be two positive numbers with $0 < \varepsilon_1 < \varepsilon_2$. Consider the conic open set of T^*X defined by

$$W_{\varepsilon_1\varepsilon_2} = \{(x;\xi) \in T^*X;\ \varepsilon_1 < |\frac{\xi_2}{\xi_1}| < \varepsilon_2 \text{ and} |\frac{\xi'}{\xi_2}| < \varepsilon_2\},$$

where $\xi' = (\xi_3, \ldots, \xi_n)$ and $|\xi'| = \sup_{i\geq 3}|\xi_i|$.

Such an open set is called a crown in T^*X.

Proposition 2 **([D]).** *Under the conditions* (H_1) *and* (H_2) , *if* X *is a sufficiently small open neighborhood of* $x = 0$, *there exist* ε_1 *and* ε_2 *with* $0 < \varepsilon_1 < \varepsilon_2$ *such that* P *is elliptic in the crown* $W_{\varepsilon_1\varepsilon_2}$.

Let $R(x, D)$ be a microdifferential operator of order m in $W_{\varepsilon_1\varepsilon_2}$. R is a formal sum satisfying a certain growth condition

$$R = \sum_{k=0}^{\infty} r_{m-k}(x;\xi)$$

where r_{m-k} is homogeneous of degree $m-k$ with respect to ξ ([K-K-K]). We have the expansion

$$r_{m-k}(x;\xi) = \sum_{l=-\infty}^{\infty} \sum_{\alpha\in\mathbf{N}^{n-2}} r_{m-k,l,\alpha}(x)\xi_1^{m-k-l-|\alpha|}\xi_2^l\xi'^\alpha.$$

We say that $R \in \mathcal{E}_K(W_{\varepsilon_1\varepsilon_2})$ if $r_{m-k,l,\alpha}(x)$ vanishes up to order $m-k-l-|\alpha|$ on $x_1 = 0$ if $m-k-l-|\alpha| > 0$. Then $\mathcal{E}_K(W_{\varepsilon_1\varepsilon_2})$ becomes an algebra.

Proposition 3. *If* j *is a positive integer, then* $D_1^{-j}\mathcal{E}_K(W_{\varepsilon_1\varepsilon_2}) \subset \mathcal{E}_K(W_{\varepsilon_1\varepsilon_2})D_1^{-j}$.

Proof. First remark that $D_1^{-j}\mathcal{E}_K(W_{\varepsilon_1\varepsilon_2}), \mathcal{E}_K(W_{\varepsilon_1\varepsilon_2})D_1^{-j} \neq \mathcal{E}_K(W_{\varepsilon_1\varepsilon_2})$, since D_1^{-1} is not invertible in $\mathcal{E}_K(W_{\varepsilon_1\varepsilon_2})$.

Now we prove the proposition. For a microdifferential operator R, we have

$$D_1 R = RD_1 + \frac{\partial R}{\partial x_1}, \quad RD_1^{-1} = D_1^{-1}R + D_1^{-1}\frac{\partial R}{\partial x_1}D_1^{-1}.$$

If $R \in \mathcal{E}_K(W_{\varepsilon_1\varepsilon_2})$, that is to say, if positive powers of D_1 appear only in the form $x_1^k D_1^k$ in R, then

$$D_1^{-1}\frac{\partial R}{\partial x_1}D_1^{-1} \in \mathcal{E}_K(W_{\varepsilon_1\varepsilon_2})D_1^{-1}.$$

Thus the case $j = 1$ is proved. The rest of the proof is completed by induction. □

Proposition 4. *Assume that $K : x_1 = 0$ is non-characteristic near $x = 0$ for a differential operator $Q(x, D)$ of order l. Then if X is a sufficiently small neighborhood of x_0, there exist positive constants ε_1 and ε_2 with $0 < \varepsilon_1 < \varepsilon_2$ such that Q is elliptic in $W_{\varepsilon_1\varepsilon_2}$. Moreover, if we define $\tilde{Q} = D_1^{-l}Q$, then $\tilde{Q}, \tilde{Q}^{-1} \in \mathcal{E}_K(W_{\varepsilon_1\varepsilon_2})$.*

Proof. We write Q in the form

$$Q(x, D) = D_1^l Q_0(x) + \sum_{j=1}^{l} D_1^{l-j} Q_j(x, D_2, D'),$$

where Q_0 is a non-vanishing function and ord $Q_j \leq j$, $[Q_j, x_1] = 0$. We have

$$\tilde{Q}(x, D) = \{1 + \sum_{j=1}^{l} D_1^{-j} Q'_j(x, D_2, D')\} Q_0(x)$$

where $Q'_j = Q_j Q_0^{-1}$. $Q'_j(x; \xi_2, \xi')$ has the form

$$Q'_j(x; \xi_2, \xi') = \sum_{l+|\alpha| \leq j} \xi_2^l \xi'^\alpha q_{j,l,\alpha}(x).$$

By choosing ε_1 and ε_2 sufficiently small, $\xi_1^{-j}\xi_2^l\xi'^\alpha$ is made arbitrarily small. Hence the ellipticity of $\tilde{Q}$ and Q follows. $\tilde{Q} \in \mathcal{E}_K(W_{\varepsilon_1\varepsilon_2})$ is obvious. Moreover, since

$$\begin{aligned}\tilde{Q}^{-1} &= Q_0(x)^{-1}[1 + \sum_{j=1}^{l} D_1^{-j} Q'_j]^{-1} \\ &= Q_0(x)^{-1}[1 + \sum_{k=1}^{\infty}(-\sum_{j=1}^{l} D_1^{-j} Q'_j)^k].\end{aligned}$$

$\tilde{Q}^{-1}$ has no positive power of D_1. This implies that $\tilde{Q}^{-1} \in \mathcal{E}_K(W_{\varepsilon_1\varepsilon_2})$. □

Proposition 5. *In the situation of Proposition 1, we have*

$$P^{-1} \in D_1^{-m+r}\mathcal{E}_K(W_{\varepsilon_1\varepsilon_2}) \subset \mathcal{E}_K(W_{\varepsilon_1\varepsilon_2})D_1^{-m+r}.$$

Proof. See the proof of Theorem 2 in [D]. □

4 Action of $\mathcal{E}_K(W_{\varepsilon_1\varepsilon_2})$

Consider the following Cauchy problem:

$$\begin{cases} D_1 g &= f \\ g|_S &= 0. \end{cases}$$

Proposition 6. *For all* $f \in \mathcal{N}^m_{q,K}$ *(resp.* $x_1^{-(q-1)/q}\mathcal{N}_{q,K}$*), there exists a unique solution* $g \in \mathcal{N}^{m+1}_{q,K}$ *(resp.* $\mathcal{N}^1_{q,K}$*).*

Proof. Let Φ be the map

$$\mathbb{C}^n_{z,x_2,x'} \to \mathbb{C}^n_x$$
$$(z, x_2, x') \mapsto (z^q, x_2, x').$$

Then we have the following isomorphisms defined by $f \mapsto f \circ \Phi$

$$\mathcal{N}^m_{q,K} \tilde{\to} \mathcal{O}^m_0$$
$$x_1^{-(q-1)/q}\mathcal{N}_{q,K} \tilde{\to} z^{-(q-1)}\mathcal{O}_0,$$

where

$$\mathcal{O}^m_0 = \{F(z, x_2, x') \in \mathcal{O}_0; F \text{ vanishes on } \tilde{S} : z = x_2 \text{ up to order } m\}.$$

Put $\tilde{f} = f \circ \Phi$ and $\tilde{g} = g \circ \Phi$. Then the Cauchy problem is

$$\begin{cases} D_z\tilde{g} &= qz^{q-1}\tilde{f} \\ \tilde{g}|_{\tilde{S}} &= 0 \end{cases}$$

where the initial hypersurface $\tilde{S}$ is $\tilde{S} = \{z = x_2\}$. □

We set $g = D_1^{-1}f$. Similarly, we can define D_2^{-1}. That is, we have linear mappings:

$$D_1^{-1}, D_2^{-1} : \mathcal{N}^m_{q,K} \to \mathcal{N}^{m+1}_{q,K}$$
$$D_1^{-1} : x_1^{-(q-1)/q}\mathcal{N}_{q,K} \to \mathcal{N}^1_{q,K}.$$

D_1^{-1} (resp.D_2^{-1}) is a right inverse of D_1 (resp. D_2).

Proposition 7 ([D]). *If R is an operator of $\mathcal{E}_K(W_{\varepsilon_1\varepsilon_2})$ of order $-l \leq 0$, then R defines a linear mapping $\mathcal{N}^m_{q,K} \to \mathcal{N}^{m+l}_{q,K}$.*

Proof. R can be written in the form

$$R = \sum_{b=0}^{\infty} \sum_{c=-\infty}^{\infty} \sum_{\alpha \in \mathbb{N}^{n-2}} r_{-l-b,c,\alpha}(x) D'^{\alpha} D_1^{-l-b-c-|\alpha|} D_2^c.$$

For $f \in \mathcal{N}^m_{q,K}$, we define Rf by

$$Rf = \sum_{b=0}^{\infty} \sum_{c=-\infty}^{\infty} \sum_{\alpha \in \mathbb{N}^{n-2}} r_{-l-b,c,\alpha}(x) (D'^{\alpha} D_1^{-l-b-c-|\alpha|} D_2^c) f.$$

Here we adopt the convention according to which integration comes before differentiation. For example, $(D_1^{-1}D_2)f$ is, by definition, $D_2(D_1^{-1}f)$. Remark that

$$D_1^{-1}D_2^{-1} = D_2^{-1}D_1^{-1}$$
$$D_j^{-1}D_k \neq D_k D_j^{-1}, \ \{j, k\} = \{1, 2\}.$$

See [D] for details, especially for the proof of the convergence of the above series. □

Proposition 8. *Assume that R is an operator of $\mathcal{E}_K(W_{\varepsilon_1\varepsilon_2})$ of order ≤ 0, and that R' is a differential operator such that $R'R$ is also a differential operator. Then for any $u \in \mathcal{N}_{q,K}$, we have*

$$(R'R)u = R'(Ru).$$

(This function does not necessarily belong to $\mathcal{N}_{q,K}$).

Proof. The convention about the order of integration and differentiation is observed. □

Proposition 9. *If K is non-characteristic for an l-th order differential operator $Q(x,D)$, then, for $m \geq l$, we have*

$$D_1^l \mathcal{N}_{q,K}^m = Q(x,D)\mathcal{N}_{q,K}^m.$$

Proof. First, we prove that $D_1^l \mathcal{N}_{q,K}^m \supset Q(x,D)\mathcal{N}_{q,K}^m$. The differential operator Q can be written in the form

$$Q(x,D) = D_1^l Q_0(x) + \sum_{j=1}^{l} D_1^{l-j} Q_j(x, D_2, D'),$$

where ord $Q_j \leq j$, $[Q_j, x_1] = 0$.

Obviously we have

$$Q_j : \mathcal{N}_{q,K}^m \to \mathcal{N}_{q,K}^{m-j}$$
$$D_1^{l-j} : \mathcal{N}_{q,K}^{m-j} \to D_1^l(D_1^{-j}\mathcal{N}_{q,K}^{m-j}) \subset D_1^l \mathcal{N}_{q,K}^m.$$

In order to prove the converse inclusion, we make use of Dunau's symbol calculus. By virtue of Proposition 4, we know that $Q = D_1^l \tilde{Q}$ and $\tilde{Q}$, $\tilde{Q}^{-1} \in \mathcal{E}_K(W_{\varepsilon_1\varepsilon_2})$.

We obtain due to Propositions 8 and 7

$$D_1^l \mathcal{N}_{q,K}^m = (Q\tilde{Q}^{-1})\mathcal{N}_{q,K}^m = Q(\tilde{Q}^{-1}\mathcal{N}_{q,K}^m) \subset Q\mathcal{N}_{q,K}^m.$$

Here remark that ord $\tilde{Q}^{-1} = 0$. □

5 Proof of the main results

We consider the problems in the situation described in the second section.

Proof of Theorem A. By virtue of Proposition 9, it is enough to prove the isomorphism:

$$P : \mathcal{N}_{q,K}^m \xrightarrow{\sim} D_1^{m-r}\mathcal{N}_{q,K}^{m-r}.$$

First we show that $P\mathcal{N}_{q,K}^m \subset D_1^{m-r}\mathcal{N}_{q,K}^{m-r}$. According to Proposition 1',

$$P = D_1^{m-r}\left\{P_1'(x, x_1D_1, D_2, D') + \sum_{l=0}^{m-r-1} D_1^{-m+r+l}P_l(x, D_2, D')\right\}$$

where ord $P_l \le m-l$ and $[x_1, P_l] = 0$.

It is easy to see that

$$P_1'(x, x_1D_1, D_2, D') : \mathcal{N}_{q,K}^m \to \mathcal{N}_{q,K}^{m-r}.$$

Moreover we have

$$\begin{aligned} P_l(x, D_2, D') &: \mathcal{N}_{q,K}^m \to \mathcal{N}_{q,K}^l \\ D_1^{-m+r+l} &: \mathcal{N}_{q,K}^l \to \mathcal{N}_{q,K}^{m-r}. \end{aligned}$$

Now the inclusion is obvious.

Secondly, P is injective since $S\backslash T$ is non-characteristic for P.

We show finally the surjectivity. When $w \in \mathcal{N}_{q,K}^{m-r}$ is given, let us solve $Pu = D_1^{m-r}w$. Recall that owing to Proposition 5, P can be decomposed as

$$P = D_1^{m-r}\tilde{P},\ \tilde{P}^{-1} \in \mathcal{E}_K(W_{\varepsilon_1\varepsilon_2}),\ \text{ord}\ \tilde{P} = r.$$

Set $u = \tilde{P}^{-1}w \in \mathcal{N}_{q,K}^m$. Then we obtain by using Proposition 8,

$$Pu = P(\tilde{P}^{-1}w) = (P\tilde{P}^{-1})w = D_1^{m-r}w. \quad \square$$

Proof of Theorem B. (B-1) is equivalent to the following assertion:

$$x_1^{-(q-1)/q}\mathcal{N}_{q,K} \subset D_1^{m-r}\mathcal{N}_{q,K}^{m-r}.$$

Since we have

$$\begin{aligned} D_1^{-1} &: x_1^{-(q-1)/q}\mathcal{N}_{q,K} \to \mathcal{N}_{q,K}^1 \\ D_1^{-(m-r-1)} &: \mathcal{N}_{q,K}^1 \to \mathcal{N}_{q,K}^{m-r}, \end{aligned}$$

the inclusion is clear.

If $r = m-1$, the converse inclusion is obvious.

(B-2) is equivalent to the following:

$$x_1^{-l/q} \notin D_1^{m-r}\mathcal{N}_{q,K}^{m-r}, \quad l \ge q,\ 1 \le r \le m-1.$$

Assume otherwise. Then there exists $u \in \mathcal{N}_{q,K}^{m-r}$ such that $D_1^{m-r}u = x_1^{-l/q}$. By an elementary computation, we see

$$D_1^{m-r-1}u = \begin{cases} \frac{q}{q-l}(x_1^{(q-l)/q} - x_2^{q-l}) & l > q \\ \log x_1 - \frac{1}{q}\log x_2 & l = q. \end{cases}$$

In either case $D_1^{m-r-1}u$ is singular on $x_2 = 0$. This contradicts the hypothesis on u. □

6 Remark on the case $r = m - 1$

This case can be treated by an elementary method: all we need is the classical Cauchy-Kowalevski theorem.

In this section, we prove, if $r = m - 1$,

$$P(x, D) : \mathcal{N}_{q,K}^m \xrightarrow{\sim} x_1^{-(q-1)/q} \mathcal{N}_{q,K},$$

where P is as in Proposition 1' (or, equivalently, Proposition 1). P is of the form

$$P(x, D) = D_1 P_1'(x, x_1 D_1, D_2, D') + P_2'(x, D_2, D')$$

where $[x_1, P_2'] = 0$.

Recall the singular change of coordinates Φ used in the proof of Proposition 6. Since

$$D_1 = \frac{1}{qz^{q-1}} D_z, \quad x_1 D_1 = \frac{1}{q} z D_z,$$

we have

$$P = \frac{1}{qz^{q-1}} D_z P_1'(z^q, x_2, x', \frac{1}{q} z D_z, D_2, D') + P_2'(z^q, x_2, x', D_2, D').$$

We want to show that:

$$P : \mathcal{O}_0^m \xrightarrow{\sim} z^{-(q-1)} \mathcal{O}_0.$$

This is equivalent to the following isomorphism:

$$\tilde{P} \underset{\text{def}}{=} qz^{q-1} P : \mathcal{O}_0^m \xrightarrow{\sim} \mathcal{O}_0.$$

Obviously,

$$\tilde{P}(z, x_2, x', D_z, D_2, D') = D_z P_1'(z^q, x_2, x', \frac{1}{q} z D_z, D_2, D') + qz^{q-1} P_2'(q^q, D_2, D').$$

Remark that $\tilde{S} : z = x_2$ is non-characteristic for $\tilde{P}$ near $(z, x_2, x') = 0$ since

$$\sigma(\tilde{P})(0; 1, -1, 0') = 1 \cdot \sigma(P_1')(0; 0, -1, 0') \neq 0.$$

Hence Cauchy-Kowalevski theorem implies that

$$\tilde{P} : \mathcal{O}_0^m \xrightarrow{\sim} \mathcal{O}_0.$$

7 The second kind of characteristic Cauchy problem

The study of this section is still in progress. Our aim is to generalize [H] by using symbol calculus. Hamada's treatment of this kind of problems is based on Leray's approach, which is very complicated.

Here, we only give an interpretation of an example in [H] from our viewpoint. We use the notation of the 2nd section with $q = 2$.

Example ([H]).
Consider

$$\begin{cases} (D_2^2 - D_1)u(x) = 0 \\ u|_S = \gamma_1 x_2^3, \quad D_1 u|_S = \gamma_2 x_2 \end{cases}$$

where

$$\gamma_1 = \sum_{m=0}^{\infty} (-1)^m \frac{\Gamma(m-\frac{3}{2})}{(2m)!}, \quad \gamma_1 = \sum_{m=0}^{\infty} (-1)^{m+1} \frac{\Gamma(m-\frac{1}{2})}{(2m)!}.$$

Then we have

$$u(x) = \sum_{m=0}^{\infty} (-1)^m \frac{\Gamma(m-\frac{3}{2})}{(2m)!} x_1^{\frac{3}{2}-m} x_2^{2m}$$

We have a completely new phenomenon in that the solution $u(x)$ has an *essential singularity*, while all the data are regular. Here we explain why an essential singularity appears, believing naively that a differential operator should be inverted.

We reduce the problem to the following one.

$$\begin{cases} (D_2^2 - D_1)u(x) = v(x), \quad v \in \mathcal{O}_0 \text{ is given}, \\ u|_S = 0, \quad D_1 u|_S = 0. \end{cases}$$

Since $D_2^2 - D_1 = D_2^2(1 - D_1 D_2^{-2})$, its inverse is calculated in terms of Neumann series:

$$(D_2^2 - D_1)^{-1} = \sum_{j=0}^{\infty} (D_1 D_2^{-2})^j D_2^{-2} = \sum_{j=0}^{\infty} D_1^j D_2^{-2j-2}.$$

In the same way as Dunau's calculus, we put $z^2 = x_1$ and set

$$\begin{aligned} u(z^2, x_2, x') &= \sum_{j=0}^{\infty} D_1^j D_2^{-2j-2} v(z^2, x_2, x') \\ &= \sum_{j=0}^{\infty} (\frac{1}{2z} D_z)^j D_2^{-2j-2} v(z^2, x_2, x') \\ &= \sum_{j=0}^{\infty} \frac{1}{(2z^2)^j} (\theta - 2(j-1)) \cdots (\theta - 2)\theta D_2^{-2j-2} v(z^2, x_2, x'), \end{aligned}$$

where $\theta = zD_z$. We can prove that it converges in $0 < |z| << 1$. The factor $\frac{1}{(2z^2)^j}$ makes the essential singularity of the solution appears.

To prove the convergence, we need the following lemma.

Lemma 10. *Let j be a positive integer. We have*

$$\sum_{k=0}^{\infty} \underbrace{(k+2(j-1))\cdots(k+2)\,k}_{j \text{ factors}}\, y^k \leq \frac{j!y^2}{(1-y)\left\{y(1-y)\right\}^j}$$

for $0 \leq y < 1$.

Proof. In fact,

$$\begin{aligned}
&\sum_{k=0}^{\infty} \underbrace{(k+2(j-1))\cdots(k+2)k}_{j \text{ factors}}\, y^k \\
&\leq \sum_{k=0}^{\infty} \underbrace{(k+2(j-1))(k+2j-3)\cdots(k+j-1)}_{j \text{ factors}}\, y^k \\
&= \frac{1}{y^{j-2}} \frac{d^j}{dy^j} \sum_{k=0}^{\infty} y^{2(j-1)+k} \\
&\leq \frac{1}{y^{j-2}} \frac{d^j}{dy^j} (1+y+y^2+\cdots) \\
&= \frac{1}{y^{j-2}} \frac{j!}{(1-y)^{j+1}} \qquad \square
\end{aligned}$$

Put $v_j(z, x_2, x') = D_2^{-2j-2} v(z^2, x_2, x')$. Then, owing to [D] Proposition 6, we have

$$|v_j(z, x_2, x')| \leq \frac{M^{j+1}}{(2j+2)!}$$

in a neighborhood of $(z, x_2, x') = (0, 0, 0)$ for some positive constant M.

We expand v_j into the form

$$v_j(z, x_2, x') = \sum_{k=0}^{\infty} F_k^{(j)}(x_2, x') z^k.$$

By using Cauchy's estimate, it is easy to see that

$$|F_k^{(j)}(x_2, x')| \leq \frac{M^{j+1}}{(2j+2)!} r^{-k}$$

for some positive constant r. This gives, for $j \geq 1$,

$$
\begin{aligned}
&|(\theta - 2(j-1)) \cdots (\theta - 2)\theta v_j| \\
&\leq \sum_{k=0}^{\infty} \frac{M^{j+1}}{(2j+2)!} r^{-k} (k + 2(j-1)) \cdots (k+2)k|z|^k \\
&\leq \frac{M^{j+1}}{(2j+2)!} \frac{j!(\frac{|z|}{r})^2}{(1 - \frac{|z|}{r}) \left\{ \frac{|z|}{r}(1 - \frac{|z|}{r}) \right\}^j} \quad \text{if} \quad |z| < r.
\end{aligned}
$$

Here we used Lemma 10 to show the last inequality. Finally we obtain

$$
\begin{aligned}
|u| \leq |D_2^{-2} v(z^2, x_2, x')| &+ \sum_{j=1}^{\infty} \left(\frac{1}{2|z^2|} \right)^j \frac{M^{j+1}}{(2j+2)!} \frac{j!(\frac{|z|}{r})^2}{(1 - \frac{|z|}{r}) \left\{ \frac{|z|}{r}(1 - \frac{|z|}{r}) \right\}^j} \\
&< \infty \quad \text{if} \quad 0 < |z| < r.
\end{aligned}
$$

This shows the convergence of u.

Remark. The generalization to

$$
D_2^m - a D_1^{\alpha_1} D_2^{\alpha_2} \cdots D_n^{\alpha_n}, \quad \alpha_1 + \cdots + \alpha_n \leq m - 1, \quad a \in \mathbb{C},
$$

is immediate.

References

[D] J.Dunau, *Un Problème de Cauchy Caractéristique*, J. Math. pures et appl. **69** (1990), 369-402.

[G-K-L] L. Gårding, T.Kotake and J.Leray, *Problème de Cauchy, I bis et VI*, Bull. Soc. Math. de France **92** (1964), 263–361.

[H] Y.Hamada, *Les singularités des solutions du problème de Cauchy à données holomorphes*, Recent developments in hyperbolic equations (Pisa, 1987) (L. Cattabriga et. al., eds.), Pitman Research Notes in Math. **183**, Longman, 1988, pp. 82–95.

[K-K-K] M.Kashiwara, T.Kawai and T.Kimura, *Foundation of Algebraic Analysis*, (in Japanese), Kinokuniya, 1980; English translation, Princeton Mathematical Series **37**, Princeton Univ. Press, 1986.

[N-S] G.Nakamura, T.Sasai, *The singularities of the solutions of the Cauchy problem for second order equations in case the initial manifold includes characteristic points*, Tôhoku Math. Journ. **28** (1976), 523-539.

[O-Y] Y.Okada, H.Yamane, *A characteristic Cauchy problem in the complex domain*, J. Math. pures et appl, (to appear).

Stokes operators for microhyperbolic equations

Keisuke Uchikoshi

Department of Mathematics, National Defense Academy, Hashirimizu, Yokosuka, Kanagawa 236, Japan

Abstract. We consider microhyperbolic equations degenerated precisely on a hyperplane, and study the propagation of the singularities.

For some class of microhyperbolic operators, we prove that the elementary solution of the Cauchy problem is the composite of holomorphic microlocal operators and quantized contact transformations. As a corollary, we give a result for the propagation of the singularity.

1. Introduction

Let $P(x, D)$ be a microdifferential operator of order m defined at $x^* = (0; 0, \cdots, 0, \sqrt{-1}) \in \mathbf{T}^*\mathbf{C}^n$ written in the form

$$(1) \qquad P(x, D) = D_1^m + \sum_{0 \leq j \leq m-1} P_j(x, D')D_1^j.$$

Here we have written $D' = (D_2, \cdots, D_n)$ as usual, and sometimes we also write as
$D'' = (D_1, \cdots, D_{n-1})$ and $D''' = (D_2, \cdots, D_{n-1})$. We assume that

$$(2) \quad \begin{cases} \text{there exist real holomorphic functions } \varphi_1(x, \xi'), \cdots, \varphi_m(x, \xi') \\ \text{homogeneous in } \xi' \text{ of degree 1, vanishing at } x^*, \text{ and the principal} \\ \text{symbol of } P \text{ is equal to } (\xi_1 + \varphi_1(x, \xi')) \cdots (\xi_1 + \varphi_m(x, \xi')). \end{cases}$$

We also assume that

$$(3) \quad \begin{cases} \text{if } i \neq j, \text{ and } (x, \xi) \text{ belongs to a small complex neighborhood of } x^*, \\ \text{then we have} \\ \qquad \varphi_i(x, \xi') = \varphi_j(x, \xi') \qquad \Longleftrightarrow \qquad x_1 = 0 \end{cases}$$

A sufficient condition for (3) is

$$(4) \quad \begin{cases} \text{for some } q_j \in \mathbf{N} \text{ and some } a_j(x, \xi') \in \mathcal{O}_{x^*} \text{ we have} \\ \varphi_j(x, \xi') = x_1^{q_j} a_j(x, \xi'),\ a_j(x^*) \neq 0\ (1 \leq j \leq m), \text{ and} \\ \qquad i \neq j \qquad \Longrightarrow \qquad (q_i, a_i(x^*)) \neq (q_j, a_j(x^*)). \end{cases}$$

(4) is not a necessary condition of (3). However, if (3) true, then using a contact transformation we may assume that (4) is true. For example, if

$$m=2,\ \varphi_1(x,\xi')=0,\ \varphi_2(x,\xi')=x_1\xi_n,$$

then (3) is true and (4) is not. Let κ be a contact transformation such that $\kappa^*(x_1)=x_1$, $\kappa^*(\xi_1)=\xi_1+x_n\xi_n$, $\kappa(x^*)=x^*$. Then we are reduced to the case

$$m=2,\ \varphi_1(x,\xi')=x_1\xi_n,\ \varphi_2(x,\xi')=2x_1\xi_n$$

satisfying (4). At first our assumption was (1), (2), and (3). But from now on we assume (4) instead of (3), because such a modification is always possible.

We consider the Cauchy problem

$$\text{(5)}\qquad \begin{cases} Pu=0, \\ D_1^{j-1}u(0,x')=v_j(x'), \quad 1\le j\le m, \end{cases}$$

where $u\in\mathcal{C}_{\mathbf{R}^n,x^*}$ and $v_j\in\mathcal{C}_{\mathbf{R}^{n-1},x^{*\prime}}$ $(x^{*\prime}=(0;0,\cdots,0,\sqrt{-1})\in\mathbf{T}^*\mathbf{C}^{n-1})$. We rewrite (5) in the following form:

$$\text{(6)}\qquad L\vec{u}=\vec{0},\ \vec{u}(0,x')=\vec{v}(x').$$

We have written

$$L(x,D)=D_1I_m+\begin{pmatrix} \varphi_1(x,D'), & -1, & & 0 \\ & \varphi_2(x,D'), & \ddots & \\ & & \ddots & -1 \\ 0 & & & \varphi_m(x,D') \end{pmatrix} + \begin{pmatrix} & 0 & \\ b_1(x,D'), & \cdots, & b_m(x,D') \end{pmatrix}$$

with some $b_j(x,D')\in\mathcal{E}_{x^*}$ of order at most $m-j$. Here $\varphi_j(x,D')$ denotes the microdifferential operator whose complete symbol is equal to $\varphi_j(x,\xi')$. Furthermore we have written

$$\vec{u}(x)=\begin{pmatrix} u_1 \\ u_2 \\ \vdots \\ u_m \end{pmatrix},\qquad \vec{v}(x')=\begin{pmatrix} \tilde{v}_1 \\ \tilde{v}_2 \\ \vdots \\ \tilde{v}_m \end{pmatrix},$$

where

$$\begin{cases} u_1=u, \\ u_{j+1}=(D_1+\varphi_j(x,D'))u_j, \quad 1\le j\le m-1, \end{cases}$$

and

$$\begin{cases} \tilde{v}_1 = v_1, \\ \tilde{v}_{j+1} = v_{j+1} + \varphi_j(0, x', D')\tilde{v}_j, \quad 1 \le j \le m-1. \end{cases}$$

According to the general theory of microhyperbolic operators in [4] , there uniquely exists an elementary solution $\tilde{E}(x, y') \in (\mathcal{C}_{\mathbf{R}^{2n-1},(x^*,-x^{*\prime})})^{m\times m}$ satisfying

$$\begin{cases} L(x, D)\tilde{E}(x, y') = O, \\ \tilde{E}(0, x', y') = \mathrm{sp}\,\delta(x' - y')I_m. \end{cases}$$

Here $A^{m\times m}$ denotes the set of $m \times m$ matrices whose components belong to a set A, and I_m denotes the unit matrix. Note that we are regarding x_1 as a parameter, because P and L are of Kowalewski type. It is easy to see that

$$E : (\mathcal{C}_{\mathbf{R}^n,x^*})^m \ni \vec{u}(x) \longmapsto \int \tilde{E}(x, y')\vec{u}(x_1, y')dy' \in (\mathcal{C}_{\mathbf{R}^n,x^*})^m$$

is well-defined, and the solution of (6) is $E\vec{v}(x')$.

We can calculate $\mathrm{supp}\,\tilde{E}$ easily. Let $j \in \{1, \cdots, m\}$. Let $\psi_j(x, \eta')$ be the solution of

$$\begin{cases} \eta_1 + \partial_{x_1}\psi_j(x, \eta') + \varphi_j(x, \eta' + \partial_{x'}\psi_j(x, \eta')) = 0, \\ \psi_j(0, x', \eta') = \sum\limits_{2\le k\le m} x_k\eta_k. \end{cases}$$

It is easy to see that if $y_k = \partial_{\eta_k}\psi_j(x, \eta')$, $\xi_k = \partial_{x_k}\psi_j(x, \eta')$, $1 \le k \le m$, then $\kappa_j : \mathbf{T}^*\mathbf{C}^n \ni (x, \xi) \longmapsto (y, \eta) \in \mathbf{T}^*\mathbf{C}^n$ is a real homogeneous symplectic transformation defined around x^*. Note that

$$\begin{cases} y_1 = x_1, \\ \eta_1 = \xi_1 + \varphi_j(x, \xi'), \\ \kappa_j(x^*) = x^*. \end{cases}$$

It is easy to see that

$$\text{(7)} \qquad \mathrm{supp}\,\tilde{E}(x, y') = \bigcup_{1\le j\le m} \{(x, y'; \xi, \eta')\infty;\ \xi_1 + \varphi_j(x, \xi') = 0,$$
$$y_k = \kappa_j^*(x_k),\ \eta_k = \kappa_j^*(\xi_k),\ 2 \le k \le m\}.$$

In the next section we shall give a more complete expression of $\tilde{E}$: It is the composite of holomorphic microlocal operators and quantized contact transformations.

2. Main results

We first discuss about a well-known result for quantized contact transformation. For each number $j \in \{1, \cdots, m\}$, let $\tilde{k}_j(x, y') \in \mathcal{C}_{\mathbf{R}^{2n-1},(x^*,-x^{*\prime})}$ be the solution of

$$\begin{cases} (D_1 + \varphi_j(x, D'))\tilde{k}_j(x, y') = 0, \\ \tilde{k}_j(0, x', y') = \operatorname{sp}\delta(x' - y'). \end{cases}$$

We can define $k_j : \mathcal{C}_{\mathbf{R}^n,x^*} \ni u(x) \longmapsto \int \tilde{k}_j(x, y')u(x_1, y')dy' \in \mathcal{C}_{\mathbf{R}^n,x^*}$, and it is a quantized contact transformation (i.e. Fourier integral operator) associated to κ_j. In fact we can calculate the complete symbol $k_j(x, \xi')$ of the operator k_j in the form

$$\exp(\psi_j'(x, \xi'))c_j(x, \xi')$$

with some elliptic amplitude function c_j of order at most 0 and the phase function $\psi_j'(x, \xi')$, defined by

$$\psi_j'(x, \xi') = \psi_j(x, \xi') - \sum_{1 \le k \le m} x_k \xi_k$$

(See [2]). Note that the kernel function $\tilde{k}_j$ is the microfunction defined by $(2\pi\sqrt{-1})^{-n+1} \int \exp((x' - y') \cdot \xi')k_j(x, \xi')d\xi'$. It is easy to see that

$$k_j x_1 = x_1 k_j, \quad k_j D_1 = (D_1 + \varphi_j(x, D'))k_j.$$

Let $K : (\mathcal{C}_{\mathbf{R}^n,x^*})^{m\times m} \longrightarrow (\mathcal{C}_{\mathbf{R}^n,x^*})^{m\times m}$ be defined by

$$K = \begin{pmatrix} k_1 & & & 0 \\ & k_2 & & \\ & & \ddots & \\ 0 & & & k_m \end{pmatrix}.$$

Our first result is the following

Theorem 1. *Let $\omega \subset \sqrt{-1}\mathbf{S}^*\mathbf{R}^{2n-1}$ be a small neighborhood of $(x^*, -x^{*\prime})$, and let $\omega^\pm = \{(x, y'; \xi, \eta')\infty \in \omega;\ \pm x_1 > 0\}$. There exist invertible matrices*

$$F^\pm(x, D'),\ G^\pm(x', D') \in (\mathcal{E}^{\mathbf{R}}_{x^*})^{m\times m}$$

defined on (a complex neighborhood of) $\omega^\pm$ such that $E = F^\pm K G^\pm$ on $\omega^\pm$ Here $\mathcal{E}^{\mathbf{R}}$ denotes the sheaf of holomorphic microlocal operators.

Holomorphic microlocal operators are some class of analytic pseudodifferential operators. They are defined by [6], and [1] gave a symbol theory for them. They have microlocal property, i.e., if $A \in \mathcal{E}^{\mathbf{R}}_{x^*}$, $u \in \mathcal{C}_{\mathbf{R}^n,x^*}$, then we have $\operatorname{supp}(Au) \subset \operatorname{supp} u$.

Example. This result is closely related to the classical theory of ordinary differential equations. For instance consider the following case:

$$L(x,D) = \begin{pmatrix} D_1 + x_1^2 D_n, & -1 \\ aD_n, & D_1 - x_1^2 D_n \end{pmatrix}, \quad a \in \mathbf{R}.$$

We define $\hat{L}(x, \frac{d}{dx_1}, \xi_n)$ by

$$\hat{L}(x, \frac{d}{dx_1}, \xi_n) = \begin{pmatrix} \frac{d}{dx_1} + x_1^2 \xi_n, & -1 \\ a\xi_n, & \frac{d}{dx_1} - x_1^2 \xi_n \end{pmatrix},$$

and let $\hat{E}(x, \xi') = \hat{E}(x_1, \xi_n)$ be the solution of

$$\hat{L}\hat{E} = O \tag{8}$$

with the Cauchy data $\hat{E}(0, \xi_n) = I_2$. This is an ordinary differential equation in x_1 with a large parameter ξ_n, investigated by [3] . In this case, the kernel function $\tilde{E}(x, y')$ is the microfunction defined by

$$(2\pi)^{n-1} \int \exp(\sqrt{-1}\zeta' \cdot (x' - y'))\hat{E}(x, \sqrt{-1}\zeta')d\zeta'$$

According to the general theory of Kashiwara and Kawai, we can obtain a microfunction in this way, but our result is more precise. Let

$$\hat{K}(x_1, \xi_n) = \begin{pmatrix} -\frac{1}{3}x_1^3 \xi_n, & 0 \\ 0, & \frac{1}{3}x_1^3 \xi_n \end{pmatrix}.$$

There exist other solutions of (8) written in the form

$$\hat{E}^{\pm}(x_1, \xi_n) = \hat{F}^{\pm}(x_1, \xi_n)\hat{K}(x_1, \xi_n)$$

on $\omega^{\pm}$. Here $\hat{F}^{\pm}$ are invertible matrices defined on $\omega^{\pm}$ satisfying

$$|\, \hat{F}^{\pm} \,|, |\, (\hat{F}^{\pm})^{-1} \,| \leq \exists C \exp(C|\, \xi_n \,|^{3/4})$$

(For an $m \times m$ matrix $A = (A_{(\mu,\nu)})$ we define $|\, A \,| = m \times \max_{\mu,\nu} |\, A_{(\mu,\nu)} \,|$).
According to an elementary theory in ordinary differential equations we have

$$\hat{E}(x_1, \xi_n) = \hat{E}^{\pm}(x_1, \xi_n)\hat{G}^{\pm}(\xi_n) \tag{9}$$

with some invertible matrix $\hat{G}^{\pm}(\xi_n)$. Substituting $x_1 = 0$ in (9), we obtain

$$\hat{E}^{\pm}(0,\xi_n)\hat{G}^{\pm} = \hat{F}^{\pm}(0,\xi_n)\hat{G}^{\pm} = I_m.$$

It follows that

$$\hat{G}^{\pm}(\xi_n) = (\hat{F}^{\pm}(0,\xi_n))^{-1},$$

and $\hat{E}(x_1,\xi_n) = \hat{F}^{\pm}(x_1,\xi_n)\hat{K}(x_1,\xi_n)\hat{G}^{\pm}(\xi_n)$. Note that in the classical theory of ordinary differential equations, it is usual to try to obtain a more precise asymptotic expansion for $\hat{F}^{\pm}$. Such a study is very difficult but it is not necessary for us (See [3]).

Note that $F^{\pm}$ (resp. $G^{\pm}$) are independent of D_1 (resp. x_1 and D_1). Therefore $G^{\pm}$ is in fact defined for any x_1. $G^{\pm}$ is the most important and we call it the *Stokes operator* .

From (7) it follows that the singularity of the solution $\vec{u}$ of (6) propagates along the union of the bicharacteristic strips of P. Let us discuss about it in detail. Let b_j be the bicharacteristic strip associated to $\xi_1 + \varphi_j(x,\xi')$ through $x^*\infty$, and let $b_j^{\pm} = \{(x,\xi)\infty \in b_j; \pm x_1 > 0\}$. It is easy to see that

(10)

$$\begin{aligned} L(x,D)\vec{u}(x) = \vec{0} \quad &\Longleftrightarrow \quad \vec{u}(x) = E\vec{v}(x') \text{ for some } \vec{v} \\ &\Longleftrightarrow \quad \vec{u}(x) = E^{+}\vec{v}^{+}(x') \text{ for some } \vec{v}^{+} \\ &\Longleftrightarrow \quad \vec{u}(x) = E^{-}\vec{v}^{-}(x') \text{ for some } \vec{v}^{-}, \end{aligned}$$

where $E^{\pm} = F^{\pm}K = E(G^{\pm})^{-1}$. Using the last statement of these equivalent conditions, the propagation in the past half space is of trivial type. If $\vec{u}(x) = {}^t(u_1, u_2, \cdots, u_m) \in (\mathcal{C}_{\mathbf{R}^n,x^*})^m$, then we define $\operatorname{supp}\vec{u} = \bigcup_{1\le j\le m} \operatorname{supp} u_j$. It follows that

$$\begin{aligned} \operatorname{supp}(E^{-}\vec{v}^{-}(x')) &= \operatorname{supp}(F^{-}K\vec{v}^{-}(x')) \\ &= \operatorname{supp}(K\vec{v}^{-}(x')) \\ &= \operatorname{supp}\begin{pmatrix} k_1 v_1^{-} \\ k_2 v_2^{-} \\ \vdots \\ k_m v_m^{-} \end{pmatrix} \\ &= \bigcup_{1\le j\le m} \operatorname{supp}(k_j v_j^{-}) \end{aligned}$$

on ω^{-}. Let $L\vec{u} = \vec{0}$. Since $\operatorname{supp}(k_j v_j^{-})$ is contained in $\{\kappa_j(x,\xi) \in \omega^{-};\ (x,\xi) \in \operatorname{supp} v^{-}\}$, we have, for instance,

(11)

$$\operatorname{supp}\vec{u} \cap \omega^{-} = b_1^{-}$$

$\Longleftrightarrow$ the components of $\vec{v}^{-}$ vanish except for v_1^{-} and $\operatorname{supp} v_1^{-} = \{x^{*\prime}\infty\}$.

More generally $\operatorname{supp}\vec{u}$ contains b_j^- if, and only if, $\operatorname{supp} v_j^-$ contains $\{x^{*\prime}\infty\}$.

On the other hand, $E^- = F^+KG^+(G^-)^{-1}$ on ω^+. In this expression E^- does not have such a diagonal property because of the additional factor $G^{+-} = G^+(G^-)^{-1}$ composed from the right, and the situation in the future half space is not so simple. In fact we have

$$\operatorname{supp}(E^-\vec{v}^-(x')) = \operatorname{supp}(F^+KG^{+-}\vec{v}^-(x')) = \operatorname{supp}(KG^{+-}\vec{v}^-(x')).$$

For instance assume the equivalent conditions in (11). Then we have

$$\begin{aligned}\operatorname{supp}(E^-\vec{v}^-(x')) &= \operatorname{supp}\begin{pmatrix} k_1G^{+-}_{(1,1)}v_1^- \\ k_2G^{+-}_{(2,1)}v_1^- \\ \vdots \\ k_mG^{+-}_{(m,1)}v_1^- \end{pmatrix} \\ &= \bigcup_{1\le j\le m} \{\kappa_j(x,\xi)\in\omega^-;\ (x,\xi)\in \operatorname{supp} G^{+-}_{(j,1)}v_1^-\}.\end{aligned}$$

This means that passing from the past to the future, the singularity may branch into different bicharacteristic strips. It is well-known in the case $m = 2$. There are also some results for the case $m = 3$, and here we only refer to a recent work [7] .

More generally, $k_iG^{+-}_{(i,j)}(x',D')k_j^{-1}$ denotes the operator transporting the singularity on
$\{(x,\xi)\in\omega^-;\xi_1+\varphi_i(x,\xi')=0\}$ into $\{(x,\xi)\in\omega^+;\xi_1+\varphi_j(x,\xi')=0\}$. We want to study how the singularity, which propagated along a simple bicharacteristic strip in the past, will branch in the future. For that purpose we need to calculate G^{+-}. This is not so easy in general, but we have the following

Theorem 2. *Let J be the set of those $j\in\{1,\cdots,m\}$ which satisfy $q_j = 1$, and let $j_0\in J$. We assume that either*

$$\text{(12)}\qquad \begin{cases}\operatorname{Im} a_{j_0}(x^*) < 0, \\ j\in J\setminus\{j_o\} \Longrightarrow \operatorname{Im} a_{j_0}(x^*) < \operatorname{Im} a_j(x^*)\end{cases}$$

or

$$\text{(13)}\qquad \begin{cases}\operatorname{Im} a_{j_0}(x^*) > 0, \\ j\in J\setminus\{j_o\} \Longrightarrow \operatorname{Im} a_{j_0}(x^*) > \operatorname{Im} a_j(x^*).\end{cases}$$

Then $G^{+-}_{(j_0,j_0)}(x',D')$ is invertible.

Corollary. *Under the above assumption we have*

$$\vec{u}\in\mathcal{C}_{\mathbf{R}^n,x^*}\setminus\{\vec{0}\},\ L\vec{u}=\vec{0},\ \operatorname{supp}\vec{u}\cap\omega^- = b_{j_0}^- \quad\Longrightarrow\quad \operatorname{supp}\vec{u}\cap\omega^+\supset b_{j_0}^+$$

for $j_0 \in J$.

Example. If $m = 4$ and

$$\varphi_1 = x_1\xi_n,\ \varphi_2 = 2x_1\xi_n,\ \varphi_3 = x_2\xi_n,\ \varphi_4 = x_3\xi_n,$$

then $J = \{1, 2\}$. If $j_0 = 2$, we have (13), and we have **Corollary** for this number.

The assumption (12) (resp. (13)) means that $\varphi_{j_0}(x, \xi')$, the characteristic root is the most negative (resp. positive), and for such a dominant root, at least some part of the singularity penetrates (from the past to the future) along the corresponding strip.

We do not know whether the assumption (12) or (13) is indispensable or not. To the contrary, if $j_0 \notin J$, the above conclusion is not always true. For instance, [5] gave the following

Example. Let $q \in \mathbf{N}$ and let

$$L = \begin{pmatrix} D_1 + c_1 x_1^q D_n, & -1 \\ d x_1^{q-1} D_n, & D_1 + c_2 x_1^q D_n \end{pmatrix}$$

with some constants $c_1, c_2 \in \mathbf{R}$ and $d \in \mathbf{C}$ satisfying $c_1 < c_2$ (therefore $m = 2$, $q_1 = q_2 = q$, and $a_j(x, \xi') = c_j \xi_n$, in our notation). In this case, we can directly calculate the fundamental solution using the hypergeometric functions, and the Stokes operator is nothing but the usual Stokes multiplier for them (They are constant operators). It follows that

(i) if $q = 1$, then $G^{+-}_{(1,1)}$ and $G^{+-}_{(2,2)}$ are always elliptic,

(ii) if $q \geq 2$ is even, then $G^{+-}_{(1,1)}$ and $G^{+-}_{(2,2)}$ are elliptic if, and only if,

$$\frac{1}{q+1}\{\frac{d}{c_2 - c_1} + \frac{1}{2}\} \notin \mathbf{Z},$$

(iii) if $q \geq 2$ is odd, then $G^{+-}_{(j,j)}$ is elliptic if, and only if,

$$\frac{1}{q+1}\{\frac{d}{c_2 - c_1} + \frac{1}{2}\} - \frac{(-1)^j}{2\pi\sqrt{-1}} \log\cos\frac{\pi}{q+1} \notin \mathbf{Z}.$$

3. Transformation by holomorphic microlocal operators

Let $q = \max(q_1, \cdots, q_m)$. Let $r > 0$, let $\theta \in [0, 2\pi]$ and let

$$
\begin{aligned}
\Omega(r) =&\{(x,\xi') \in \mathbf{C}^n \times \mathbf{C}^{n-1};\ |\,x\,| < r,\\
&|\,\xi''\,| < r\mathrm{Im}\ \xi_n,\ |\,\mathrm{Re}\ \xi_n\,| < r\mathrm{Im}\ \xi_n\},\\
\Omega_i(r) =&\{(x,\xi') \in \Omega(r);\ i+1 < r\mathrm{Im}\ \xi_n\},\\
\Omega^\theta(r) =&\{(x,\xi') \in \Omega(r);\ x_1 \neq 0,\ |\,\arg x_1 - \theta\,| < r\},\\
\Omega'^{\theta}_i(r) =&\{(x,\xi') \in \Omega_i(r);\ |\,x_1\,| > r(\mathrm{Im}\ \xi_n)^{-1/mq},\ |\,\arg x_1 - \theta\,| < r\},\\
\Omega''_i(r) =&\{(x,\xi') \in \Omega_i(r);\ |\,x_1\,| < 2r(\mathrm{Im}\ \xi_n)^{-1/mq}\},
\end{aligned}
$$

and

$$\Omega^\theta_i(r) = \Omega'^{\theta}_i(r) \cup \Omega''_i(r).$$

We define

$$
\bar{M}(x,D') = \begin{pmatrix} \varphi_1(x,D'), & -1, & & 0 \\ & \varphi_2(x,D'), & \ddots & \\ & & \ddots & -1 \\ 0 & & & \varphi_m(x,D') \end{pmatrix}
$$

and $M(x,D) = D_1 I_m + \bar{M}(x,D')$.

Then we have the following

Proposition 1. *Let $\theta \in [0, 2\pi]$. Then there exist*

$$F^\theta(x,D'),\ \tilde{G}^\theta(x,D') \in (\mathcal{E}^{\mathbf{R}}(\Omega^\theta(r)))^{m\times m}$$

such that

$$
\begin{cases} L(x,D)F^\theta(x,D') = F^\theta(x,D')M(x,D), \\ F^\theta(x,D')\tilde{G}^\theta(x,D') = \tilde{G}^\theta(x,D')F^\theta(x,D') = Id. \end{cases}
$$

on $\Omega^\theta(r)$.

To prove **Proposition 1**, we need to prepare a symbol theory for holomorphic microlocal operators containing x_1 as a parameter. It is the same as [1], but we resume the result for the sake of the readers' convenience. We denote by $\mathcal{S}(\Omega^\theta(r))$ the set of formal series $f = \sum_{i\in\mathbf{Z}_+} f_i(x,\xi')$ satisfying

(i) $f_i \in \mathcal{O}(\Omega^\theta_i)$, $i \in \mathbf{Z}_+$,

(ii) there exists some $R \in (0,1)$, and for any $\varepsilon > 0$ there exists some $C_\varepsilon > 0$ such that

$$|\,f_i(x,\xi')\,| \leq C_\varepsilon R^i \exp(\varepsilon|\,\mathrm{Im}\ \xi_n\,|) \quad \text{on } \Omega^\theta_i(r),$$

for each i. If $f = \sum_i f_i$, we define a formal series $f^\sharp = \sum_{i \in \mathbf{Z}_+} f_i^\sharp$ by $f_i^\sharp = \sum_{0 \le j \le i} f_j$, and $\mathcal{N}(\Omega^\theta(r))$ by

$$\mathcal{N}(\Omega^\theta(r)) = \{f \in \mathcal{S}(\Omega^\theta(r));\ f^\sharp \in \mathcal{S}(\Omega^\theta(r))\}.$$

We identify a function f_0 with $f_0 + 0 + 0 + \cdots \in \mathcal{S}(\Omega^\theta(r))$, if it satisfies (i) and (ii) for $i = 0$.

Finally we define

$$\mathcal{S}^\theta = \varinjlim_{r>0} \mathcal{S}(\Omega^\theta(r)), \quad \mathcal{N}^\theta = \varinjlim_{r>0} \mathcal{N}(\Omega^\theta(r)).$$

Let $f = \sum_i f_i(x, \xi') \in \mathcal{S}(\Omega^\theta(r))$. We define $\mathcal{F}(f)(x, y')$ by

$$\mathcal{F}(f)(x, y') = (2\pi\sqrt{-1})^{n-1} \int_{\Delta_i} e^{\xi' \cdot (x' - y')} f_i(x, \xi') d\xi',$$

where $\Delta_i = \{\xi' \in \sqrt{-1}\mathbf{R}^{n-1};\ |\ \xi''\ | < r\mathrm{Im}\ \xi_n,\ i + 1 < r\mathrm{Im}\ \xi_n\}$. Then we have the following

Lemma 1. *Let $f = \sum_i f_i(x, \xi') \in \mathcal{S}(\Omega^\theta(r))$. $\mathcal{F}(f)$ is holomorphic on*

$$\{(x, y') \in \mathbf{C}^n \times \mathbf{C}^{n-1};\ |\ (x, y')\ | < r',$$
$$\mathrm{Re}(\xi' \cdot (x' - y')) < 0 \text{ for any } \xi' \in \Delta_0\},$$

for $0 < r' << r$, and it defines the kernel function of a holomorphic microlocal operator on $\Omega^\theta(r')$ (We denote this operator by $\mathcal{M}(f)(x, D')$). If $f \in \mathcal{N}^\theta$, then $\mathcal{M}(f) = 0$.

Lemma 2. *Let $f = \sum_i f_i(x, \xi')$, $g = \sum_i g_i(x, \xi') \in \mathcal{S}^\theta$. We can define $f \circ g \in \mathcal{S}^\theta$ by*

$$(f \circ g)_i = \sum_{j+k+|\alpha'|=i} \frac{1}{\alpha'!} \partial_{\xi'}^{\alpha'} f_j \partial_{x'}^{\alpha'} g_k,$$

and we have

$$\mathcal{M}(f)(x, D')\mathcal{M}(g)(x, D') = \mathcal{M}(f \circ g)(x, D').$$

Remark. (i) We write $\sum_i f_i = \sum_i g_i$ if, and only if, $f_i = g_i$ for any i. This does not simply mean that the sums of these two series are the same.

(ii) Let $\sum f_i \in \mathcal{S}(\Omega^\theta(r))$, let $g_0 = 0$ and $g_i = f_{i-1}$ for $i \ge 1$. Then we have $\sum(f_i - g_i) \in \mathcal{N}(\Omega^\theta(r))$.

(iii) A function f_0 belongs to $\mathcal{N}(\Omega^\theta(r))$ if, and only if, it is exponentially decreasing.

(iv) If $f \in \mathcal{S}^\theta$, then $f(0,x',\xi') \in \mathcal{S}^\theta$ is well-defined.

(v) For a formal series $f = \sum_i f_i$ we define $\partial_{x_1} f = \sum_i \partial_{x_1} f_i$.

If $A(x,D) = \sum_\alpha A_\alpha D^\alpha$, then we define $\sigma_i(A)(x,\xi) = \sum_{|\alpha|=i} A_\alpha \xi^\alpha$. The complete symbols of $\bar{L}(x,D') = L(x,D) - D_1 I_m$ and $\bar{M}(x,D')$ are graded as follows:

$$\bar{L}_i = \delta_{i,0} \begin{pmatrix} \varphi_1(x,\xi'), & -1, & & 0 \\ & \varphi_2(x,\xi'), & \ddots & \\ & & \ddots & -1 \\ 0 & & & \varphi_m(x,\xi') \end{pmatrix} + \begin{pmatrix} & & 0 & \\ \sigma_{m-1-i}(b_1)(x,\xi'), & & \cdots & \sigma_{-i}(b_m)(x,\xi') \end{pmatrix},$$

and

$$\bar{M}_i = \delta_{i,0} \bar{M}(x,\xi').$$

Now **Proposition 1** is a corollary of the following

Proposition 2. *Let $\theta \in [0,2\pi]$. Then there exist $F^\theta(x,\xi')$, $\tilde{G}^\theta(x,\xi') \in (\mathcal{S}^\theta)^{m\times m}$ such that*

$$(14) \qquad \begin{cases} \partial_{x_1} F^\theta(x,\xi') + \bar{L}(x,\xi') \circ F^\theta(x,\xi') = F^\theta(x,\xi') \circ \bar{M}(x,\xi'), \\ F^\theta(x,\xi') \circ \tilde{G}^\theta(x,\xi') = \tilde{G}^\theta(x,\xi') \circ F^\theta(x,\xi') = I_m. \end{cases}$$

In the next section, we shall give a sketch of the proof.

Now we can prove **Theorem 1** as follows. Let $0 \le \theta_1 \le \theta_2 \le \cdots \le \theta_\ell \le 2\pi$ be such that, **Proposition 1** is true for each θ_j with $r = r_j$ respectively, and $[0,2\pi] \subset \bigcup_{1\le j\le \ell} (\theta_j - r_j, \theta_j + r_j)$. Note that we can always choose such numbers. Let $K = K(x,\xi')$ be the complete symbol of the Fourier integral operator K, and let $G^{\theta_i}(x',\xi') = \tilde{G}^{\theta_i}(0,x',\xi')$. We may assume that $\partial_{x_1} + \bar{M} \circ K = 0$, $K(0,x',\xi')$ without any modulo classes. Then we have $(F^{\theta_i} \circ K \circ G^{\theta_i}(x',\xi'))_k = (F^{\theta_j} \circ K \circ G^{\theta_j}(x',\xi'))_k$ on $\Omega_k^{\theta_i}(r_i) \cap \Omega_k^{\theta_j}(r_j)$. In fact, we have

$$
\begin{aligned}
&\partial_{x_1}(F^{\theta_i} \circ K \circ G^{\theta_i}(x', \xi')) + \bar{L} \circ (F^{\theta_i} \circ K \circ G^{\theta_i}(x', \xi')) \\
=&((\partial_{x_1} + \bar{L}\circ)F^{\theta_i}) \circ K \circ G^{\theta_i}(x', \xi')) + F^{\theta_i} \circ \partial_{x_1}(K \circ G^{\theta_i}(x', \xi')) \\
=&F^{\theta_i} \circ \bar{M} \circ K \circ G^{\theta_i}(x', \xi') + F^{\theta_i} \circ \partial_{x_1}(K \circ G^{\theta_i}(x', \xi')) \\
=&F^{\theta_i} \circ ((\partial_{x_1} + \bar{M}\circ)K) \circ G^{\theta_i}(x', \xi') \\
=&O,
\end{aligned}
$$

and

$$[F^{\theta_i} \circ K \circ G^{\theta_i}(x', \xi')]_{x_1=0} = F^{\theta_i}(0, x', \xi') \circ I_m \circ \tilde{G}^{\theta_i}(0, x', \xi') = I_m.$$

It follows that

$$
\begin{cases}
\partial_{x_1}(F^{\theta_i} \circ K \circ G^{\theta_i}(x', \xi')) + \bar{L} \circ (F^{\theta_i} \circ K \circ G^{\theta_i}(x', \xi')) = O, \\
[F^{\theta_i} \circ K \circ G^{\theta_i}(x', \xi')]_{x_1=0} = I_m
\end{cases}
$$

It is easy to see that such a symbol is unique, and we obtain

$$(F^{\theta_i} \circ K \circ G^{\theta_i}(x', \xi'))_k = (F^{\theta_j} \circ K \circ G^{\theta_j}(x', \xi'))_k$$

on $\Omega_k^{\theta_i}(r_i) \cap \Omega_k^{\theta_j}(r_j)$. Therefore $\{(F^{\theta_i} \circ K(x, \xi') \circ G^{\theta_i}(x', \xi'))_k;\ 1 \leq i \leq \ell\}$ defines a function on $\bigcup_{1 \leq j \leq \ell} \Omega_k^{\theta_j}(r_j)$. Decreasing $r > 0$ if necessary, the last set contains $\Omega_k(r)$. Since $0 \in (\theta_j - r, \theta_j + r)$ and $\pi \in (\theta_k - r, \theta_k + r)$ for some j and k, we obtain **Theorem 1**.

4. Study of ordinary differential equations

In this section we give a sketch of the proof of **Proposition 2**. For the sake of simplicity, we assume that $L = L(x_1, \xi')$, and that $\bar{L}_i = O$ if $i \neq 0$ (Of course these assumptions are not necessary). Without loss of generality, we may assume that the characteristic roots are arranged in the following way:

$$(1 \leq) q_1 \leq q_2 \leq \cdots \leq q_m,$$

and if $q_i = q_j$, $i < j$, then we have either

$$\operatorname{Im} a_i(x^*) < 0 < \operatorname{Im} a_j(x^*)$$

or

$$\operatorname{Im} a_i(x^*) \cdot \operatorname{Im} a_j(x^*) > 0, \ |\operatorname{Im} a_i(x^*)| > |\operatorname{Im} a_j(x^*)|.$$

Let $q = \max_{1 \le j \le m} q_j$. Therefore $q = q_m$ under the above assumption. Considering an arbitrary θ we omit the index θ and 0 in F_0^θ, $\bar{L}_0^\theta$, and $\bar{M}_0^\theta$. We can rewrite (15) in the form

$$\partial_{x_1} F + \bar{L} F - F \bar{M} = O. \tag{15}$$

Let $\theta'^{(j)}$, $\theta''^{(j)} \in \mathbf{R}$, $(1 \le j \le m)$ be numbers with the properties

(i) $\theta \in (\theta'^{(m)}, \theta''^{(m)}) \subset (\theta'^{(m-1)}, \theta''^{(m-1)}) \subset \cdots \subset (\theta'^{(1)}, \theta''^{(1)})$,

(ii) $\dfrac{\pi}{q_j + 1} < \theta''^{(j)} - \theta'^{(j)} < \dfrac{2\pi}{2q_j + 1}$, $1 \le j \le m$,

(iii) there uniquely exists some $k_j \in \mathbf{Z} \cap (\frac{q_j+1}{\pi}\theta'^{(j)}, \frac{q_j+1}{\pi}\theta''^{(j)})$.

Note that for any θ we can always choose such numbers. In fact it is trivial if $m = 1$. Assume that $m_0 > 2$ and that we can choose the above numbers if $m = m_0 - 1$. Let $m = m_0$. If $q_m = q_{m-1}$, then we may take $\theta'^{(m)} = \theta'^{(m-1)}$, $\theta''^{(m)} = \theta''^{(m-1)}$. If $q_m > q_{m-1}$, then $\dfrac{\pi}{q_m + 1} < \theta''^{(m-1)} - \theta'^{(m-1)}$, and we can choose a subset $(\theta'^{(m)}, \theta''^{(m)})$ of $(\theta'^{(m-1)}, \theta''^{(m-1)})$ such that (i) – (iii) are true.

It is easy to see that in each interval $[\theta'^{(j)}, \theta''^{(j)}]$ the function $\pm\sin((q_j + 1)\tau)$ attains its maximum (resp. minimum) at a uniquely determined point $\tau = \theta^{(j),+}$ (resp. $\theta^{(j),-}$).

As in [3] , to obtain an asymptotic expansion of the solution of (15), we had better consider the following two cases separately:

(i) $|x_1| \le \exists\text{constant}(\operatorname{Im} \xi_n)^{-1/mq}$,

(ii) $|x_1| \ge \exists\text{constant}(\operatorname{Im} \xi_n)^{-1/mq}$.

Let $1 << 1/r << C_1 << \cdots << C_m$. We define

$$\begin{aligned}
&\Omega'^{(j),\theta}(r, C_j) \\
&= \{(x, \xi') \in \mathbf{C}^n \times \mathbf{C}^{n-1}; \\
&\quad (-1)^{q_j+1} \operatorname{Re}(x_1^{q_j+1}) + \frac{1}{r}|\operatorname{Im}(x_1^{q_j+1})| > C_j^{2(q_j+1)} (\operatorname{Im} \xi_n)^{-(q_j+1)/(mq)}, \\
&\quad |\operatorname{Re}(x_1^{q_j+1})| + \frac{1}{r}|\operatorname{Im}(x_1^{q_j+1})| < C_j^{q_j+1}, \ \arg x_1 \in (\theta'^{(j)}, \theta''^{(j)}), \\
&\quad C_j |x'| < 1, \ |\xi'''| < r\operatorname{Im} \xi_n, \ |\operatorname{Re} \xi_n| < r\operatorname{Im} \xi_n\}
\end{aligned}$$

for $1 \le j \le m$. Note that we may assume

$$\Omega'^{(m),\theta}(r, C_m) \subset \Omega'^{(m-1),\theta}(r, C_{m-1}) \subset \cdots \subset \Omega'^{(1),\theta}(r, C_1).$$

We define $\Phi_{j,k}(x_1,t,\xi')$, $(x_1,t,\xi') \in \mathbf{C} \times \mathbf{C} \times \mathbf{C}^{n-1}$ by

$$\begin{cases} \partial_{x_1}\Phi_{j,k}(x_1,t,\xi') + \varphi_j(x_1,\xi') - \varphi_k(x_1,\xi') = 0, \\ \Phi_{j,k}(x_1,x_1,\xi') = 0 \end{cases}$$

for $1 \le j,k \le m$. Let $x_1^{(j),\pm}$ be the point on the boundary of $\Omega'^{(j),\theta}$ defined by

$$\begin{cases} |\operatorname{Re}((x_1^{(j),\pm})^{q_j+1})| + \dfrac{1}{r}|\operatorname{Im}((x_1^{(j),\pm})^{q_j+1})| = C_j^{q_j+1}, \\ \arg x_1^{(j),\pm} = \theta^{(j),\pm}. \end{cases}$$

Let x_1 be an arbitrary point such that $(x,\xi') \in \Omega'^{(j),\theta}(r,C_j)$ for some (x',ξ'). It is easy to see that we can connect $x_1^{(j),\pm}$ and x_1 by a continuous curve $\gamma^{(j),\pm}(x_1)$ with length at most $C_j^{-1/2}$ such that

(i) if $(x,\xi') \in \Omega'^{(j),\theta}(r,C_j)$ and $t \in \gamma^{(j),\pm}(x_1) \setminus \{x_1^{(j),\pm}\}$, then we have $(t,x',\xi') \in \Omega'^{(j),\theta}(r,C_j)$,

(ii) if $(x,\xi') \in \Omega'^{(j),\theta}(r,C_j)$, $t \in \gamma^{(j),\pm}(x_1) \setminus \{x_1^{(j),\pm}\}$, and $\pm j \le \pm k$, then $\pm \operatorname{Re}\Phi_{j,k}(x_1,t,\xi') \le 0$.

As $\gamma^{(j),\pm}(x_1)$, we may in fact take the union of at most two line segments.

We can find a solution $F \in (\mathcal{O}(\Omega'^{(m),\theta}(r,C_m)))^{m\times m}$ of (15) in several steps. Let $\pm \operatorname{Im} a_1(x^*) > 0$. At first we consider

$$(16) \quad \partial_{x_1}F'^{(1)}(x,\xi') + \bar{L}(x,\xi')F'^{(1)}(x,\xi') - F'^{(1)}(x,\xi')\bar{M}'^{(1)}(x,\xi') = O,$$

where

$$F'^{(1)}(x,\xi') = \begin{pmatrix} * & & & 0 \\ * & 1, & & \\ \vdots & & \ddots & \\ * & 0 & & 1 \end{pmatrix}, \quad \bar{M}'^{(1)}(x,\xi') = \begin{pmatrix} \varphi_1, & \\ 0, & \\ \vdots & * \\ 0, & \end{pmatrix}.$$

Calculating the $(j,1)$-component of (16), it follows that

(17)

$$\partial x_1 F'^{(1)}_{(j,1)}(x,\xi') + \sum_{1\le k\le m} \bar{L}_{(j,k)}(x,\xi')F'^{(1)}_{(k,1)}(x,\xi') - F'^{(1)}_{(j,1)}(x,\xi')\varphi_1(x,\xi') = 0.$$

If $k \ge 2$, calculating the $(1,k)$-component of (16) we obtain

$$(18) \quad \bar{L}_{(1,k)}(x,\xi') - F'^{(1)}_{(1,1)}(x,\xi')\bar{M}'^{(1)}_{(1,k)}(x,\xi') = 0,$$

and if $j,k \ge 2$ we obtain

$$(19) \quad \bar{L}_{(j,k)}(x,\xi') - F'^{(1)}_{(j,1)}(x,\xi')\bar{M}'^{(1)}_{(1,k)}(x,\xi') + \bar{M}'^{(1)}_{(j,k)}(x,\xi') = 0.$$

Note that the other components of the left hand side of (16) are always equal to 0. If $F'^{(1)}(x,\xi')$ satisfies (17) and $F'^{(1)}_{(1,1)}(x,\xi')$ is invertible, $\bar{M}'^{(1)}(x,\xi')$ is automatically defined by (18) and (19). Let us solve (17) by successive approximation on $\Omega'^{(1),\theta}(r,C_1)$:

$$(20)\qquad \partial x_1 F'^{(1,i)}_{(j,1)}(x,\xi') + (\bar{L}_{(j,j)}(x,\xi') - \varphi_1(x,\xi'))F'^{(1,i)}_{(j,1)}(x,\xi') = -(\sum_{k\neq j}\bar{L}_{(j,k)}(x,\xi') - \varphi_j(x,\xi'))F'^{(1,i-1)}_{(k,1)}(x,\xi'),\ i\in \mathbf{Z}_+$$

Here $F'^{(1,-1)}(x,\xi') = 0$ (We want to obtain a matrix $F'^{(a)}(x,\xi')$ in the form

$$\sum_{b\in \mathbf{Z}_+} F'^{(a,b)}(x,\xi'),$$

and $F'^{(a,b)}_{(c,d)}(x,\xi')$ denotes the (c,d)-component of $F'^{(a,b)}(x,\xi')$).

We can solve (20) by

$$F'^{(1,i)}_{(j,1)}(x,\xi') = -\int_{\gamma^{(1),\mp}(x_1)} \exp(\Phi_{j,1}(x_1,t,\xi'))\sum_{k\neq j}\bar{L}_{(j,k)}(t,\xi')F'^{(1,i-1)}_{(k,1)}(t,x',\xi')dt + \delta_{i,0}\delta_{j,1}$$

on $\Omega'^{(1),\theta}(r,C_1)$. Let $C >> 1$. It is easy to see that

$$|\ F'^{(1,i)}_{(j,1)}(x,\xi') - \delta_{i,0}\delta_{j,1}\ | \leq (C\rho^{(1),\mp}(x_1)(\mathrm{Im}\ \xi_n)^{\frac{m-1}{m}})^{i+1}(\mathrm{Im}\ \xi_n)^{\frac{(m-1)(j-1)}{m}}$$

on $\Omega'^{(1),\theta}(r,C_1)$, where $\rho^{(1),\mp}(x_1)$ denotes the length of $\gamma^{(1),\mp}(x_1)$. We may assume that $\rho^{(1),\mp}(x_1) << 1/C$ and it follows that

$$F'^{(1)}(x,\xi') = \sum_i F'^{(1,i)}(x,\xi')$$

is convergent. More precise calculation shows that

$$|\ F'^{(1,i)}_{(j,1)}(x,\xi') - \delta_{i,0}\delta_{j,1}\ | << (\mathrm{Im}\ \xi_n)^{\frac{(m-1)(j-1)}{m}},$$

and it follows that $F'^{(1)}(x,\xi') \in (\mathcal{O}(\Omega'^{(1),\theta}(r,C_1)))^{m\times m}$ is invertible.

We next consider

$$(21)\quad \partial_{x_1}F''^{(1)}(x,\xi') + \bar{M}'^{(1)}(x,\xi')F''^{(1)}(x,\xi') - F''^{(1)}(x,\xi')\bar{M}''^{(1)}(x,\xi') = O,$$

where

$$F''^{(1)}(x,\xi') = \begin{pmatrix} 1, & *, & \cdots & * \\ & 1, & & 0 \\ & & \ddots & \\ 0 & & & 1 \end{pmatrix},\quad \bar{M}''^{(1)}(x,\xi') = \begin{pmatrix} \varphi_1, & 0, & \cdots & 0 \\ 0, & & & \\ \vdots & & * & \\ 0, & & & \end{pmatrix}.$$

We can solve this equation in $\Omega'^{(1),\theta}(r, C_1)$ similarly. Calculating the $(1, j)$-component of (20) for $j \geq 2$, it follows that

$$
\begin{aligned}
&\partial x_1 F''^{(1)}_{(1,j)}(x,\xi') + \varphi_1(x,\xi')F''^{(1)}_{(1,j)}(x,\xi') \\
(22) \qquad & - \sum_{1\leq k\leq m} F''^{(1)}_{(1,k)}(x,\xi')\bar{M}''^{(1)}_{(k,j)}(x,\xi') = 0.
\end{aligned}
$$

If we can solve these equations, $\bar{M}''^{(1)}(x,\xi')$ is given by

$$\bar{M}''^{(1)}_{(i,j)}(x,\xi') = \bar{M}'^{(1)}_{(i,j)}(x,\xi'), \qquad i, j \geq 2.$$

We can also solve (22) by successive approximation as above, and the solution is given by $F''^{(1)}_{(1,j)}(x,\xi') = \sum_{i\in\mathbf{Z}_+} F''^{(1,i)}_{(1,j)}(x,\xi')$, where

$$
\begin{aligned}
&F''^{(1,i)}_{(1,j)}(x,\xi') \\
&= - \int_{\gamma^{(1),\pm}(x_1)} \exp(\Phi_{j,1}(x_1,t,\xi')) \sum_{k\neq j} F''^{(1,i-1)}_{(1,k)}(t,x',\xi')\bar{M}''^{(1)}_{(k,j)}(t,\xi')dt + \delta_{i,0}\delta_{j,1}.
\end{aligned}
$$

Let $F^{(1)}(x,\xi') = F'^{(1)}(x,\xi')F''^{(1)}(x,\xi')$. Then $F^{(1)}(x,\xi')$ is invertible and

$$\partial_{x_1}F^{(1)}(x,\xi') + \bar{L}(x,\xi')F^{(1)}(x,\xi') - F^{(1)}(x,\xi')\bar{M}''^{(1)}(x,\xi') = O.$$

Let us show that in this way we can inductively construct an invertible $F^{(j)}(x,\xi')$
$\in (\mathcal{O}(\Omega'^{(j),\theta}(r,C_j)))^{m\times m}$ satisfying

$$\partial_{x_1}F^{(j)}(x,\xi') + \bar{L}(x,\xi')F^{(j)}(x,\xi') - F^{(j)}(x,\xi')\bar{M}''^{(j)}(x,\xi') = O,$$

on $\Omega'^{(j),\theta}(r,C_j)$, where

$$
\bar{M}''^{(j)}(x,\xi') = \left(\begin{array}{ccc|c}
\varphi_1, & & & 0 \\
& \ddots & & \\
& & \varphi_j, & \\
\hline
0 & & & *
\end{array}\right)
\begin{array}{c} \uparrow \\ j \\ \downarrow \\ \uparrow \\ m-j \\ \downarrow \end{array} .
$$

$$\leftarrow \; j \; \rightarrow \leftarrow \; m-j \; \rightarrow$$

Assume that $1 \leq j_0 \leq m-1$ and that this is true for $1 \leq j \leq j_0$. Let us consider the case $j = j_0+1$. Let $\pm \operatorname{Im} a_j(x^*) > 0$. We first look for $F'^{(j)}(x,\xi')$ satisfying

(23)

$$\partial_{x_1}F'^{(j)}(x,\xi') + \bar{M}''^{(j-1)}(x,\xi')F'^{(j)}(x,\xi') - F'^{(j)}(x,\xi')\bar{M}'^{(j)}(x,\xi') = O,$$

where

$$
F'^{(j)}(x,\xi') = \begin{pmatrix} 1, & & & & & & 0 \\ & \ddots & & & & & \\ & & 1, & & & & \\ & & & * & & & \\ & & & * & 1, & & \\ & & & \vdots & & \ddots & \\ 0 & & & \underset{\substack{\frown \\ j}}{*} & 0 & & 1 \end{pmatrix},
$$

and

$$
\bar{M}'^{(j)}(x,\xi') = \begin{pmatrix} \varphi_1, & & & & 0 \\ & \ddots & & & \\ & & \varphi_j, & \multicolumn{2}{|c}{} \\ & & & \multicolumn{2}{|c}{*} \\ 0 & & & \multicolumn{2}{|c}{} \end{pmatrix} \begin{matrix} \uparrow \\ j-1 \\ \downarrow \\ \uparrow \\ m-j+1 \\ \downarrow \end{matrix} .
$$

$$
\leftarrow \quad j \quad \rightarrow\leftarrow \quad m-j \quad \rightarrow
$$

We need to note that $\Omega'^{(j-1),\theta}(r, C_{j-1}) \subset \Omega'^{(j),\theta}(r, C_j)$, and $\bar{M}''^{(j-1)}(x,\xi')$ is defined on $\Omega'^{(j-1),\theta}(r, C_{j-1})$. Therefore this equation is well-defined on $\Omega'^{(j),\theta}(r, C_j)$. Calculating the (k,j)-component of (23) for $j \le k \le m$, it follows that

$$
\partial_{x_1} F'^{(j)}_{(k,j)}(x,\xi') + \sum_{j \le \ell \le m} \bar{M}'^{(j)}_{(k,\ell)}(x,\xi') F'^{(j)}_{(\ell,j)}(x,\xi') - F'^{(j)}_{(k,j)}(x,\xi')\varphi_j(x,\xi') = 0.
$$

We can similarly calculate $F'^{(j)}(x,\xi') = \sum_i F'^{(j,i)}(x,\xi')$, where

$$
\begin{aligned}
&F'^{(j,i)}_{(k,j)}(x,\xi') \\
&= - \int_{\gamma^{(j),\mp}(x_1)} \exp(\Phi_{k,j}(x_1,t,\xi')) \sum_{\ell \ne k} \bar{M}''^{(j-1)}_{(k,\ell)}(t,\xi') F'^{(j,i-1)}_{(\ell,j)}(t,x',\xi')dt + \delta_{i,0}\delta_{j,k}.
\end{aligned}
$$

Here note that if $k \ge j$, and $t \in \gamma^{(j),\mp}$, then $\operatorname{Re}\Phi_{k,j}(x_1,t,\xi') \le 0$, and thus the sum is convergent. We can also prove that $F'^{(j)}(x,\xi')$ is invertible, and as for $\bar{M}'^{(j)}(x,\xi')$, we only need to set

$$
\begin{aligned}
\bar{M}'^{(j)}_{(k,\ell)}(x,\xi') &= ((F'^{(j)}(x,\xi'))^{-1}\bar{M}''^{(j-1)}(x,\xi'))_{(k,\ell)}, \\
&\qquad j \le k \le m,\ j+1 \le \ell \le m.
\end{aligned}
$$

We next consider

$$\partial_{x_1} F''^{(j)}(x,\xi') + \bar{M}'^{(j)}(x,\xi')F''^{(j)}(x,\xi') - F''^{(j)}(x,\xi')\bar{M}''^{(j)}(x,\xi') = O,$$

where

$$F''^{(j)}(x,\xi') = \begin{pmatrix} 1, & & & & & \\ & \ddots & & & & 0 \\ & & 1, & * & \cdots & * \\ & & & 1, & & 0 \\ & & & & \ddots & \\ 0 & & & & & 1 \end{pmatrix} (j\ .$$

We can similarly calculate $F''^{(j)}(x,\xi') = \sum_i F''^{(j,i)}(x,\xi')$, where

$$F''^{(j,i)}_{(j,k)}(x,\xi')$$

$$= - \int_{\gamma^{(j)},\pm(x_1)} \exp(\Phi_{j,k}(x_1,t,\xi')) \sum_{\ell \neq j} \bar{M}'^{(j)}_{(j,\ell)}(t,\xi')F''^{(j,i-1)}_{(\ell,k)}(t,x',\xi')dt + \delta_{i,0}\delta_{j,k}$$

for $j+1 \le k \le m$, and $\bar{M}''^{(j)}_{(k,\ell)}(x,\xi') = \bar{M}'^{(j)}_{(k,\ell)}(x,\xi')$ for $j+1 \le k,\ell \le m$. Now we obtain (23) for

$$F^{(j)}(x,\xi') = F^{(j-1)}(x,\xi')F'^{(j)}(x,\xi')F''^{(j)}(x,\xi').$$

Note that we have

$$\bar{M}''^{(m-1)}(x,\xi') = \begin{pmatrix} \varphi_1 & & 0 \\ & \ddots & \\ 0 & & \varphi_m \end{pmatrix}.$$

We have obtained an invertible $F = F'^{(m)}(x,\xi') \in (\mathcal{O}(\Omega'^{(m),\theta}(r,C_m))^{m\times m}$ satisfying (13).

In the notation of the previous section, we have $\Omega'^{(m),\theta}(r,C_m) \subset \Omega'^{\theta}_0(r')$ if $0 < r' << 1$, and thus we need to estimate this matrix also on $\Omega''^{\theta}_0(r')$. But this is very easy. Let $(x,\xi') \in \Omega''^{\theta}_0(r')$. For this (x',ξ') we take a complex number x_1^o satisfying $(x_1^o,x',\xi') \in \Omega'^{\theta}_0(r') \cap \Omega''^{\theta}_0(r')$. For instance, we may take $x_1^o = Re^{\sqrt{-1}\theta}$, where R is a number satisfying $r'(\mathrm{Im}\ \xi_n)^{-1/mq} < R < 2r'(\mathrm{Im}\ \xi_n)^{-1/mq}$. $F' = F$ is a solution of

$$\partial_{x_1} F'(x,\xi') + \bar{L}(x,\xi')F'(x,\xi') - F'(x,\xi')\bar{M}(x,\xi') = O,$$
$$F'(x_1^o,x',\xi') = F(x_1^o,x',\xi')$$

on $\Omega'^{\theta}_0(r')$. It can be extended to a whole neighborhood of $x_1 = x_1^o$, and it is easy to see that

$$| F_{(j,k)}(x,\xi') | \le C(\mathrm{Im}\ \xi_n)^{\frac{(m-1)(j-k)}{m}} \exp\{C| x_1 - x_1^o | \mathrm{Im}\ \xi_n\} + | F_{(j,k)}(0,x',\xi') |$$

for some $C > 0$. However we have $|\, x_1 - x_1^o \,| \leq C(\mathrm{Im}\ \xi_n)^{-1/mq}$ on $\Omega''^{\theta}_0(r')$, and it follows that

$$|\, F_{(j,k)}(x, \xi') \,| \leq 2C(\mathrm{Im}\ \xi_n)^{\frac{(m-1)(j-k)}{m}} \exp\{C^2(\mathrm{Im}\ \xi_n)^{\frac{mq-1}{mq}}\}$$

there (and also on $\Omega'^{\theta}_0(r')$).

Therefore we obtain **Proposition 2**. In this article we do not discuss about the calculation of the Stokes operator any more.

References

[1] T. Aoki, *Symbols and formal symbols of pseudodifferential operators*, Advanced Studies in Pure Mathematics **4** (1984), 181–208.
[2] L. Hörmander, *Fourier integral operators I*, Acta Math. **127** (1971), 79–183.
[3] M. Iwano and Y. Shibuya, *Reduction of the order of a linear ordinary differential equation containing a small parameter*, Kōdai Math. Report **15** (1963), 1–28.
[4] M. Kashiwara and T. Kawai, *Microhyperbolic pseudodifferential operators I*, J. Math. Soc. Japan **27** (1975), 359–404.
[5] S. Nakane, *Propagation of singularities and uniqueness in the Cauchy problem at a class of doubly characteristic points*, Comm. Partial Differential Equations **6** (1981), 917–927.
[6] M. Sato, T. Kawai, and M. Kashiwara, *Microfunctions and pseudo-differential equations*, Lecture Notes in Math. **287** (H. Komatsu, ed.), Springer, 1973, pp. 265–529.
[7] H. Yamane, *Branching of singularities for some second or third order microhyperbolic operators*, J. Math. Sci. Univ. Tokyo **2** (1995), 671-749.

Part II
Mathematical Physics

Instanton-type formal solutions to the second Painlevé equations with a large parameter

Dedicated to Professor H. Komatsu on his sixtieth birthday

Takashi Aoki

Department of Mathematics and Physics, Kinki University, Higashi-Osaka, Osaka 577, Japan

Introduction

The purpose of this article is to give some calculation concerned with a part of the paper [AKT3]. In the paper, we have constructed families of formal solutions having two arbitrary constants to the Painlevé equations with a large parameter and discussed formal reduction of the second Painlevé equations to the first Painlevé equation (with a large parameter). First few terms of the formal solutions are given in the appendix of the paper for the cases of the first and the second Painlevé equations, but detailed calculation has not been given there. Thus, we present here the construction of formal solutions to the second Painlevé equations in detail. For the first Painlevé case, see [A].

The contents of this article are based on the collaboration with Professor T. Kawai and Professor Y. Takei. The author would thank them for their many valuable suggestions and stimulating discussions with them.

1 Formal power series solutions to the second Painlevé equations

Let us consider the second Painlevé equation with a large parameter η:

$$(P_{\mathrm{II}}): \quad \frac{d^2\lambda}{dt^2} = \eta^2(2\lambda^3 + t\lambda + \alpha),$$

where α is a non-zero complex number. Standard description of the Painlevé equations does not have such a large parameter η. See [KT2] for the origin of the large parameter η. We look for a formal solution to (P_{II}) which is expanded in a series of negative powers of η. Put the expression

$$\lambda = \lambda(t,\eta) = \sum_{j=0}^{\infty} \eta^{-j}\lambda_j(t) \tag{1}$$

into (P_{II}) and compare the coefficients of the powers of η of both sides. Then we find the following recursive relations for $\lambda_j = \lambda_j(t)$:

$$\lambda_0 = \lambda_0(t) \text{ is a root of the cubic equation } 2\lambda_0^3 + t\lambda_0 + \alpha = 0, \tag{2}$$

$$\lambda_{j+2} = \frac{1}{\Delta}(\lambda_j'' - 2 \sum_{\substack{j_1+j_2+j_3=j+2 \\ j_1\cdot j_2\cdot j_3<j+2}} \lambda_{j_1}\lambda_{j_2}\lambda_{j_3}) \quad (j \geq -1). \tag{3}$$

Here we set $\Delta = 6\lambda_0^2 + t$. Conversely, if $\{\lambda_j\}$ satisfies (2) and (3), then the formal series $\lambda = \lambda(t, \eta)$ defined by (1) satisfies (P_{II}). Hence we have

Proposition 1 ([KT2]). *There exists a formal solution λ to (P_{II}) of the form*

$$\lambda = \sum_{j=0}^{\infty} \eta^{-j}\lambda_j(t)$$

for which the leading term λ_0 is a root of the cubic equation $2\lambda_0^3 + t\lambda_0 + \alpha = 0$. The construction is done in a recursive manner and each λ_j is uniquely determined once the branch of the root λ_0 is fixed. Furthermore, λ_{2j-1} vanishes identically for every $j = 1, 2, 3, \cdots$.

Remark. In [KT2], the construction is done by using the Hamiltonian systems of [O]. This solution is denoted by λ_{II}.

It is easy to see that λ_0' and Δ' can be written in terms of λ_0 and Δ:

$$\lambda_0' = -\frac{\lambda_0}{\Delta}, \tag{4}$$

$$\Delta' = 1 - 12\frac{\lambda_0^2}{\Delta}. \tag{5}$$

By using these relations, one can observe that λ_{2j} $(j = 1, 2, \cdots)$ is written in the form

$$\lambda_{2j} = \frac{\lambda_0}{\Delta^{3j}} \sum_{k=0}^{2j-1} c_{j,k} \left(\frac{\lambda_0^2}{\Delta}\right)^k,$$

where $c_{j,k}$ are integers. The first few terms of λ_{2j} are as follows:

$$\lambda_2 = \frac{\lambda_0}{\Delta^3}\left(2 - 12\frac{\lambda_0^2}{\Delta}\right),$$

$$\lambda_4 = \frac{\lambda_0}{\Delta^6}\left(40 - 1656\frac{\lambda_0^2}{\Delta} + 15984\left(\frac{\lambda_0^2}{\Delta}\right)^2 - 42336\left(\frac{\lambda_0^2}{\Delta}\right)^3\right),$$

$$\lambda_6 = \frac{\lambda_0}{\Delta^9}\left(2240 - 238336\frac{\lambda_0^2}{\Delta} + 6768576\left(\frac{\lambda_0^2}{\Delta}\right)^2 - 75689856\left(\frac{\lambda_0^2}{\Delta}\right)^3 \right.$$
$$\left. + 360187776\left(\frac{\lambda_0^2}{\Delta}\right)^4 - 609638400\left(\frac{\lambda_0^2}{\Delta}\right)^5\right).$$

2 Instanton-type formal solutions to the second Painlevé equations

One can regard the formal solution constructed in the preceding section as an analogue of a WKB solution in the exact WKB analysis of Schrödinger equations (cf. [AKT1], [AKT2]). For a Schrödinger equation, we can write down the general (formal) solution as a linear combination of the WKB solutions of the equation. But the Painlevé equations are nonlinear and the structure of the set of all of their solutions is not simple. We want to know the form of the general solutions to the second Painlevé equations. Thus, we will construct a family of formal solutions that have two free parameters.

We employ the method of multiple-scale analysis (cf. [BO]). It is [JK] that first used this method in the analysis of the first and the second Painlevé equations. We note that our procedure can be applied not only to the first and the second Painlevé equations but also to the other Painlevé equations (P_J) $(J = \mathrm{III}, \cdots, \mathrm{IV})$ (cf. [AKT3]) and we can deal with all the terms of formal solutions.

First we take the following change of unknown function in (P_{II}):

$$\lambda = \lambda_0 + \eta^{-\frac{1}{2}} \Lambda,$$

where λ_0 is the same as in the preceding section. Then we see that Λ satisfies the equation

$$\frac{d^2 \Lambda}{dt^2} = \eta^2 \cdot \Delta \Lambda + \eta^{\frac{3}{2}} \cdot 6\lambda_0 \Lambda^2 + \eta \cdot 2\Lambda^3 - \eta^{\frac{1}{2}} \lambda_0''. \tag{6}$$

We next introduce a new independent variable τ and a function $T = T(t, \eta)$ by

$$T(t,\eta) = \eta \int \sqrt{\Delta} dt.$$

We look for a solution $\Lambda = \Lambda(t, \eta)$ to (6) in the form

$$\Lambda(t,\eta) = L(t,\tau,\eta)|_{\tau = T(t,\eta)}.$$

Since $\frac{d}{dt} = \frac{\partial}{\partial t} + \frac{\partial T}{\partial t}\frac{\partial}{\partial \tau}$, $L(t, \tau, \eta)$ should satisfy, as a function of (t, τ), the following equation:

$$\frac{\partial^2 L}{\partial \tau^2} - L = \eta^{-\frac{1}{2}} \frac{6\lambda_0}{\Delta} L^2 + \eta^{-1} \left(\frac{2}{\Delta} L^3 - \frac{2}{\sqrt{\Delta}} \frac{\partial^2 L}{\partial t \partial \tau} - \frac{\sqrt{\Delta}'}{\Delta} \frac{\partial L}{\partial \tau} \right) \tag{7}$$
$$- \eta^{\frac{3}{2}} \frac{\lambda_0''}{\Delta} - \eta^{-2} \frac{1}{\Delta} \frac{\partial L}{\partial t^2}.$$

Suppose that L has an expansion of the form

$$L = \sum_{k=0}^{\infty} \eta^{-\frac{k}{2}} L_{\frac{k}{2}}(t,\tau).$$

Put this into (7) and compare the coefficients of the both sides. Then we have the following series of differential equations for $\{L_{\frac{k}{2}}\}$:

$$\frac{\partial^2 L_0}{\partial \tau^2} - L_0 = 0, \tag{8}$$

$$\frac{\partial^2 L_{\frac{1}{2}}}{\partial \tau^2} - L_{\frac{1}{2}} = \frac{6\lambda_0}{\Delta} L_0^2, \tag{9}$$

$$\frac{\partial^2 L_1}{\partial \tau^2} - L_1 = \frac{6\lambda_0}{\Delta} \cdot 2L_0 L_{\frac{1}{2}} + \frac{2}{\Delta} L_0^3 - \frac{2}{\sqrt{\Delta}} \frac{\partial L_0}{\partial t \partial \tau} - \frac{\sqrt{\Delta}'}{\Delta} \frac{\partial L_0}{\partial \tau}, \tag{10}$$

$$\begin{aligned} \frac{\partial^2 L_{\frac{3}{2}}}{\partial \tau^2} - L_{\frac{3}{2}} = \frac{6\lambda_0}{\Delta}(2L_0 L_1 + L_{\frac{1}{2}}^2) + \frac{2}{\Delta} \cdot 3L_0^2 L_{\frac{1}{2}} - \frac{2}{\sqrt{\Delta}} \frac{\partial L_{\frac{1}{2}}}{\partial t \partial \tau} \\ - \frac{\sqrt{\Delta}'}{\Delta} \frac{\partial L_{\frac{1}{2}}}{\partial \tau} - \frac{\lambda_0''}{\Delta}, \end{aligned} \tag{11}$$

$$\begin{aligned} \frac{\partial^2 L_{\frac{k}{2}}}{\partial \tau^2} - L_{\frac{k}{2}} = \frac{6\lambda_0}{\Delta} \sum_{k_1+k_2=k-1} L_{\frac{k_1}{2}} L_{\frac{k_2}{2}} + \frac{2}{\Delta} \sum_{k_1+k_2+k_3=k-2} L_{\frac{k_1}{2}} L_{\frac{k_2}{2}} L_{\frac{k_3}{2}} \\ - \frac{2}{\sqrt{\Delta}} \frac{\partial L_{\frac{k-2}{2}}}{\partial t \partial \tau} - \frac{\sqrt{\Delta}'}{\Delta} \frac{\partial L_{\frac{k-2}{2}}}{\partial \tau} - \frac{1}{\Delta} \frac{\partial^2 L_{\frac{k-4}{2}}}{\partial t^2} \quad (k \geq 4). \end{aligned} \tag{12}$$

If we have a sequence of functions $L_{\frac{k}{2}}(t, \tau)$ satisfying (8)-(12), we find a formal solution

$$\Lambda = \sum_{k=0}^{\infty} \eta^{-\frac{k}{2}} L_{\frac{k}{2}}(t, \tau)|_{\tau = T(t)}$$

to (6). We can solve (8) easily:

$$L_0 = a_1^{(0)} e^{\tau} + a_{-1}^{(0)} e^{-\tau},$$

where $a_{\pm 1}^{(0)} = a_{\pm 1}^{(0)}(t)$ are arbitrary functions of t. They will be determined soon. If we regard them as being given, the right-hand side of (9) is known and we find a solution $L_{\frac{1}{2}}$ to (9) of the form

$$L_{\frac{1}{2}} = a_2^{(\frac{1}{2})} e^{2\tau} + a_0^{(\frac{1}{2})} + a_{-2}^{(\frac{1}{2})} e^{-2\tau},$$

where

$$\begin{cases} a_2^{(\frac{1}{2})} = \dfrac{2\lambda_0^2}{\Delta} a_1^{(0)2}, \\ a_0^{(\frac{1}{2})} = -\dfrac{12\lambda_0}{\Delta} a_1^{(0)} a_{-1}^{(0)}, \\ a_{-2}^{(\frac{1}{2})} = \dfrac{2\lambda_0^2}{\Delta} a_{-1}^{(0)2}. \end{cases}$$

Then the right-hand side of (10) can be written in terms of $a_{\pm 1}^{(0)}$, λ_0 and Δ. We require here the condition of non-secularity (cf. [BO]) on the right-hand side:

the coefficients of $e^{\pm\tau}$ must vanish. This yields a system of ordinary differential equations for $a_{\pm 1}^{(0)}$:

$$\begin{cases} \dfrac{\partial a_1^{(0)}}{\partial t} + \dfrac{(\sqrt{\Delta})'}{2\sqrt{\Delta}} a_1^{(0)} - \dfrac{3}{\sqrt{\Delta}} \left(1 - 20\dfrac{\lambda_0^2}{\Delta}\right) a_1^{(0)2} a_{-1}^{(0)} = 0, \\ \dfrac{\partial a_{-1}^{(0)}}{\partial t} + \dfrac{(\sqrt{\Delta})'}{2\sqrt{\Delta}} a_{-1}^{(0)} + \dfrac{3}{\sqrt{\Delta}} \left(1 - 20\dfrac{\lambda_0^2}{\Delta}\right) a_1^{(0)} a_{-1}^{(0)2} = 0. \end{cases} \tag{13}$$

Multiplying the first equation by $a_{-1}^{(0)}$ and the second by $a_1^{(0)}$, and adding them up, we find the following simple equation for $a_1^{(0)}a_{-1}^{(0)}$:

$$\frac{\partial}{\partial t}(a_1^{(0)}a_{-1}^{(0)}) + \frac{(\sqrt{\Delta})'}{\sqrt{\Delta}} a_1^{(0)}a_{-1}^{(0)} = 0.$$

Hence we have

$$a_1^{(0)}a_{-1}^{(0)} = C\Delta^{-\frac{1}{2}}. \tag{14}$$

Here C is a constant. Putting this into (13) and using (4) and (5), we have

$$\begin{cases} \dfrac{\partial a_1^{(0)}}{\partial t} + \left(\left(\dfrac{1}{4} - 5C\right)\dfrac{\Delta'}{\Delta} - 2C\dfrac{\lambda_0'}{\lambda_0}\right) a_1^{(0)} = 0, \\ \dfrac{\partial a_{-1}^{(0)}}{\partial t} + \left(\left(\dfrac{1}{4} + 5C\right)\dfrac{\Delta'}{\Delta} + 2C\dfrac{\lambda_0'}{\lambda_0}\right) a_{-1}^{(0)} = 0. \end{cases} \tag{15}$$

Thus we see that $a_1^{(0)}$ and $a_{-1}^{(0)}$ are written in the form

$$\begin{cases} a_1^{(0)} = c_+\Delta^{5C-\frac{1}{4}}\lambda_0^{2C}, \\ a_{-1}^{(0)} = c_-\Delta^{-5C-\frac{1}{4}}\lambda_0^{-2C}, \end{cases} \tag{16}$$

where c_+ and c_- are constants. It follows from (14) that the relation

$$c_+c_- = C$$

holds. Hence we have determined $a_\pm^{(0)}$:

$$\begin{cases} a_1^{(0)} = c_+\Delta^{5c_+c_- -\frac{1}{4}}\lambda_0^{2c_+c_-}, \\ a_{-1}^{(0)} = c_-\Delta^{-5c_+c_- -\frac{1}{4}}\lambda_0^{-2c_+c_-}. \end{cases} \tag{17}$$

Thus we get L_0 and $L_{\frac{1}{2}}$:

$$L_0 = \frac{c_+}{\Delta^{\frac{1}{4}}}\theta(t)^{c_+c_-}e^{\tau} + \frac{c_-}{\Delta^{\frac{1}{4}}}\theta(t)^{-c_+c_-}e^{-\tau},$$

$$L_{\frac{1}{2}} = \frac{2c_+^2\lambda_0}{\Delta^{\frac{3}{2}}}(\theta(t)^{c_+c_-}e^{\tau})^2 - \frac{12c_+c_-\lambda_0}{\Delta^{\frac{3}{2}}} + \frac{2c_-^2\lambda_0}{\Delta^{\frac{3}{2}}}(\theta(t)^{c_+c_-}e^{\tau})^{-2},$$

where we set $\theta(t) = \Delta^5\lambda_0^2$.

Suppose that we have constructed $L_{\frac{k}{2}}$ for $k = 0, 1, \cdots, 2j-1$ in the form

$$L_{\frac{k}{2}} = \sum_{l=0}^{k+1} b_{k+1-2l}^{(\frac{k}{2})}(t)(\theta(t)^{c_+c_-}e^{\tau})^{k+1-2l} \tag{18}$$

and that $b_{\pm 1}^{(j-1)}$ are determined by the non-secularity condition for the equation for L_j: the coefficients of $e^{\pm\tau}$ in the right-hand side are equal to zero. Suppose further that each $b_l^{(\frac{k}{2})}(t)$ has the form

$$b_l^{(\frac{k}{2})}(t) = \frac{g_l^{(\frac{k}{2})}(r)}{\Delta^{\frac{1}{4}+\frac{3}{4}k}}. \tag{19}$$

Here, for even k (hence l is odd), $g_l^{(\frac{k}{2})}(r)$ is a power series of $r = r(t) = \dfrac{\lambda_0^2}{\Delta}$ and for odd k (hence l is even), $g_l^{(\frac{k}{2})}(r)/\sqrt{r}$ is a power series of r and the power series converges if $|r| < \frac{1}{4}$ in each case. We note that the form of (18) is based on the observation that the coefficient of $e^{l\tau}$ in $L_{\frac{k}{2}}$ always has a factor $\theta(t)^{c_+c_-l}$; we factor this term out and let $b_l^{(\frac{k}{2})}$ be the quotient.

Under the above assumption of induction, we see that the right-hand side of the equation for L_j is known and that it does not have any even powers of $e^{\pm\tau}$. Hence there is a solution L_j to the equation of the form

$$L_j = \sum_{l=0}^{2j+1} b_{2j+1-2l}^{(j)}(t)(\theta(t)^{c_+c_-}e^{\tau})^{2j+1-2l},$$

where $b_{2j+1-2l}^{(j)}(t)$ are uniquely determined in the form (19) except for $b_{\pm 1}^{(j)}(t)$ which will be determined soon. Now the right-hand side of the equation for $L_{j+\frac{1}{2}}$ is written in terms of known functions and $b_{\pm 1}^{(j)}(t)$. The coefficients of the odd powers of $e^{\pm\tau}$ of the right-hand side of the equation are equal to zero and hence it is natural to require that $L_{j+\frac{1}{2}}$ has the form

$$L_{j+\frac{1}{2}} = \sum_{l=0}^{2j+2} b_{2j+2-2l}^{(j+\frac{1}{2})}(t)(\theta(t)^{c_+c_-}e^{\tau})^{2j+2-2l}. \tag{20}$$

If we suppose this form, the coefficients $b_{2j+2-2l}^{(j+\frac{1}{2})}(t)$ are uniquely determined. Then the right-hand side of the equation for L_{j+1} is written in terms of known functions and $b_{\pm 1}^{(j)}(t)$. Now we impose the condition of non-secularity for the equation: the coefficients of $e^{\pm\tau}$ in the right-hand side of the equation for L_{j+1} should vanish. This yields a system of inhomogeneous linear differential equations of first order for $b_{\pm 1}^{(j)}(t)$ of the form:

$$\left(\frac{d}{dt} + A\right)\begin{pmatrix} b_1^{(j)} \\ b_{-1}^{(j)} \end{pmatrix} = \begin{pmatrix} f_1^{(j)} \\ f_{-1}^{(j)} \end{pmatrix}, \tag{21}$$

where we set

$$A = \begin{pmatrix} c_{11} & c_{12} \\ c_{21} & c_{22} \end{pmatrix}$$

with

$$c_{11} = \frac{1}{4\Delta}\left\{(1 - 12c_+c_-) - 12(1 - 20c_+c_-)\frac{\lambda_0^2}{\Delta}\right\},$$

$$c_{12} = -\frac{3}{\Delta}c_+^2\left(1 - 20\frac{\lambda_0^2}{\Delta}\right),$$

$$c_{21} = \frac{3}{\Delta}c_-^2\left(1 - 20\frac{\lambda_0^2}{\Delta}\right),$$

$$c_{22} = \frac{1}{4\Delta}\left\{(1 + 12c_+c_-) - 12(1 + 20c_+c_-)\frac{\lambda_0^2}{\Delta}\right\}$$

and where $f_{\pm 1}^{(j)}$ are known functions. We see that $f_{\pm 1}^{(j)}$ have the form

$$f_{\pm 1}^{(j)} = \frac{\tilde{f}_{\pm 1}^{(j)}}{\Delta^{\frac{5}{4}+\frac{3}{2}j}}$$

with some convergent series $\tilde{f}_{\pm 1}^{(j)}$ of r that are convergent in $|r| < \frac{1}{4}$. Thus we can determine $b_{\pm 1}^{(j)}(t)$ uniquely under the assumption that they have the form

$$b_{\pm 1}^{(j)}(t) = \frac{g_{\pm 1}^{(j)}(r)}{\Delta^{\frac{1}{4}+\frac{3}{2}j}} \tag{22}$$

with some convergent series $g_{\pm 1}^{(j)}(r)$. Hence we have L_j and $L_{j+\frac{1}{2}}$. Setting $\tau = T(t,\eta)$, we obtain the following

Theorem 2. *There is a family of formal solutions to* (P_{II}) *of the form*

$$\lambda = \lambda_{c_+,c_-}(t,\eta) = \lambda_0(t) + \sum_{k=0}^{\infty}\sum_{l=0}^{k+1} b_{k+1-2l}^{(\frac{k}{2})}(t)(\theta(t)^{c_+c_-}e^{T(t,\eta)})^{k+1-2l},$$

where $b_{k+1-2l}^{(\frac{k}{2})}(t)$, $\theta(t)$ *and* $T(t,\eta)$ *are constructed as above and* c_+ *and* c_- *are arbitrary constants.*

We call these formal solutions $\lambda_{c_+,c_-}(t,\eta)$ instanton-type solutions. This terminology was first used in [KT1] to the Painlevé equations. They dealt with the case where one of $c_\pm$ equals zero. By the uniqueness (up to the choice of the branch of λ_0) of λ_{II} constructed in the first section, we see that $\lambda_{0,0} = \lambda_{\mathrm{II}}$ holds.

We note that we have not discussed the ambiguity of $b_{\pm 1}^{(j)}(t)$ that comes from null solutions of (21). Regarding the structure of the null solutions, see [AKT3]. First few terms of $b_l^{(\frac{k}{2})}(t)$ are as follows:

$$b_1^{(0)} = \frac{c_+}{\Delta^{\frac{1}{4}}},$$

$$b_{-1}^{(0)} = \frac{c_-}{\Delta^{\frac{1}{4}}},$$

$$b_2^{(\frac{1}{2})} = \frac{2c_+{}^2\sqrt{r(t)}}{\Delta},$$
$$b_0^{(\frac{1}{2})} = -\frac{12\,c_+\,c_-\sqrt{r(t)}}{\Delta},$$
$$b_{-2}^{(\frac{1}{2})} = \frac{2c_-{}^2\sqrt{r(t)}}{\Delta},$$

$$b_3^{(1)} = \frac{1}{\Delta^{\frac{7}{4}}}\left(\frac{c_+{}^3}{4} + 3\,c_+{}^3\,r(t)\right),$$
$$\begin{aligned} b_1^{(1)} = \frac{1}{\Delta^{\frac{7}{4}}}\Bigg(&\frac{3\,c_+}{16}\left(1 - 8\,c_+\,c_- + 24\,c_+{}^2\,c_-{}^2\right)\\ &+\frac{c_+}{4}\left(-15 + 88\,c_+\,c_- - 1128\,c_+{}^2\,c_-{}^2\right)r(t)\\ &+\frac{c_+ - 48\,c_+{}^3\,c_-{}^2}{12\,(-1+4\,r(t))}\Bigg), \end{aligned}$$
$$\begin{aligned} b_{-1}^{(1)} = \frac{1}{\Delta^{\frac{7}{4}}}\Bigg(&\frac{-3\,c_-}{16}\left(1 + 8\,c_+\,c_- + 24\,c_+{}^2\,c_-{}^2\right)\\ &+\frac{c_-}{4}\left(15 + 88\,c_+\,c_- + 1128\,c_+{}^2\,c_-{}^2\right)r(t)\\ &+\frac{-c_- + 48\,c_+{}^2\,c_-{}^3}{12\,(-1+4\,r(t))}\Bigg), \end{aligned}$$
$$b_{-3}^{(1)} = \frac{1}{\Delta^{\frac{7}{4}}}\left(\frac{c_-{}^3}{4} + 3\,c_-{}^3\,r(t)\right),$$

$$b_4^{(\frac{3}{2})} = \frac{\sqrt{r(t)}}{\Delta^{\frac{5}{2}}}c_+{}^4(1+4\,r(t)),$$
$$\begin{aligned} b_2^{(\frac{3}{2})} = \frac{\sqrt{r(t)}}{\Delta^{\frac{5}{2}}}\Bigg(&\frac{c_+{}^2}{4}\left(27 - 148\,c_+\,c_- + 72\,c_+{}^2\,c_-{}^2\right)\\ &+c_+{}^2\left(-55 + 324\,c_+\,c_- - 1128\,c_+{}^2\,c_-{}^2\right)r(t)\\ &+\frac{1}{3(-1+4\,r(t))}\left(c_+{}^2 - 48\,c_+{}^4\,c_-{}^2\right)\Bigg), \end{aligned}$$
$$b_0^{(\frac{3}{2})} = \frac{2\sqrt{r(t)}}{\Delta^{\frac{5}{2}}}\left(1 + 78\,c_+{}^2\,c_-{}^2\right) - 12\left(1 + 120\,c_+{}^2\,c_-{}^2\right)r(t),$$
$$\begin{aligned} b_{-2}^{(\frac{3}{2})} = -\frac{\sqrt{r(t)}}{\Delta^{\frac{5}{2}}}\Bigg(&\frac{c_-{}^2}{4}\left(27 + 148\,c_+\,c_- + 72\,c_+{}^2\,c_-{}^2\right)\\ &+c_-{}^2\left(55 + 324\,c_+\,c_- + 1128\,c_+{}^2\,c_-{}^2\right)r(t)\\ &+\frac{1}{3\,(-1+4\,r(t))}\left(-c_-{}^2 + 48\,c_+{}^2\,c_-{}^4\right)\Bigg), \end{aligned}$$
$$b_{-4}^{(\frac{3}{2})} = \frac{\sqrt{r(t)}}{\Delta^{\frac{5}{2}}}c_-{}^4(1+4\,r(t)).$$

Remarks. 1 By the concrete forms of $b_l^{(\frac{k}{2})}$ given above, we can guess the general form of $b_l^{(\frac{k}{2})}$ as follows:

$$b_l^{(\frac{k}{2})}(t) = \frac{\tilde{g}_l^{(\frac{k}{2})}(r)}{\Delta^{\frac{1}{4}+\frac{3}{4}k}(1-4r)^{\frac{k+1}{2}-\frac{|l|}{2}}}. \tag{23}$$

Here, for even k (hence l is odd), $\tilde{g}_l^{(\frac{k}{2})}(r)$ is a polynomial of r of degree $k+\frac{1}{2}-\frac{|l|}{2}$ and for odd k (hence l is even), $\tilde{g}_l^{(\frac{k}{2})}(r)/\sqrt{r}$ is a polynomial of r of degree $k-\frac{|l|}{2}$.
2 We have not discussed the convergence of the instanton-type formal solutions. Actually they are divergent series in generic case, but we conjecture that they are Borel summable in some sense. In the case where one of $c_\pm$ vanishes, [C] discusses the Borel summability of these types of formal solutions.
3 The instanton-type solutions will play a rôle when we establish general connection formulas for Painlevé equations. See [AKT3] and [T] for this problem.

References

[A] T. Aoki, *Multiple-scale analysis for Painlevé transcendents with a large parameter*, to appear in the Proceedings of the Workshop "Singularities and PDE's" held at Stefan Banach International Mathematical Center.

[AKT1] T.Aoki, T.Kawai and Y.Takei, *The Bender-Wu analysis and the Voros theory*, ICM-90 Satellite Conf. Proc. "Special Functions" (M.Kashiwara and T.Miwa, eds.) Springer-Verlag, 1991, pp. 1-29.

[AKT2] T.Aoki, T.Kawai and Y.Takei, *Algebraic analysis of singular perturbations -On exact WKB analysis*, to appear in Sugaku Expositions, AMS.

[AKT3] T.Aoki, T.Kawai and Y.Takei, *WKB analysis of Painlevé transcendents with a large parameter, II*, (to appear).

[BO] C. M. Bender and S. T. Orszag, Advanced Mathematical Methods for Scientists and Engineers, McGraw-Hill, 1978.

[C] O. Costin, *Exponential asymptotics, transseries, and generalized Borel summation for analytic, nonlinear, rank-one system of ordinary differential equations*, International Mathematics Research Notices, **8** (1995), 376-417.

[JK] N. Joshi and M. Kruskal, *Connection results for the first Painlevé equation*, "Painlevé transcendents" (D. Levi and P. Winternitz, eds.), Plenum Press, New York, 1992, pp. 61-79.

[KT1] T. Kawai and Y. Takei, *WKB analysis and deformation of Schrödinger equations*, RIMS Kôkyûroku **854**, Kyoto Univ., 1993, pp. 22-42.

[KT2] T. Kawai and Y. Takei, *WKB analysis of Painlevé transcendents with a large parameter, I*, RIMS preprint No. 1007, 1995.

[O] K. Okamoto, Isomonodromic deformation and Painlevé equations, and Garnier systems, J. Fac. Sci., Univ. Tokyo, Sect. IA, **33** (1986), 575-618.

[T] Y. Takei, On the connection formula for the first Painlevé equation, (to appear).

Pseudodifferential and Fourier integral operators in scattering theory

Dedicated to Professor Komatsu on his sixtieth birthday

Christian Gérard

Centre de Mathématiques URA 169 CNRS, Ecole Polytechnique, 91128 Palaiseau Cedex, France

1 Introduction

This text is an abridged version of a talk given at the joint CNRS-JSPS meeting "New Trends in Microlocal Analysis" in Tokyo in Spetember 1995. We describe in this talk some new results obtained jointly with Jan Dereziński about the wave operators for Schrödinger hamiltonians

$$H = \frac{1}{2}D^2 + V(x), \tag{1.1}$$

for short-range and long-range potentials V. (These results will appear in a book published by Springer Verlag in the collection Texts and Monographs in Physics).

We will denote by H_0 the free hamiltonian $\frac{1}{2}D^2$. The potentials arising in (1.1) fall naturally into two classes: the short-range potentials where roughly speaking

$$|V(x)| \leq C\langle x\rangle^{-\mu},\ \mu > 1, \tag{1.2}$$

and the long-range potentials where

$$|\partial_x^\alpha V(x)| \leq C\langle x\rangle^{-|\alpha|-\mu},\ \mu > 0, |\alpha| \leq 2. \tag{1.3}$$

For short-range potentials the wave operators are defined as

$$\mathrm{s-}\lim_{t\to\infty} \mathrm{e}^{itH}\mathrm{e}^{-itH_0} =: \Omega_{\mathrm{sr}}^+,$$

and one has

$$\mathrm{Ran}\Omega_{\mathrm{sr}}^+ = \mathcal{H}_{\mathrm{c}}(H). \tag{1.4}$$

This property of the wave operator goes under the name of asymptotic completeness.

For long-range potentials, the wave operators are defined in terms of a modified free evolution

$$\Omega_{\mathrm{lr}}^+ := \mathrm{s-}\lim_{t\to+\infty} \mathrm{e}^{itH}\mathrm{e}^{-iS(t,D)}, \tag{1.5}$$

where $S(t,\xi)$ is a solution of the Hamilton-Jacobi equation:

$$\partial_t S(t,\xi) = \frac{1}{2}\xi^2 + V(\partial_\xi S(t,\xi)).$$

Property (1.4) also holds for long-range potentials under hypothesis (1.3).

In this lecture we will describe some results about the nature of the operators $\Omega_{\rm sr}^+$ and $\Omega_{\rm lr}^+$. In particular we would like to know to what extent these operators can be represented as pseudodifferential or Fourier integral operators. Clearly to obtain this kind of results it is necessary to assume a smoothness condition on the potential. We will hence consider the following *smooth short-range condition*

$$\begin{aligned} &\lim_{|x|\to\infty} V(x) = 0, \\ &\textstyle\int_0^\infty \langle R\rangle^{|\alpha|} \sup_{|x|\geq R} |\partial_x^\alpha V(x)| \mathrm{d}R < \infty, \ |\alpha| \geq 0, \end{aligned} \tag{1.6}$$

or the *smooth long-range condition*:

$$\begin{aligned} &\lim_{|x|\to\infty} V(x) = 0, \\ &\textstyle\int_0^\infty \langle R\rangle^{|\alpha|-1} \sup_{|x|\geq R} |\partial_x^\alpha V(x)| \mathrm{d}R < \infty, \ |\alpha| \geq 1. \end{aligned} \tag{1.7}$$

The wave operators for long-range potentials are not invariantly defined. They depend on the choice of a solution of the Hamilton-Jacobi equation. It turns out that another equivalent of modified wave operators, introduced by Isozaki and Kitada is more useful.

The Isozaki-Kitada construction is based on a time-independent modifier which is a Fourier integral operator $J_{\rm lr}^+$ defined by

$$J_{\rm lr}^+\phi(x) := (2\pi)^{-n} \int \mathrm{e}^{i\Phi_{\rm lr}^+(x,\xi) - i\langle y,\xi\rangle} q^+(x,\xi)\phi(y)dy\xi, \tag{1.8}$$

associated with a solution $\Phi_{\rm lr}^+(x,\xi)$ of the eikonal equation:

$$\frac{1}{2}\xi^2 = \frac{1}{2}(\nabla_x \Phi_{\rm lr}^+(x,\xi))^2 + V(x).$$

Here $q^+(x,\xi)$ is a cutoff equal to 1 in an appropriate outgoing region defined as:

$$\Gamma_{\epsilon,\sigma}^+ := \{(x,\xi) | |\xi| \geq \epsilon, \langle x,\xi\rangle \geq \sigma|x||\xi|\}.$$

For a correct choice of the phase function $\Phi_{\rm lr}$ the wave operator (1.5) can also be recovered as the following limit:

$$\Omega_{\rm lr}^+ 1_{[\epsilon,+\infty)}(H_0) = \mathrm{s-}\lim_{t\to+\infty} e^{itH} J e^{-itH_0} 1_{[\epsilon,+\infty)}(H_0). \tag{1.9}$$

The representation (1.9) will be useful in the sequel. We denote by $S(1,g_1)$ the standard symbol class defined by the conditions

$$|\partial_x^\alpha \partial_\xi^\beta a(x,\xi)| \leq C_{\alpha,\beta} \langle x\rangle^{-|\alpha|}, \ |\alpha|, |\beta| \geq 0.$$

We will say that a symbol a belongs to $S^{-\infty}$ in outside a subset Γ of $T^*\mathbf{R}^n$ if:

$$|\partial_x^\alpha \partial_\xi^\beta a(x,\xi)| \leq C_{\alpha,\beta,N} \langle x\rangle^{-|N|}, \ \forall N, |\alpha|, |\beta| \geq 0, \ (x,\xi) \notin \Gamma.$$

A Fourier integral operator of the form (1.8) for an amplitude function $c \in S(1,g_1)$ will be denoted by $J(\Phi_{\rm lr}^+, c)$.

2 Results

Let us now describe the results. We prove that if we multiply the wave operator with a pseudodifferential cutoff supported in an outgoing region and with the energy bounded away from zero, then we obtain a pseudodifferential operator in the short-range case and a Fourier integral operator in the long-range case. The evolution of a state initially localized in an outgoing region is well approximated by classical mechanics, which is the intuitive reason for such results.

Theorem 2.1. *Let $\epsilon > \epsilon_0 > 0$ and $\sigma > \sigma_0 > -1$. Let $p_+(x,\xi) \in S(1,g_1)$ such that $p_+ \in S(\langle x\rangle^{-\infty})$ outside $\Gamma^+_{\epsilon,\sigma}$ and $\chi \in C_0^\infty(\mathbf{R}\backslash\{0\})$.*
1) Assume the smooth short-range condition (1.6). Then there exist $a_1, a_2 \in S(1,g_1)$ with $a_1, a_2 \in S(\langle x\rangle^{-\infty})$ outside $\Gamma^+_{\epsilon_0,\sigma_0}$ such that:

$$\Omega^+_{\mathrm{sr}}\chi(H_0)p_+(x,D) = a_1(x,D),$$

$$\Omega^{+*}_{\mathrm{sr}}\chi(H)p_+(x,D) = a_2(x,D).$$

2) Assume the smooth long-range condition (1.7). Then there exists $c_1, c_2 \in S(1,g_1)$ such that $c_1, c_2 \in S(\langle x\rangle^{-\infty})$ outside $\Gamma^+_{\epsilon_0,\sigma_0}$ and $r_{-\infty,1}, r_{-\infty,2} \in S(\langle x\rangle^{-\infty})$ such that

$$\Omega^+_{\mathrm{lr}}\chi(H_0)p_+(x,D) = J(\Phi^+_{\mathrm{lr}}, c_1) + r_{-\infty,1}(x,D),$$

$$\Omega^{+*}_{\mathrm{lr}}\chi(H)p_+(x,D) = J(\Phi^+_{\mathrm{lr}}, c_2)^* + r_{-\infty,2}(x,D).$$

On infrared singularities

Takahiro Kawai[1] *and Henry P. Stapp*[2]

[1] Research Institute for Mathematical Sciences, Kyoto University, Kyoto 606-01, Japan
[2] Lawrence Berkeley Laboratory, University of California, Berkeley, CA 94720, USA

The traditional separation of infrared divergent part of the S-matrix from a finite remainder ([YFS], [GY]) is effective only at points where the S-matrix is non-singular, as was pointed out in [S2]. This limitation is due primarily to the approximation

$$e^{ikx} \sim 1 \qquad (|k| \ll 1), \tag{1}$$

which is used to replace

$$\delta^4(p+k) = \int \frac{d^4x}{(2\pi)^4} e^{i(p+k)x} \qquad \text{by} \qquad \delta^4(p) = \int \frac{d^4x}{(2\pi)^4} e^{ipx}$$

in the demonstration that the infrared divergent terms originating from real photons are cancelled by those originating from virtual photons (e.g. [GY] (3.20) ff). An approximation of this sort seems to be necessary, if we treat the separation of the infrared divergent parts in momentum space. (Cf. Problem A below.) However, the separation can be neatly done in coordinate space, even at singular points of the S-matrix. ([S2]) In view of the fact that a point x in the coordinate space represents the cotangential component of the singularity spectrum of a function on the momentum space (e.g. [KS1],[Sa]) the recipe of Stapp [S2] may be regarded as the microlocalization of the traditional separation of infrared divergences. The core-spirit of microlocal analysis (e.g. [K^3]) is to make use of both p-variables and x-variables in the analysis. In fact, to study the infrared finiteness of the remainder terms (the Q-coupling part in the sense of [S2] and [KS4]) p-variables play a central role, while the cancellation of infrared divergent terms (the C-coupling part in the sense of [S2]) proceeds in coordinate space.

The purpose of this report is to list the mathematical difficulties that are encountered if one tries to stick to momentum space. As we emphasized above, one can circumvent these problems by studying the problem on the coordinate space. Still, we think they are interesting mathematical problems.

Problem A. To study the cancellation mechanism for the infrared divergence in the transition probability, let us consider its kernel function, which can be represented by a bubble diagram function. (Cf. e.g. [KS1],[I],[S1]) Besides the ordinary symbols used in the above cited references, we use the symbol

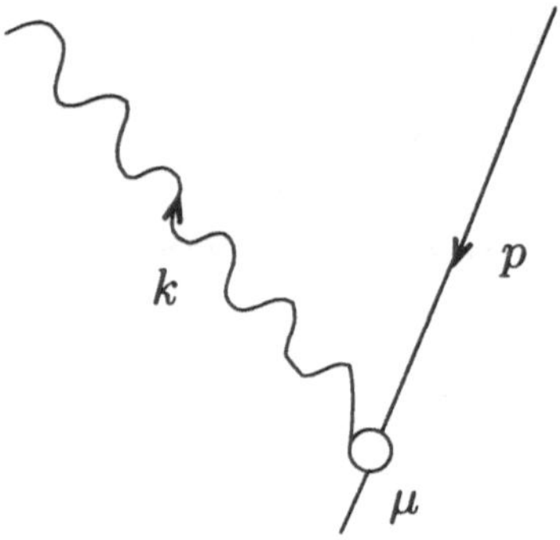

$$(\mu = 0, 1, 2, 3)$$

to designate the coupling that appears after one uses the Ward identity on the C-coupling factor, namely a coupling that in the limit $|k| \to 0$, loosely speaking, tends to

$$\frac{ip_\mu}{2pk} \tag{2}$$

where p and k are the charged-particle and photon momenta, respectively. (See [KS4] p.2489 for the precise definition.)

If we start with the simplest Feynman diagram D_0 in Figure 1, then the kernel function of the lowest order photon contribution to the probability is a sum of several terms like D_1, D_2, D_3, etc. specified below. Here and in what follows a solid line represents a charged particle with mass $m > 0$ and a dotted line represents a hard photon. The factor 1/2 in D_2 and D_3 arises from the decomposition of the photon propagator into the sum of its real part and its pure imaginary part:

$$\frac{1}{k^2 + i0} = \text{p.v.}\frac{1}{k^2} - \pi i \delta(k^2), \tag{3}$$

where $\text{p.v.}\frac{1}{k^2} = \frac{1}{2}\left(\frac{1}{k^2 + i0} + \frac{1}{k^2 - i0}\right)$. (See [KS2] for the definition (at $k = 0$) of $\delta(k^2)$ etc.) In the following we omit the index μ to simplify the notation.

Figure 1.

D_0 :

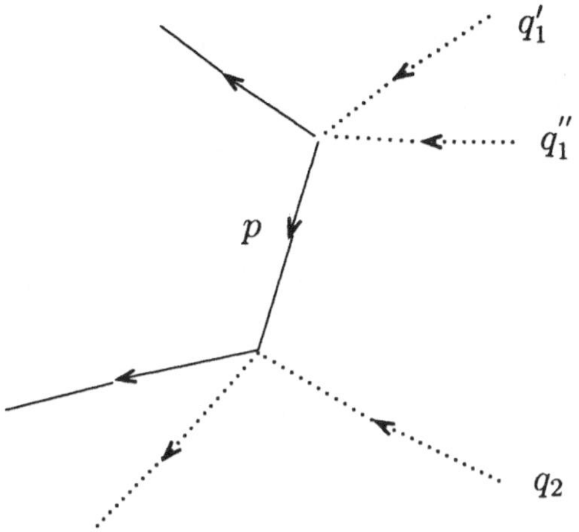

D_1 :

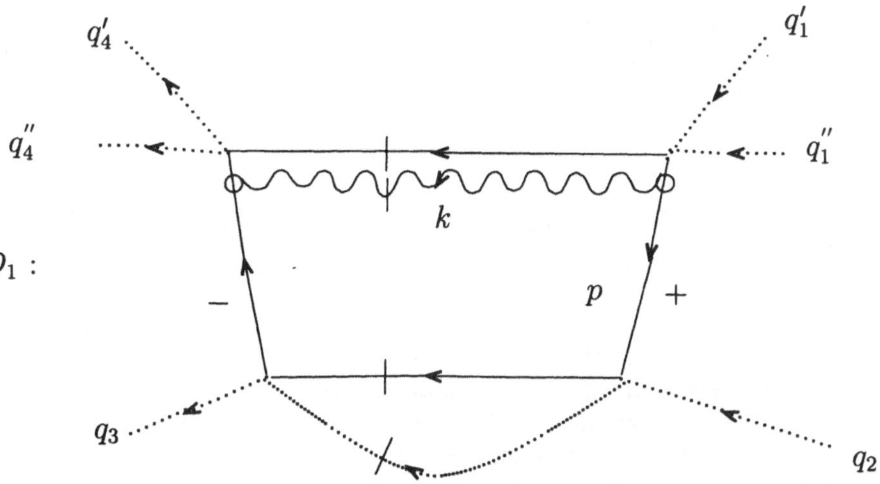

D_2 :

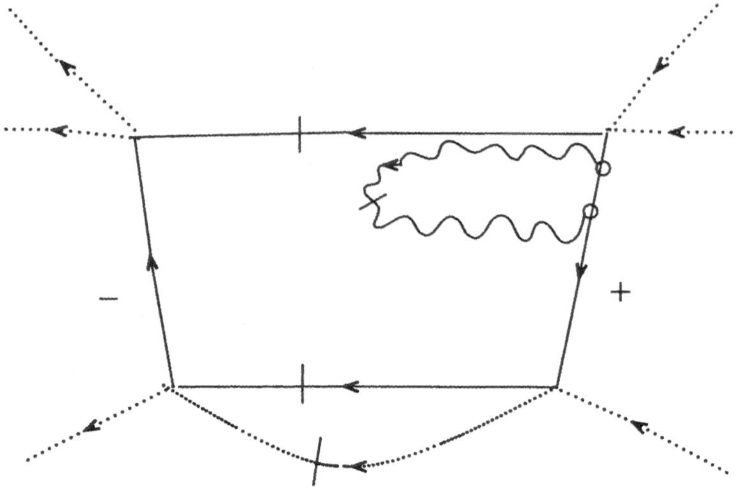

$\times\left(-\frac{1}{2}\right)$

D_3 :

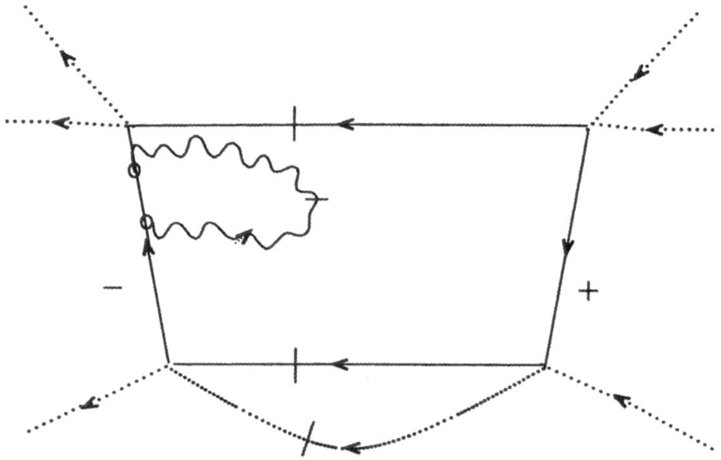

$\times\left(-\frac{1}{2}\right)$

The infrared divergent part of the contribution D_1 from a real photon is expected to be cancelled by those of the contributions D_2 and D_3 from a virtual photon. To discuss the cancellation, the reasonable way might be to look at the integrand of each contribution and to try to locate the singularities at the same place. In order to do so, one should shift both $q_1' + q_1''$ and $q_4' + q_4''$ by k in D_1; otherwise stated, if we would keep the external momenta in diagram D_1 to be the same as those in diagrams D_2 and D_3, their integrands would be singular at different places except when $k = 0$. The cancellation argument would then fail unless we employ the approximation (1) in the form of the replacement of the argument of the delta function. But this "approximation" is not mathematically justifiable when the shift in the argument is from a singular point (e.g., a pole) to a non-singular point.

Problem B. Even before considering the infrared divergence due to $\delta(k^2)$ coupled with two factors like (2), the bubble diagram functions associated with D_1 etc. do not determine a well-defined kernel function of the probability. To make this statement precise let us consider the bubble diagram function F_Δ associated with the diagram Δ given below. Then the general rule in microlocal analysis (cf. e.g. [K³]) does not allow us to legitimately restrict F_Δ to the submanifold $\{q_2 = q_3\}$ $(= \{q_2 = q_3, q_1' + q_1'' = q_4' + q_4''\})$, whereas the kernel function of the transition probability should be evaluated on the submanifold $\{q_2 = q_3, q_1' + q_1'' = q_4' + q_4''\}$.

Figure 2.

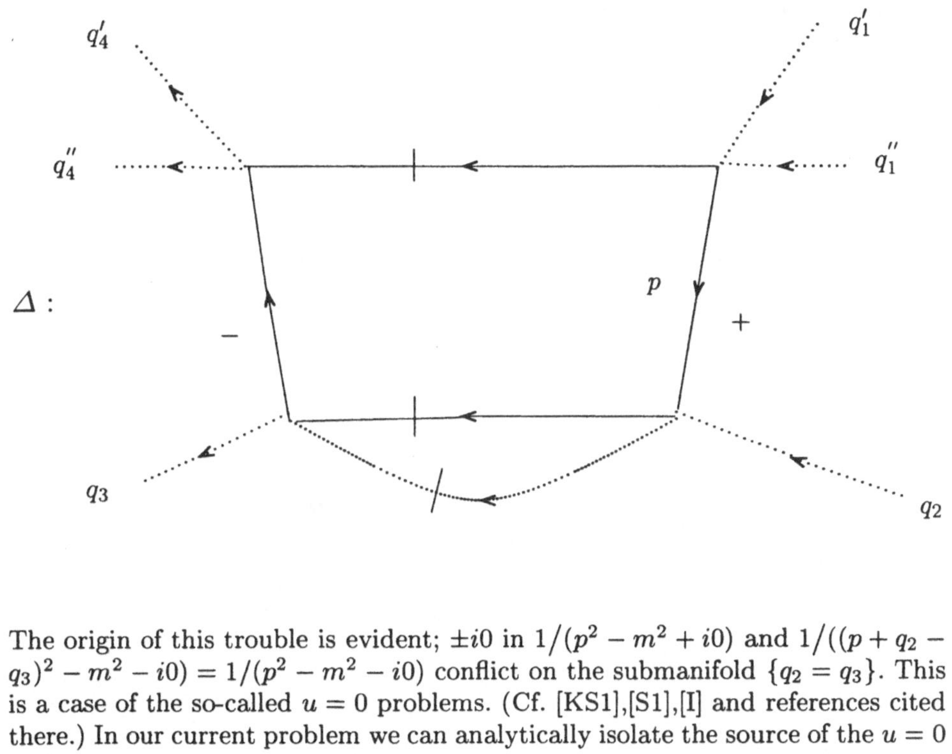

The origin of this trouble is evident; $\pm i0$ in $1/(p^2 - m^2 + i0)$ and $1/((p + q_2 - q_3)^2 - m^2 - i0) = 1/(p^2 - m^2 - i0)$ conflict on the submanifold $\{q_2 = q_3\}$. This is a case of the so-called $u = 0$ problems. (Cf. [KS1],[S1],[I] and references cited there.) In our current problem we can analytically isolate the source of the $u = 0$ problem:
if we write

$$\frac{1}{p^2 - m^2 + i0} = -2\pi i\delta(p^2 - m^2) + \frac{1}{p^2 - m^2 - i0} \tag{4}$$

then we find that the trouble originates from the following factor

$$\frac{-2\pi i\delta(p^2 - m^2)}{(p + q_2 - q_3)^2 - m^2 - i0} = \frac{-2\pi i\delta(p^2 - m^2)}{2p(q_2 - q_3) + (q_2 - q_3)^2 - i0}. \tag{5}$$

Because $q_2 - q_3 = -(q_1' + q_1'') + (q_4' + q_4'')$ holds by virtue of the over-all conservation of energy-momentum vectors, this factor is kept intact if we shift both $q_1' + q_1''$ and $q_4' + q_4''$ by k (as we did in Problem A). Thus one might hope the quotient of bubble diagram functions $(D_1^0 + D_2^0 + D_3^0)/\Delta^0$ with D_1^0 etc. being defined by replacing $(p^2 - m^2 + i0)^{-1}$ by $-2\pi i \delta(p^2 - m^2)$ in D_1 etc., would be well-defined (and rather tame) because of the cancellation of the common troublesome factor $2p(q_2 - q_3) + (q_2 - q_3)^2$ in the quotient. Although we have not yet succeeded in doing the actual computation of the quotient, we believe the treatment of the $u = 0$ problem of this sort should be interesting and worth doing.

Problem C. If we combine some contributions from virtual photons (like D_2 and D_4 given below) to eliminate the infrared divergence, we then find that their sum $D_2 + D_4$ is infrared convergent (i.e. determines a well-defined function) but that it presents a singularity $(\varphi + i0)^{3/2} \log(\varphi + i0)$ at points where grad $\varphi \neq 0$; this singularity is stronger than the singularity $(\varphi + i0)^{3/2}$ for the bubble diagram function associated with Δ. Here φ is a defining function of the Landau-Nakanishi surface for the diagram Δ. (Cf. [KS2] and [KS3].)

Figure 3.

D_4 :

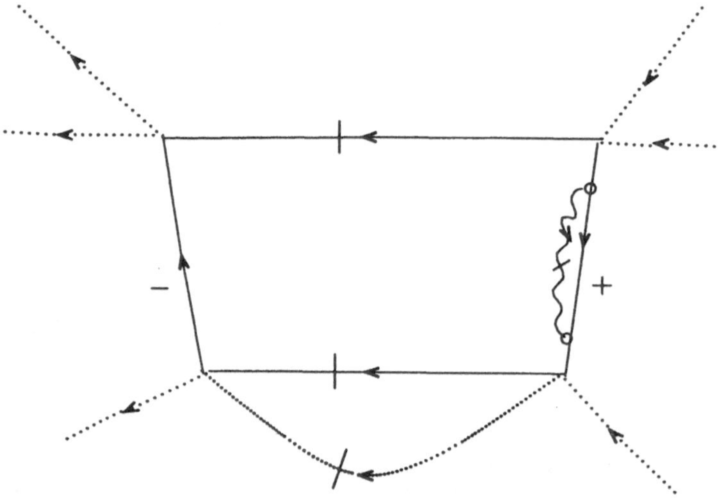

$\times(-1)$.

This stronger singularity should be cancelled by the contributions (again suitably combined to kill the infrared divergences) from the real photon, because it is a perturbative counterpart in the momentum space of the claim of unitarity of the operator $U(L(x))$ (acting on the photon space) established by [S2] in coordinate space. Probably this approach would be one way to bypass Problem A without using the approximation (1). We have not tried this approach, mainly because the Landau-Nakanishi geometry along $q_2 = q_3$ seems to be rather complicated.

We believe, however, that a study of the geometry near $q_2 = q_3$ would be an interesting mathematical subject.

References

[GY] G. Grammer and D.R. Yennie, Phys. Rev. **D8**, 4332 (1973).

[I] D. Iagolnitzer, The S-matrix, North Holland, Amsterdam, 1978.

[K^3] M. Kashiwara, T. Kawai and T. Kimura, Foundations of Algebraic Analysis, Princeton University Press, 1986.

[KS1] T. Kawai and H.P. Stapp, in Lecture Notes in Physics, No.39, ed. H. Araki, Springer, 1975.

[KS2] ———————, in Algebraic Analysis (M. Kashiwara and T. Kawai, eds.), vol.1, Academic Press, New York, 1988, pp. 309-330.

[KS3] ——————— , Ann. Inst. Fourier (Grenoble), **43**, 1301 (1993).

[KS4] ——————— , Phys. Rev. **D52**, 2484, 2508 and 2517 (1995).

[Sa] M. Sato, in Lect. Notes in Physics, No.39, ed. H. Araki, Springer, 1975.

[S1] H.P. Stapp, in Structural Analysis of Collision Amplitudes, ed. R. Balian and D. Iagolnitzer, North-Holland, New York, 1976.

[S2] ——————— , Phys. Rev. **28D**, 1386 (1983).

[YFS] D. Yennie, S. Frautschi, and H. Suura, Ann. Phys. (N.Y.) **13**, 379 (1961).

The Navier-Stokes equation with distributions as initial data and application to self-similar solutions

Hideo Kozono [1] and *Masao Yamazaki* [2]

[1] Graduate School of Mathematics, Nagoya University, Chikusa-ku, Nagoya, Aichi 464-01, Japan
[2] Department of Mathematics, Hitotsubashi University, Kunitachi, Tokyo 186, Japan

Abstract. In this paper we construct new function spaces of the Besov type from the Morrey spaces, and show that the Navier-Stokes equation and semilinear heat equations have unique time-global solutions with a bound for small initial data, which are not necessarily Radon measures, in some of the above function spaces. Thus we obtain some self-similar solutions of these equations provided that the initial data is a homogeneous function.

We also constructed a local version of these function spaces, and give conditions on the initial data in these spaces sufficient for the unique existence of time-local solutions of the above equations with a bound near $t = 0$.

Next, as a generalization of the above result on the Navier-Stokes equation, we consider the stationary Navier-Stokes equation with an external force, and give a condition on the external force sufficient for the existence of small solutions belonging to appropriate Morrey spaces. Further, the stability of the above stationary solution, in Morrey spaces larger than or equal to the original Morrey spaces, is shown under an appropriate condition. The stability of the above stationary solution in new function spaces introduced above is also verified. Also in this case, we obtain some self-similar solutions provided that the external force and the initial data are homogeneous.

Part of the results of this paper is proved in [KY1] and [KY2].

1 Introduction

In this paper we consider the Cauchy problem for the Navier-Stokes equation on $\mathbb{R}^n$ for $n \geq 2$ of the following form:

$$\text{(1.1)} \qquad \frac{\partial u}{\partial t} - \Delta_x u + (u \cdot \nabla_x)u + \nabla_x \pi = 0 \qquad \text{in} \quad]0,\infty[\times \mathbb{R}^n,$$

$$\text{(1.2)} \qquad \nabla_x \cdot u = 0 \qquad \text{in} \quad]0,\infty[\times \mathbb{R}^n,$$

$$\text{(1.3)} \qquad u(0,x) = a(x) \qquad \text{on} \quad \mathbb{R}^n,$$

Suppose throughout this paper that the initial data $a(x)$ satisfies $\nabla_x \cdot a(x) = 0$.

We also consider the Cauchy problem for semilinear heat equations on $\mathbb{R}^n$ of the following form:

$$\frac{\partial u}{\partial t}(t,x) = \Delta u(t,x) + f\big(u(t,x)\big) \quad \text{in} \quad]0,\infty[\times\mathbb{R}^n, \tag{1.4}$$

$$u(0,x) = a(x) \quad \text{on} \quad \mathbb{R}^n, \tag{1.5}$$

where $f(\sigma)$ is a locally Lipschitz continuous function on C satisfying

$$|f(\sigma) - f(\rho)| \le C|\sigma - \rho|\left(1 + |\sigma|^{\gamma-1} + |\rho|^{\gamma-1}\right), \tag{1.6}$$

or, more strongly,

$$f(0) = 0 \quad \text{and} \quad |f(\sigma) - f(\rho)| \le C|\sigma - \rho|\left(|\sigma|^{\gamma-1} + |\rho|^{\gamma-1}\right) \tag{1.7}$$

for every σ, $\rho \in C$ with some constants $\gamma > 1$ and $C > 0$. The estimates (1.6) and (1.7) yield the inequalities $|f(\sigma)| \le C\left(|\sigma|^\gamma + 1\right)$ and $|f(\sigma)| \le C|\sigma|^\gamma$ respectively, possibly with a larger constant C. A typical example of $f(\sigma)$ satisfying (1.7) is $f(\sigma) = \pm|\sigma|^{\gamma-1}\sigma$.

The purpose of this paper is to construct new function spaces in the same way as the Besov spaces, based on the Morrey spaces in place of the standard L^p-spaces, and to show that, if the initial data $a(x)$ belongs to some function spaces above and its norm is sufficiently small, then the Cauchy problems (1.1)–(1.3) and (1.4)–(1.5) admit unique time-global strong solutions with a bound near $t = 0$, provided that the function $f(\sigma)$ in (1.4) satisfies (1.7) with some constant $\gamma > 1 + 2/n$. For the Navier-Stokes equation, the spaces where initial data can be taken are strictly larger than those in Kato [K2]. Moreover, we can take distributions other than Radon measures as initial data. It is also to be noted that these function spaces contain non-trivial homogeneous functions, with which as initial data the equations above have self-similar solutions.

We also introduce a local version of the above function spaces, and show that the above Cauchy problems, with $f(\sigma)$ satisfying (1.6) in (1.4), admit unique time-local strong solutions with a bound near $t = 0$ for initial data in the local function spaces under an additional assumption. For the Navier-Stokes equation, the spaces where initial data can be taken are strictly larger than those in Taylor [Ta], Kato [K2] and Federbush [Fe].

Further, in the case $n \ge 3$, we consider the stationary Navier-Stokes equation on $\mathbb{R}^n$, and give a condition on the external force sufficient for the unique existence of small stationary solutions belonging to appropriate Morrey spaces. Then we show the stability of the above stationary solution by showing the global solvability of the Cauchy problem for the nonstationary Navier-Stokes equation with the same external force with initial data sufficiently close to the stationary solution. Then the global solvability of (1.1)–(1.3) can be regarded as a special case of this problem with the null stationary solution.

Note that the equations (1.1)–(1.2) and (1.4) have the homogeneity property as follows: Suppose that $\big(u(t,x), p(t,x)\big)$ is a solution of (1.1)–(1.2), and put $u_\lambda(t,x) = \lambda u(\lambda^2 t, \lambda x)$ and $p_\lambda(t,x) = \lambda^2 p(\lambda^2 t, \lambda x)$ for every $\lambda > 0$. Then we have

$$\begin{aligned}&\frac{\partial u_\lambda}{\partial t}(t,x) - \Delta_x u_\lambda(t,x) + \big(u_\lambda(t,x)\cdot\nabla_x\big)u_\lambda(t,x) + \nabla_x p_\lambda(t,x)\\ &= \lambda^3\left\{\frac{\partial u}{\partial t}(t,x) - \Delta_x u(t,x) + \big(u(t,x)\cdot\nabla_x\big)u(t,x) + \nabla_x p(t,x)\right\} = 0;\end{aligned}$$

that is, the function $u_\lambda(t,x)$ is also a solution of (1.1)–(1.2). Furthermore, we have the equality $u_\lambda(0,x) = \lambda u(0,\lambda x)$. In the same way, let $u(t,x)$ be a solution of (1.4) with $f(\sigma) = \pm|\sigma|^{\gamma-1}\sigma$, and put $u_\lambda(t,x) = \lambda^{2/(\gamma-1)}u(\lambda^2 t, \lambda x)$ for every $\lambda > 0$. Then we have

$$\frac{\partial u_\lambda}{\partial t}(t,x) - \Delta_x u_\lambda(t,x) + f\big(u_\lambda(t,x)\big) = \lambda^{2\gamma/(\gamma-1)} f\big(u(t,x)\big) = 0;$$

that is, the function $u_\lambda(t,x)$ is also a solution of (1.4). Furthermore, we have the equality $u_\lambda(0,x) = \lambda^{2/(\gamma-1)}u(0,\lambda x)$.

It follows that the condition on the smallness of the initial data sufficient for the global solvability of (1.1)–(1.3) [resp. (1.4)–(1.5)] must be measured by the norm invariant under the scaling $a_\lambda(x) = \lambda a(\lambda x)$. [resp. under the scaling $a_\lambda(x) = \lambda^{2/(\gamma-1)}a(\lambda x)$.] Furthermore, if the initial data $a(x)$ is invariant under the scaling above, then the unique solution becomes self-similar.

We briefly review some previous researches on these equations on $\mathbb{R}^n$. Historically, the Cauchy problems with measures as initial data are considered first for (1.4)–(1.5). Brézis and Friedman [BF] showed that the Cauchy problem (1.4)–(1.5) admits a solution on $]0,T] \times \mathbb{R}^n$ for $f(\sigma) = -|\sigma|^{\gamma-1}\sigma$ and $a(x) = \delta(x)$ if and only if $\gamma < 1 + 2/n$. Here (1.5) is interpreted that $u(t,\cdot)$ approaches to $\delta(x)$ in the sense of distributions as $t \to +0$. (The condition is independent of the choice of $T > 0$, since blow-up in finite time does not take place with this $f(\sigma)$.) The condition $\gamma < 1 + 2/n$ follows from the result of Haraux and Weissler [HW] on the existence of a non-trivial solution for the Cauchy problem (1.4)–(1.5) with $a(x) \equiv 0$ for $f(\sigma) = |\sigma|^{\gamma-1}\sigma$ with $\gamma > 1 + 2/n$. Baras and Pierre [BP] obtained characterizations in terms of various capacities of the Radon measures with which as the initial data (1.4)–(1.5) is solvable with $f(\sigma) = -|\sigma|^{\gamma-1}\sigma$. Niwa [N] introduced spaces of measures of the Morrey type, and obtained a sufficient condition for the local well-posedness and the global well-posedness of the Cauchy problem (1.4)–(1.5) with initial data in these spaces for $f(\sigma) = \pm|\sigma|^{\gamma-1}\sigma$.

For the Navier-Stokes equation, Cottet [Co], Giga, Miyakawa and Osada [GMO] and Kato [K3] showed the global existence in the case $n = 2$ and $\nabla \times a(x)$ is a finite Radon measure. Miyakawa and Yamada [MY] and Michaux and Rakotoson [MR] treated the same case on bounded domains. In the case $n = 3$, Giga and Miyakawa [GM] proved the global existence for initial data $a(x)$ such that $\nabla \times a(x)$ belongs to a space of measures of the Morrey type and is sufficiently small. Taylor [Ta] proved the time-local existence of the solutions of the Navier-Stokes equations on $\mathbb{R}^n$ with locally integrable functions in the Morrey spaces as the initial data satisfying some conditions. He also proved the time-global existence of the solutions of the Navier-Stokes equations on compact manifolds with locally integrable functions in the Morrey spaces with small norm as the initial data, together with results on more general parabolic equations with initial data in spaces of measures of the Morrey type. Kato [K2] proved the time-global existence of the solutions with initial data in spaces of measures of the Morrey type with small norm, together with the time-local existence under somewhat weaker conditions. Federbush [Fe] treated initial data in the spaces of Morrey type somewhat smaller than the spaces considered in [Ta] and [K2].

On the other hand, there are other results on the Navier-Stokes equation in function spaces larger than L^p. Kobayashi and Muramatu [KM] obtained the time-local existence of a strong solution in a general domain $\Omega \subset \mathbb{R}^n$ with initial data in some abstract Besov spaces, together with a uniqueness result. Grubb [G] treated non-homogeneous boundary value problems in the framework of the Besov spaces. Kato and Ponce [KP] solved the equation in the Sobolev spaces of negative order. Cannone and Meyer [CM] recently treated the same problems in general function spaces defined through the Littlewood-Paley dyadic decomposition. In many cases such function spaces give a more natural framework to consider the Navier-Stokes equations. For example, recent results by Borchers and Miyakawa [BM] and Kozono and Yamazaki [KY3] imply that the space $L^{n,\infty}$ is more suitable than L^n for the treatment of the Navier-Stokes exterior problem.

Furthermore, the results above are related to some results on self-similar solutions of the Navier-Stokes equation based on these results. In the 2-dimensional case, Giga and Kambe [GK] showed that, if the initial vorticity is a finite Radon measure with the mass M sufficiently small, then there exists a unique global solution of (1.1)–(1.3), which approaches asymptotically as $t \to \infty$ to the self-similar solution with initial vorticity $M\delta$. Moreover, the vorticity is equal to the solution of the heat equation with initial data $M\delta$. This problem was studied further by Carpio [Car]. In the 3-dimensional case, Giga and Miyakawa [GM] showed that, if the initial vorticity $\omega(x)$ is small in the space of measures of Morrey type and invariant under the scaling $\lambda^2\omega(\lambda x)$, then the solution becomes self-similar. Then Carpio [Car] showed that, if the initial vorticity $\omega(x)$ is small in the same space of measures and if the rescaled function $\lambda^2\omega(\lambda x)$ converges to some Radon measure μ in the weak-$*$ topology of Radon measures as $\lambda \to \infty$, then the unique global solution of (1.1)–(1.3) approaches asymptotically as $t \to \infty$ to the solution of (1.1)–(1.3) with initial vorticity μ. Cannone and Planchon [CP] showed that, if the initial data is homogeneous of degree -1 and is sufficiently small in the homogeneous Besov space $\dot{B}^0_{3,\infty}$, then there exists a unique global solution of (1.1)–(1.3), which is self-similar. Cannone [Can] relaxed the above result by showing that the conclusion remains valid provided that the initial data is sufficiently small in the space $\dot{B}^{3/p-1}_{p,\infty}$, which is strictly larger than the space $\dot{B}^{0}_{3,\infty}$ in view of the embedding theorem of Sobolev type. Note that the initial data of the self-similar solution is homogeneous of order -1, and hence it cannot be an element of the standard L^p-space for any p.

In this paper we unify and generalize their results in the case $\Omega = \mathbb{R}^n$, by introducing new function spaces which contain the Morrey spaces as well as the Besov spaces as proper subspaces. Moreover, our spaces contain measures other than Radon measures, and thus we obtain some self-similar solutions not given in the papers cited above. Furthermore, we give self-similar solutions of (1.4)–(1.5) and the Navier-Stokes equation with external force homogeneous of degree -3.

2 The Morrey spaces

In this section we recall the definition and some property of the Morrey spaces.

Definition 2.1. For p and q satisfying $1 \le q \le p < \infty$, the *Morrey space* $\mathcal{M}_{p,q} = \mathcal{M}_{p,q}(\mathbb{R}^n)$ and the *local Morrey space* $M_{p,q} = M_{p,q}(\mathbb{R}^n)$ are defined as the sets of functions $u(x) \in L^q_{\mathrm{loc}}(\mathbb{R}^n)$ such that

$$\|u\,|\mathcal{M}_{p,q}\| = \sup_{x_0\in\mathbb{R}^n} \sup_{R>0} R^{n/p-n/q} \,\|u\,|L^q\,(B(x_0,R))\,\| < \infty$$

and

$$\|u\,|M_{p,q}\| = \sup_{x_0\in\mathbb{R}^n} \sup_{0<R\le 1} R^{n/p-n/q} \,\|u\,|L^q\,(B(x_0,R))\,\| < \infty$$

respectively, where $B(x_0, R)$ denotes the closed ball in $\mathbb{R}^n$ with center x_0 and radius R.

Next, for p satisfying $1 \le p < \infty$, the spaces $\mathcal{M}_p$ and M_p are defined as the set of Radon measures μ on $\mathbb{R}^n$ such that

$$\|\mu\,|\mathcal{M}_p\| = \sup_{x_0\in\mathbb{R}^n} \sup_{R>0} R^{n/p-n}|\mu|\,(B(x_0,R)) < \infty$$

and

$$\|\mu\,|M_p\| = \sup_{x_0\in\mathbb{R}^n} \sup_{0<R\le 1} R^{n/p-n}|\mu|\,(B(x_0,R)) < \infty$$

respectively, where $|\mu|$ denotes the total variation of μ.

Roughly speaking, the elements of the space $\mathcal{M}_{p,q}$ have local singularities like those of the elements of L^q, and decay rate like that of the elements of L^p. It is easy to see the relation $\mathcal{M}_{p,q} \subset M_{p,q}$. Further, the spaces $\mathcal{M}_{p,1}$ and $M_{p,1}$ can be regarded as closed subspaces of $\mathcal{M}_p$ and M_p respectively.

See Peetre [P] and the papers cited therein for the functions spaces generalizing the Morrey spaces. These spaces are denoted as $L^{q,n-nq/p}$ in Campanato [Cam] and as $\mathcal{E}^{q,-n/p}$ in Peetre [P]. (The latter notation is theoretically superior, since the number $-n/p$ is exactly the degree of homogeneity of the space.) We employ notations similar to the ones employed in Giga and Miyakawa [GM] and Taylor [Ta].

Then these spaces have the following property.

Proposition 2.2. *Suppose that the integers m and n satisfy $1 \leq m < n$, and put $x' = (x_1, \cdots, x_m)$ for $x = (x_1, \cdots, x_n) \in \mathbb{R}^n$. Let $u(x') \in \mathcal{M}_{p,q}(\mathbb{R}^m)$ be a function on $\mathbb{R}^m$, and define a function $v(x)$ on $\mathbb{R}^n$ by $v(x) = u(x')$. Then we have $v(x) \in \mathcal{M}_{np/m,q}(\mathbb{R}^n)$.*

Proposition 2.3. *Suppose that $1 \leq r < q < p < \infty$. Then we have the inclusion relations*

$$\mathcal{M}_{p,p} = L^p \subset L^{p,\infty} \subset \mathcal{M}_{p,q} \subset \mathcal{M}_{p,r} \text{ and } M_{p,p} = L^p_{\mathrm{unif}} \subset M_{p,q} \subset M_{p,r} .$$

3 Function spaces of Besov type based on the Morrey spaces

In order to introduce our function spaces, we first introduce the Littlewood-Paley dyadic decomposition. Let $\chi(t)$ be a monotone-decreasing smooth function on $[0, \infty)$ such that $\chi(t) \equiv 1$ for $t \leq 4/3$ and $\chi(t) \equiv 0$ for $t \geq 3/2$. Then, putting $\Phi(\xi) = \chi(|\xi|)$ and $\varphi_j(\xi) = \Phi(2^{-j}\xi) - \Phi(2^{1-j}\xi)$ for $\xi \in \mathbb{R}^n$ and $j \in \mathbb{Z}$, we have $\varphi_j(\xi) \in C_0^\infty(\mathbb{R}^n)$, $0 \leq \varphi_j(\xi) \leq 1$, $\operatorname{supp}\varphi_j \subset \{\xi \mid 2^{j-1} < |\xi| < 2^{j+1}\}$, $\varphi_j(\xi) = \varphi_0(2^{-j}\xi)$, $\sum_{j=-\infty}^{\infty} \varphi_j(\xi) \equiv 1$ and $\Phi(\xi) + \sum_{j=1}^{\infty} \varphi_j(\xi) \equiv 1$.

Next, let $\mathcal{S}'$ and $\mathcal{P}$ denote the set of tempered distributions on $\mathbb{R}^n$ and the set of polynomials with n variables respectively, and put $f(D)u = \mathcal{F}^{-1}\left[f(\xi)\mathcal{F}[u](\xi)\right]$ for $u \in \mathcal{S}'$.

Then, for p and q as above and for $r \in [1, \infty]$ and $s \in \mathbb{R}$, we define the function spaces $\mathcal{M}^s_{p,q}$, $\mathcal{N}^s_{p,q,r}$, $M^s_{p,q}$, $N^s_{p,q,r}$ as follows: (See Bergh and Löfström [BL] or Triebel [Tr].)

Definition 3.1. Let $\mathcal{M}^s_{p,q}$ and $\mathcal{N}^s_{p,q,r}$ denote the sets of $u \in \mathcal{S}'/\mathcal{P}$ such that

$$\left\|u \,\middle|\mathcal{M}^s_{p,q}\right\| = \left\|(-\Delta_x)^{s/2}u\middle|\, \mathcal{M}_{p,q}\right\| < \infty$$

and

$$\left\|u \,\middle|\mathcal{N}^s_{p,q,r}\right\| = \left\|\left\{2^{sj}\left\|\varphi_j(D)u\right|\mathcal{M}_{p,q}\right\|\right\}_{j\in\mathbb{Z}}\middle|\,\ell^r(\mathbb{Z})\right\| < \infty$$

respectively. Next, let $M^s_{p,q}$ and $N^s_{p,q,r}$ denote the sets of $u \in \mathcal{S}'$ such that

$$\|u\,|M^s_{p,q}\| = \left\|(1-\Delta_x)^{s/2}u\right| M_{p,q}\Big\| < \infty$$

and

$$\|u\,|N^s_{p,q,r}\| = \|\Phi(D)u\,|M_{p,q}\| + \left\|\left\{2^{sj}\,\|\varphi_j(D)u|\,M_{p,q}\|\right\}_{j=1}^{\infty}\right|\ell^r(\mathbb{N}_+)\Big\| < \infty$$

respectively.

Then these spaces enjoy the following propositions.

Proposition 3.2. *We have the following inclusion relations and equalities:* $\mathcal{N}^s_{p,q,1} \subset \mathcal{M}^s_{p,q} \subset \mathcal{N}^s_{p,q,\infty}$, $\mathcal{N}^0_{p,1,1} \subset \mathcal{M}_{p,1} \subset \mathcal{M}_p \subset \mathcal{N}^0_{p,1,\infty}$, $\mathcal{M}^s_{p,p} = \dot{H}^s_p$, $\mathcal{N}^s_{p,p,r} = \dot{B}^s_{p,r}$, $N^s_{p,q,1} \subset M^s_{p,q} \subset N^s_{p,q,\infty}$, $N^0_{p,1,1} \subset M_{p,1} \subset M_p \subset N^0_{p,1,\infty}$, $M^s_{p,p} = H^s_p$ *and* $N^s_{p,p,r} = B^s_{p,r}$. *Here* $\dot{H}^s_p$ *and* $\dot{B}^s_{p,r}$ *stand for the homogeneous Sobolev spaces and the homogeneous Besov spaces, and* H^s_p *and* $B^s_{p,r}$ *stand for the inhomogeneous Sobolev spaces and the homogeneous Besov spaces respectively.*

Proposition 3.3. *Suppose that* $1 \le q < p < \infty$ *and* $0 < \theta < 1$. *Then we have* $\mathcal{N}^s_{p,q,r} \subset \mathcal{N}^{s-(1-\theta)n/p}_{p/\theta,q/\theta,r} \subset \dot{B}^{s-n/p}_{\infty,r}$ *and* $N^s_{p,q,r} \subset N^{s-(1-\theta)n/p}_{p/\theta,q/\theta,r} \subset B^{s-n/p}_{\infty,r}$. *Furthermore, if* $q > 1$, *we have*

$$\mathcal{M}^s_{p,q} \subset \mathcal{M}^{s-(1-\theta)n/p}_{p/\theta,q/\theta} \text{ and } M^s_{p,q} \subset M^{s-(1-\theta)n/p}_{p/\theta,q/\theta}.$$

Proposition 3.4. *Suppose that* $s_0 \ne s_1$ *and* $0 < \theta < 1$, *and put*

$$s = (1-\theta)s_0 + \theta s_1.$$

Then the spaces $\mathcal{N}^s_{p,q,r}$ *and* $N^s_{p,q,r}$ *coincide with the real interpolation spaces* $\left(\mathcal{M}^{s_0}_{p,q}, \mathcal{M}^{s_1}_{p,q}\right)_{\theta,r}$ *and* $\left(M^{s_0}_{p,q}, M^{s_1}_{p,q}\right)_{\theta,r}$ *respectively, and the spaces* $\mathcal{M}^s_{p,q}$ *and* $M^s_{p,q}$ *coincide with the complex interpolation spaces*

$$\left[\mathcal{M}^{s_0}_{p,q}, \mathcal{M}^{s_1}_{p,q}\right]_\theta \text{ and } \left[M^{s_0}_{p,q}, M^{s_1}_{p,q}\right]_\theta$$

respectively. In particular, the spaces $\mathcal{N}^s_{p,q,r}$ *and* $N^s_{p,q,r}$ *are independent of the choice of the Littlewood-Paley decomposition.*

Proposition 3.5. *Suppose that the integers* m *and* n *satisfy* $1 \le m < n$, *and put* $x' = (x_1, \cdots, x_m)$ *for* $x = (x_1, \cdots, x_n) \in \mathbb{R}^n$. *Let* $u(x') \in \mathcal{N}^s_{p,q,r}(\mathbb{R}^m)$ *[resp.* $N^s_{p,q,r}(\mathbb{R}^m)$ *] be a function on* $\mathbb{R}^m$, *and define a function* $v(x)$ *on* $\mathbb{R}^n$ *by* $v(x) = u(x')$. *Then we have* $v(x) \in \mathcal{N}^s_{np/m,q,r}(\mathbb{R}^n)$. *[resp.* $N^s_{np/m,q,r}(\mathbb{R}^n)$. *]*

Proposition 3.6. *Let σ be a real number, and suppose that a function $P(\xi) \in C^{[n/2]+1}(\mathbb{R}^n \setminus \{0\})$ [resp. $P(\xi) \in C^{[n/2]+1}(\mathbb{R}^n)$] satisfies the estimate*

$$\left|\frac{\partial^{|\alpha|}P}{\partial\xi^\alpha}(\xi)\right| \leq C|\xi|^{\sigma-|\alpha|} \left[\text{resp. } \left|\frac{\partial^{|\alpha|}P}{\partial\xi^\alpha}(\xi)\right| \leq C\left(1+|\xi|^2\right)^{(\sigma-|\alpha|)/2}\right]$$

for every $\alpha \in \mathbf{N}^n$ such that $|\alpha| \leq [n/2]+1$. Then the operator $P(D)$ is bounded from $\mathcal{N}^s_{p,q,r}$ to $\mathcal{N}^{s-\sigma}_{p,q,r}$ [resp. from $N^s_{p,q,r}$ to $N^{s-\sigma}_{p,q,r}$] for every p, q, r and s. Moreover, if $1 < q \leq p$, then the operator $P(D)$ is bounded from $\mathcal{M}^s_{p,q}$ to $\mathcal{M}^{s-\sigma}_{p,q}$ [resp. from $N^s_{p,q,r}$ to $N^{s-\sigma}_{p,q,r}$] for every r and s.

Proposition 3.7. *If $s < n/p$, then we can choose a canonical representative in $\mathcal{S}'$ for every element of $\mathcal{M}^s_{p,q}$ and $\mathcal{N}^s_{p,q,r}$ modulo $\mathcal{P}$. Hence in this case we can (and do) regard these spaces as subspaces of $\mathcal{S}'$.*

This fact can be proved in the same way as in the case of homogeneous Besov spaces. (See Bourdaud [Bo] for example.)

Remark 3.8. If $q < p$ or $r = \infty$, the Laplacian regarded as a closed operator on the spaces $\mathcal{M}^s_{p,q}$, $M^s_{p,q}$, $\mathcal{N}^s_{p,q,r}$ or $N^s_{p,q,r}$ is not densely defined. It follows that the Gauss semigroup fails to be a C^0-semigroup on these spaces, although it is uniformly bounded and satisfies the semigroup property. It follows that, in the consideration of the Cauchy problems (1.1)–(1.3) and (1.4)–(1.5), the initial conditions (1.3) and (1.5) cannot be satisfied in the usual sense, and have to be considered in weaker sense accordingly.

Example 3.9. Proposition 3.6 implies that $\mathrm{p.v.}\,\dfrac{1}{x} \in \mathcal{N}^0_{1,1,\infty}(\mathbb{R})$, since it is the Hilbert transform of the measure $\delta(x) \in \dot{B}^0_{1,\infty}(\mathbb{R}) = \mathcal{N}^0_{1,1,\infty}(\mathbb{R})$. (Note that $\mathrm{p.v.}\,\dfrac{1}{x}$ is not a Radon measure.) Hence Propositions 3.5 and 3.3 imply $\mathrm{p.v.}\,\dfrac{1}{x_1} \in \mathcal{N}^0_{n,1,\infty}(\mathbb{R}^n) \subset \mathcal{N}^{n/p-1}_{p,p/n,\infty}(\mathbb{R}^n)$ for every $p \geq n$.

Example 3.10. Suppose that $1 \leq q < p < \infty$. Then direct calculation and Proposition 3.2 imply $u(x) = \prod_{j=1}^n (x_j)_+^{-1/p} \in \mathcal{M}_{p,q} \subset \mathcal{N}^0_{p,q,\infty}$. It follows from Proposition 3.6 that

$$\frac{\partial u}{\partial x_1}(x) = -\frac{1}{p}\,\mathrm{p.f.}(x_1)_+^{-1-1/p}\prod_{j=2}^n (x_j)_+^{-1/p} \in \mathcal{N}^{-1}_{p,q,\infty}.$$

4 Results on (1.1)–(1.3)

We have the following three theorems for the Cauchy problem (1.1)–(1.3).

Theorem 4.1. (uniqueness) *Suppose that $n \geq 2$, and that p and q satisfy the conditions $1 \leq q \leq p < \infty$ and that $n \leq p$. Then there exists a constant*

M_1 such that, for every $T \in]0,+\infty]$ and every $a(x) \in S'$ such that $\nabla \cdot a = 0$, there exists at most one solution of (1.1)–(1.2) *on* $]0,T[\times \mathbb{R}^n_x$ *satisfying the following:*

(1) *For every $T' \in]0,T[$, we have* $\sup_{0<t\le T'} \left\| u(t,\cdot) \left| N^{n/p-1}_{p,q,\infty} \right.\right\| < \infty$ *and* $\sup_{0<t\le T'} t^{1/2-n/4p} \left\| u(t,\cdot) \left| M_{2p,2q} \right.\right\| < \infty$.
(2) $\lim_{T'\to+0} \sup_{0<t\le T'} t^{1/2-n/4p} \left\| u(t,\cdot) \left| M_{2p,2q} \right.\right\| < M_1$.
(3) *$u(t,\cdot) \to a$ in the sense of tempered distributions as $t \to +0$.*

Theorem 4.2. (existence of global solutions) *Let the numbers n, p, q and M_1 be the same as in Theorem 4.1. Then there exist a positive constant δ_1 and a continuous, strictly increasing function $\omega_1(\delta)$ on $[0,\delta_1]$ satisfying $\omega_1(0) = 0$ and $\omega_1(\delta_1) \le M_1$ such that, for every $a(x) \in \mathcal{N}^{n/p-1}_{p,q,\infty}$ satisfying $\nabla_x \cdot a(x) = 0$ and $\delta = \left\| a(x) \left| \mathcal{N}^{n/p-1}_{p,q,\infty} \right.\right\| \le \delta_0$, there exists a solution $u(t,x)$ of* (1.1)–(1.2) *on $]0,\infty[\times \mathbb{R}^n$ satisfying the following conditions:*

(1) $u(t,x) \in C^\infty(]0,\infty[\times \mathbb{R}^n)$.
(2) $\sup_{t>0} \left\| u(t,\cdot) \left| \mathcal{N}^{n/p-1}_{p,q,\infty} \right.\right\| < \infty$.
(3) $\sup_{t>0} t^{1/2-n/4p} \left\| u(t,\cdot) \left| \mathcal{M}_{2p,2q} \right.\right\| \le \omega_1(\delta)$.
(4) *$u(t,\cdot) \to a$ in the weak-$*$ topology of the space $\dot{B}^{-1}_{\infty,\infty}$ as $t \to +0$.*

Theorem 4.3. (existence of local solutions) *Let the numbers n, p, q and M_1 be the same as in Theorem 4.1. Then there exist a positive constant δ_1' such that, for every $a(x) \in N^{n/p-1}_{p,q,\infty}$ satisfying $\nabla_x \cdot a(x) = 0$ and*

$$\limsup_{j\to\infty} 2^{(n/p-1)j} \left\| \varphi_j(D) u \left| M_{p,q} \right.\right\| < \delta_1',$$

there exist a positive constant T and a solution $u(t,x)$ of (1.1)–(1.2) *on $]0,T[\times \mathbb{R}^n$ satisfying the following conditions:*

(1) $u(t,x) \in C^\infty(]0,T[\times \mathbb{R}^n)$.
(2) $\sup_{0<t<T} \left\| u(t,\cdot) \left| N^{n/p-1}_{p,q,\infty} \right.\right\| < \infty$.
(3) $\sup_{0<t<T} t^{1/2-n/4p} \left\| u(t,\cdot) \left| M_{2p,2q} \right.\right\| < M_1$.
(4) *$u(t,\cdot) \to a$ in the weak-$*$ topology of the space $B^{-1}_{\infty,\infty}$ as $t \to +0$.*

Example 4.4. The function $a(x) = c\left(0,\cdots,0,\mathrm{p.v.}\,\dfrac{1}{x_1}\right)$ enjoys the assumption of Theorem 2 provided that $|c|$ is sufficiently small, in view of Example 3.9. Since this function is homogeneous of degree -1, it follows that the unique solution $u(t,x)$ with $a(x)$ as initial data is self-similar; namely, it satisfies the equality $u(t,x) = t^{-1/2}u\left(1,t^{-1/2}x\right)$. In this case we see by the uniqueness that $u(t,x)$ depends only on t and x_1. Substituting this fact into the equation and making use of the uniqueness again, we see that $u(t,x) = c\,(0,\cdots,0,v(t,x_1))$, where $v(t,x_1)$ is the solution of the heat equation with $\mathrm{p.v.}\,\dfrac{1}{x_1}$ as initial data.

Remark 4.5. Theorem II of Kato [K2] asserts the time-global existence for initial data in $\mathcal{M}_n$ with small norm. This case may be covered by Theorems 4.1 and 4.2, since Propositions 3.2 and 3.3 imply the inclusion relation $\mathcal{M}_n \subset \mathcal{N}^0_{n,1,\infty} \subset \mathcal{N}^{n/p-1}_{p,p/n,\infty}$ for every $p \geq n$.

Remark 4.6. Suppose that the initial vorticity $\nabla \times a$ belongs to $\mathcal{M}_{n/2}$ with small norm. This case may be covered by Theorems 4.1 and 4.2, since Proposition 3.2 implies $\nabla \times a \in \mathcal{N}^0_{n/2,1,\infty}$ for this a. From this fact we can deduce by using Propositions 3.3 and 3.6 and the Biot-Savard law coupled with the equality $\nabla \cdot a = 0$ that $a \in \mathcal{N}^1_{n/2,1,\infty} \subset \mathcal{N}^{n/p-1}_{p,2p/n,\infty}$ holds for every $p \geq n$. Hence Theorems 4.1 and 4.2 generalize the results of Cottet [Co], Giga, Miyakawa and Osada [GMO] and Kato [K3] in the case $n = 2$, and the result of Giga and Miyakawa [GM] in the case $n = 3$, except the existence of the solution with large data in the case $n = 2$.

Remark 4.7. In the theorems above, the solution $u(t, x)$ may not be strongly continuous at $t = 0$ in $N^{n/p-1}_{p,q,\infty}$ or $\mathcal{N}^{n/p-1}_{p,q,\infty}$. This fact is mainly due to the fact that the Laplacian is not densely defined in these spaces, which is closely related to the non-reflexivity of these spaces. Hence, in order to ensure the uniqueness, we need a condition on some sort of smallness near $t = 0$ like (2) in Theorem 4.1. Conditions of this type was employed by Giga, Miyakawa and Osada [GMO] and Kato [K3]. In Giga and Miyakawa [GM], Taylor [Ta] and Kato [K2], the uniqueness is verified under a somewhat stronger assumption like

$$\sup_{0<t\leq T'} t^{1/2-n/4p} \left\| u(t,\cdot) \,|M_{2p,2q} \right\| < M_1,$$

although this assumption may not be explicitly stated in some of the above papers. It is still an open question whether conditions of this type is really necessary for the uniqueness. (See Kato [K3].) However, conditions of this type is really essential for semilinear heat equation (1.4). (See Remark 5.7 in Section 5.)

It is also to be noted that, if the initial value $a(x)$ belongs to the standard L^n-space, then Kato [K1] proved the unique existence of a time-local solution, and also the global solvability provided the L^n-norm of $a(x)$ is small enough. Since the Laplacian in L^n is densely defined, the solution $u(t, x)$ satisfies

$$\lim_{t\to+0} t^{1/2-n/4p} \left\| u(t,\cdot) \,|L^{2p} \right\| = 0$$

in this case.

5 Results on (1.4)–(1.5)

We have the following three theorems for the Cauchy problem (1.4)–(1.5).

Theorem 5.1. (uniqueness) *Suppose that the function $f(\sigma)$ satisfies the condition (1.6) with some constant $\gamma > 1$, and let p, q and s be real numbers such that $\gamma \le q \le p$, $n(\gamma - 1) < 2p$, $-2/\gamma < s < 0$ and $s \ge n/p - 2/(\gamma - 1)$. Then there exists a constant M_2 such that, for every $T \in]0, +\infty]$ and every $a(x) \in \mathcal{S}'$, there exists at most one solution $u(t,x)$ of (1.4) on $]0,T[\times \mathbb{R}$ such that the following hold:*

1. *For every $T' \in]0,T[$, we have $\sup_{0<t\le T'} \left\| u(t,\cdot) \left| N^s_{p,q,\infty} \right. \right\| < \infty$ and $\sup_{0<t\le T'} t^{-s/2} \left\| u(t,\cdot) \,|M_{p,q} \right\| < \infty$.*
2. *$\limsup_{T'\to +0} \sup_{0<t\le T'} t^{-s/2} \left\| u(t,\cdot) \,|M_{p,q} \right\| < M_2$.*
3. *$u(t,\cdot) \to a$ in the sense of tempered distributions as $t \to +0$.*

Theorem 5.2. (existence of global solutions) *Suppose that $f(\sigma)$ satisfies (1.7) with some constant $\gamma > 1 + 2/n$, and let p and q be real numbers such that $\gamma \le q \le p$ and that $n(\gamma - 1) < 2p < n\gamma(\gamma - 1)$. Then there exist a positive number δ_2 and a continuous, strictly increasing function $\omega_2(\delta)$ on $[0, \delta_2]$ satisfying $\omega(0) = 0$ and $\omega_2(\delta_2) \le M_2$ such that, for every $a(x) \in \mathcal{N}^{n/p-2/(\gamma-1)}_{p,q,\infty}$ satisfying $\delta = \left\| a(x) \left| \mathcal{N}^{n/p-2/(\gamma-1)}_{p,q,\infty} \right. \right\| \le \delta_2$, there exists a solution $u(t,x)$ of (1.4) on $]0,\infty[\times \mathbb{R}^n$ satisfying the following:*

1. *$u(t,x) \in C^1(]0,\infty[\times \mathbb{R}^n)$.*
2. *$\sup_{t>0} \left\| u(t,\cdot) \left| \mathcal{N}^{n/p-2/(\gamma-1)}_{p,q,\infty} \right. \right\| < \infty$.*
3. *$\sup_{t>0} t^{1/(\gamma-1)-n/2p} \left\| u(t,\cdot) \,|\mathcal{M}_{p,q} \right\| \le \omega_2(\delta)$.*
4. *$u(t,\cdot) \to a$ in the weak-$*$ topology of the space $\dot{B}^{-2/(\gamma-1)}_{\infty,\infty}$ as $t \to +0$.*

Theorem 5.3. (existence of local solutions) *Let the function $f(\sigma)$ and the constants γ, p and q be the same as in Theorem 5.1. Then there exists a positive number δ_2' such that, for every $a(x) \in N^s_{p,q,\infty}$ satisfying*

$$\limsup_{j\to\infty} 2^{sj} \left\| \varphi_j(D)u \,|M_{p,q} \right\| < \delta_2', \tag{5.1}$$

there exist a positive constant T and a solution $u(t,x)$ of (1.4) on $]0,T[\times \mathbb{R}^n$ satisfying the following:

1. *$u(t,x) \in C^1(]0,T[\times \mathbb{R}^n)$.*
2. *$\sup_{0<t<T} \left\| u(t,\cdot) \left| \mathcal{N}^s_{p,q,\infty} \right. \right\| < \infty$.*
3. *$\sup_{t>0} t^{-s/2} \left\| u(t,\cdot) \,|\mathcal{M}_{p,q} \right\| < M_2$.*
4. *$u(t,\cdot) \to a$ in the weak-$*$ topology of the space $B^{s-n/p}_{\infty,\infty}$ as $t \to +0$.*

Example 5.4. If $n \ge 3$, we have

$$1 + \frac{2}{n} < \rho(n) = \frac{3 - n + \sqrt{n^2 - 2n + 9}}{2} < 2.$$

Hence, for every $\gamma \in]\rho(n), 2[$, we have $1+2/n < \gamma$. Putting $p = n(\gamma-1)/(3-\gamma)$, we have $n/p - 2/(\gamma - 1) = -1$. The inequality $\gamma^2 - (3 - n)\gamma - n > 0$ implies

$\gamma < p$. Further, the inequality $1 < \gamma < 2$ implies $(3-\gamma)\gamma - 2 > 0$, and hence

$$n(\gamma-1) < \frac{2n(\gamma-1)}{3-\gamma} = 2p < n\gamma(\gamma-1).$$

It follows from Example 3.10 that the function $a(x) = c(\partial u/\partial x_1)(x)$ can be taken as the initial data of Theorem 5.2 with $q = \gamma$, provided that $|c|$ is sufficiently small. Moreover, since the function $a(x)$ is homogeneous of degree $-1 - n/p = -2/(\gamma-1)$, it follows that the solution $u(t,x)$ is self-similar; namely, it satisfies the equality $u(t,x) = t^{-1/(\gamma-1)} u\left(1, t^{-1/(\gamma-1)}x\right)$.

Remark 5.5. The condition $\gamma > 1 + 2/n$, which is necessary and sufficient for the existence of nontrivial global nonnegative solution of (1.4)–(1.5) with $f(\sigma) = |\sigma|^{\gamma-1}\sigma$, (see Fujita [Fu] and Weissler [W],) is equivalent to the inequality $2\gamma < n\gamma(\gamma-1)$. Hence it is necessary and sufficient for the existence of p and q satisfying the assumption of Theorem 5.2.

Remark 5.6. The results of Weissler [W] implies the existence of a function $a_0(x) \in \mathcal{N}^{n/p-2/(\gamma-1)}_{p,q,\infty}$ such that, for sufficiently large constant c, the Cauchy problem (1.4)–(1.5) with $f(\sigma) = |\sigma|^{\gamma-1}\sigma$ admits no solution for initial data $a(x) = ca_0(x)$, even locally in time. This implies that the assumption (5.1) is necessary even for the local solvability.

Remark 5.7. As we have seen in Remark 4.7, we need some condition on the smallness at $t = 0$ like (2) in Theorem 5.1 for the uniqueness. In this case conditions of this type is really necessary: Haraux and Weissler [HW] showed that, if γ satisfies $1 + 2/n < \gamma < n/(n-2)$, then the equation (1.4) with $f(\sigma) = |\sigma|^{\gamma-1}\sigma$ possesses a non-trivial self-similar solution $u(t,x) = t^{1/(\gamma-1)}u(1, x/\sqrt{t})$ which approaches 0 in L^p for some range of p as $t \to +0$. For this $u(t,x)$ we have $\left\|u(t,\cdot)\,|\mathcal{N}^s_{p,q,\infty}\right\| = Ct^{n/2p-s/2+1/(\gamma-1)}$. It follows that, $u(t,\cdot) \to 0$ strongly in $\mathcal{N}^s_{p,q,\infty}$ if and only if $s < n/p - 2/(\gamma-1)$, and $u(t,\cdot) \to 0$ in the weak-$*$ topology of $\mathcal{N}^{n/p-2/(\gamma-1)}_{p,q,\infty}$. Hence, in order to ensure the uniqueness, we must impose some condition to exclude this self-similar solution.

6 The unique existence and the stability of small stationary solutions

In this section we assume that $n \geq 3$, and fix a number r such that $2 < r \leq n$. Then we consider the following stationary Navier-Stokes equation with an external force $f(x)$ in $\mathbb{R}^n$:

$$-\Delta_x w(x) + \left(w(x)\cdot\nabla_x\right)w(x) + \nabla_x \pi(x) = f(x), \tag{6.1}$$

$$\nabla_x \cdot w(x) = 0, \tag{6.2}$$

and give a condition on $f(x)$ sufficient for the unique existence of a small solution of (6.1)–(6.2) in suitable Morrey spaces. We also study the stability of the above stationary solution.

For the unique existence of small solutions of (6.1)–(6.2), we have the following theorem.

Theorem 6.1. (unique existence of stationary solutions) *There exist a positive number δ_3 and a continuous, strictly monotone-increasing function $\omega_3(\delta)$ on $[0, \delta_3]$ satisfying $\omega(0) = 0$ such that the following hold:*

(1) *For every $f(x) \in (\mathcal{D}')^n$, there exists at most one solution $w(x)$ of (6.1)–(6.2) in $\mathcal{M}_{n,r}$ satisfying the condition $\|w \,|\mathcal{M}_{n,r}\| < \omega_3(\delta_3)$.*
(2) *For every $f(x) \in (\mathcal{M}^{-2}_{n,r})^n$ satisfying $\delta = \|f \,|\mathcal{M}^{-2}_{n,r}\| < \delta_3$, there exists a solution $w(x) \in (\mathcal{M}_{n,r})^n$ of (6.1)–(6.2) with $\|u \,|\mathcal{M}_{n,r}\| \le \omega_3(\delta)$.*

Remark 6.2. In particular, we obtain a sufficient condition on the existence of a small stationary solution in the standard space L^n with external force $f = \nabla F$ with F small in $L^{n/2}$, by choosing $r = n$ in the above theorem and applying the Sobolev embedding theorem $\dot{H}^{-1}_{n/2} \subset \dot{H}^{-2}_n$. This corresponds to Theorem A of Kozono and Sohr [KS]. Although only the case $n = 3$ is treated there, it is easy to see that the same argument works for general $n \ge 3$, since the stationary Stokes equation is uniquely solvable in the space $\dot{H}^1_{n/2}$ in $\mathbb{R}^n$ for $n \ge 3$.

Remark 6.3. Suppose that $2 \le r < n$. Then $M_{n,r}$ contains the weak-L^n space $L^{n,\infty}$. It follows that the function space $M^{-2}_{n,r}$ contains nontrivial elements which is homogeneous of degree -3 (such as $|x|^{-3}$) in the case $r < n$ and $n \ge 4$. Choosing one of these functions as $f(x)$ in Theorem 6.1 and observing the uniqueness, we see that the stationary solution $w(x)$ is homogeneous of degree -1.

We can also treat solutions $w(x)$ which does not necessarily satisfy $w(x) \to 0$ as $|x| \to \infty$, as we can see in the following example.

Example 6.4. Suppose that $n \ge 4$, and let r be an integer satisfying $2 < r < n$. Next, let $\bar{w} = \big(w_1(x_1, \cdots, x_r), \cdots, w_r(x_1, \cdots, x_r)\big)$ be a sufficiently small function in $L^r(\mathbb{R}^r)$ satisfying (6.2) on $\mathbb{R}^r$, and put

$$\bar{f} = -\Delta_x \bar{w} + \nabla_x(\bar{w} \otimes \bar{w})$$

on $\mathbb{R}^r$. Then $w = (\bar{w}, 0)$ and $f = (\bar{f}, 0)$ satisfies (6.1)–(6.2) on $\mathbb{R}^n$. Moreover, we have $w \in \mathcal{M}_{n,r}$ and $f \in \mathcal{M}^{-2}_{n,r}$ on $\mathbb{R}^n$.

In order to verify the stability of the above stationary solution, we consider the following nonstationary Navier-Stokes equation on $\mathbb{R}^n$ with the same external force as above:

$$\frac{\partial v}{\partial t} - \Delta_x v + (v \cdot \nabla_x) v + \nabla_x q = f(x), \quad \text{in} \quad]0, \infty[\times \mathbb{R}^n, \tag{6.3}$$

(6.4) $$\nabla_x \cdot v = 0 \quad \text{in} \quad]0,\infty[\times \mathbb{R}^n,$$

(6.5) $$v(0,x) = a(x) \quad \text{on} \quad \mathbb{R}^n$$

for initial data $a(x)$ close enough to the stationary solution.

Then we have the following result on the stability of the above stationary solution in the Sobolev type spaces.

Theorem 6.5. (stability of stationary solutions) *Suppose that the constants p, q and σ_0 satisfy $n/2 < p < \infty$, $1 < q \le pr/n$ and $n/2p < \sigma_0 < \min\{1, n/p\}$. Further, let $w(x)$ be the solution of (6.1)–(6.2) given in Theorem 6.1, (2). Then there exists a positive number $\delta_4 \le \delta_3$ such that, for every $f(x) \in \left(\mathcal{M}_{n,r}^{-2}\right)^n$ satisfying $\|f\,|\mathcal{M}_{n,r}^{-2}\| < \delta_4$, there exist positive numbers ε_1 and L_1 such that, for every $a(x) \in \left(\mathcal{M}_{p,q}^{n/p-1}\right)^n$ satisfying*

$$\nabla_x \cdot a(x) = 0 \, and \, \varepsilon = \left\| a(x) - w(x) \left| \mathcal{M}_{p,q}^{n/p-1} \right. \right\| < \varepsilon_1,$$

there uniquely exists a time-global solution $v(t,x)$ of (6.3)–(6.4) satisfying the conditions

(6.6) $$\sup_{0<t\le T} t^{1/2-n/4p} \left\| v(t,\cdot) - w \left| \mathcal{M}_{p,q}^{n/2p} \right. \right\| < \infty \quad \text{for every } T > 0,$$

$$\sup_{0<t\le T} \left\| v(t,\cdot) - w \left| \mathcal{M}_{p,q}^{n/p-1} \right. \right\| < \infty \quad \text{for every } T > 0,$$

$$\limsup_{t\to 0} t^{1/2-n/4p} \left\| v(t,\cdot) - w \left| \mathcal{M}_{p,q}^{n/2p} \right. \right\| < L_1$$

and the initial condition (6.5) in the following sense: For every s such that $-1 \le s \le n/p - 1$ we have

(6.7) $$\sup_{0<t\le T} t^{s/2+1/2-n/2p} \left\| v(t,\cdot) - a \left| \mathcal{M}_{p,q}^{s} \right. \right\| < \infty \quad \text{for every } T > 0.$$

Furthermore, for every σ such that $n/p - 1 \le \sigma \le \sigma_0$, there exists a continuous, strictly monotone-increasing function $\psi_{1,\sigma}(\varepsilon)$ on $[0, \varepsilon_1]$ satisfying $\psi_{1,\sigma}(0) = 0$ such that the estimate

(6.8) $$\sup_{t>0} t^{\sigma/2+1/2-n/2p} \left\| v(t,\cdot) - w \left| \mathcal{M}_{p,q}^{\sigma} \right. \right\| \le \psi_{1,\sigma}(\varepsilon)$$

holds if $\varepsilon < \varepsilon_1$.

Remark 6.6. Suppose that $p \ge n$. Then Propositions 2.3 and 3.3 imply $w(x) \in \mathcal{M}_{n,r} \subset \mathcal{M}_{n,nq/p} \subset \mathcal{M}_{p,q}^{n/p-1}$. Hence (6.8) with $\sigma = n/p - 1$ implies the stability of $w(x)$ in $\mathcal{M}_{p,q}^{n/p-1}$. In particular, putting $p = n$ and $q = r$, we obtain the stability of $w(x)$ in $\mathcal{M}_{n,r}$. On the other hand, (6.8) with $\sigma > n/p - 1$ gives the decay order of the above solution.

Finally, we have the following result on the stability of the above stationary solution in the Besov type spaces.

Theorem 6.7. (stability of stationary solutions) *Let* p, q, σ_0, $f(x)$ *and* $w(x)$ *be the same as in Theorem 6.5. Then there exist positive numbers* ε_2 *and* L_2 *such that, for every* $a(x) \in \left(\mathcal{N}_{p,q,\infty}^{n/p-1}\right)^n$ *satisfying* $\nabla_x \cdot a(x) = 0$ *and* $\varepsilon = \left\| a(x) - w(x) \left| \mathcal{M}_{p,q}^{n/p-1} \right\| < \varepsilon_2\right.$, *there uniquely exists a time-global solution* $v(t,x)$ *of* (6.3)–(6.4) *satisfying the conditions* (6.6),

$$\sup_{0<t\leq T} \left\| v(t,\cdot) - w \left| \mathcal{N}_{p,q,\infty}^{n/p-1} \right\| < \infty \quad \text{for every } T>0,\right.$$

$$\limsup_{t\to 0} t^{1/2-n/4p} \left\| v(t,\cdot) - w \left| \mathcal{M}_{p,q}^{n/2p} \right\| < L_2\right.$$

and the initial condition (6.5) *in the sense that* (6.7) *holds for every* s *such that* $-1 \leq s < n/p-1$. *Furthermore, for every* σ *such that* $n/p-1 < \sigma \leq \sigma_0$, *there exists a continuous, strictly monotone-increasing function* $\psi_{2,\sigma}(\varepsilon)$ *on* $[0,\varepsilon_2]$ *satisfying* $\psi_{2,\sigma}(0) = 0$ *such that the estimate*

$$\sup_{t>0} t^{\sigma/2+1/2-n/2p} \left\| v(t,\cdot) - w \left| \mathcal{M}_{p,q}^{\sigma} \right\| \leq \psi_{2,\sigma}(\varepsilon)\right. \tag{6.9}$$

holds if $\varepsilon < \varepsilon_2$. *Moreover, there exists a continuous, strictly monotone-increasing function* $\psi_3(\varepsilon)$ *on* $[0,\varepsilon_2]$ *satisfying* $\psi_3(0) = 0$ *such that the estimate*

$$\sup_{t>0} \left\| v(t,\cdot) - w \left| \mathcal{N}_{p,q,\infty}^{n/p-1} \right\| \leq \psi_3(\varepsilon)\right. \tag{6.10}$$

holds if $\varepsilon < \varepsilon_2$.

Remark 6.8. Suppose that $p \geq n$. Then Remark 6.6 and Proposition 3.2 imply $w(x) \in \mathcal{N}_{p,q,\infty}^{n/p-1}$. Hence (6.10) implies the stability of $w(x)$ in $\mathcal{N}_{p,q,\infty}^{n/p-1}$. On the other hand, (6.9) gives the decay order of the above solution.

Remark 6.9. In the case $n \geq 3$, Theorem 6.7 is a generalization of Theorem 4.2 to the case $w(x) \neq 0$. In particular, Theorem 6.7 admits some distributions other than Radon measures as perturbation.

Remark 6.10. If the stationary solution and the initial data are both homogeneous of degree -1, then the perturbation $u(t,x) = v(t,x) - w(x)$ is self-similar; namely, it satisfies the equality $u(t,x) = t^{-1/2} u\left(1, t^{-1/2}x\right)$.

References

[BP] P. Baras et M. Pierre, *Problèmes paraboliques semi-linéaires avec données mesures*, Applicable Anal. **18** (1984), 111–149.

[BL] J. Bergh and J. Löfström, *Interpolation Spaces*, Springer, Berlin, 1976.

[BM] W. Borchers and T. Miyakawa, *On stability of exterior stationary Navier-Stokes flows*, Acta. Math. **174** (1995), 311–382.

[Bo] G. Bourdaud, *Réalisations des espaces de Besov homogènes*, Ark. Mat. **26** (1988), 41–54.

[BF] H. Brézis and A. Friedman, *Nonlinear parabolic equations involving measures as initial conditions*, J. Math. Pures Appl. **62(9)** (1983), 73–97.

[Cam] S. Campanato, *Proprietà di una famiglia di spazi funzionali*, Ann. Scuola Norm. Sup. Pisa **18** (1964), 137–160.

[Can] M. Cannone, *A generalization of a theorem by Kato on Navier-Stokes equations*, (preprint).

[CM] M. Cannone and Y. Meyer, *Littlewood-Paley decomposition and Navier-Stokes equations*, Methods Appl. Anal. **2** (1995), 307–319.

[CP] M. Cannone and F. Planchon, *Self-similar solutions for Navier-Stokes equations in* $\mathbb{R}^3$, Comm. Partial Differential Equations (to appear).

[Car] A. Carpio, *Comportement asymptotique des solutions des équations du tourbillon en dimensions 2 et 3*, C. R. Acad. Sci. Paris, Sér. I **316** (1993), 1289–1294.

[Co] G. Cottet, *Équations de Navier-Stokes dans le plan avec tourbillon initial mesure*, C. R. Acad. Sci. Paris, Sér. I **303** (1986), 105–108.

[Fe] P. Federbush, *Navier and Stokes meet the wavelet*, Comm. Math. Phys. **155** (1993), 219–248.

[Fu] H. Fujita, *On the blowing up of solutions of the Cauchy problem for* $u_t = \Delta u + u^{1+\alpha}$, J. Fac. Sci. Univ. Tokyo, I **13** (1966), 109–124.

[GK] Y. Giga and T. Kambe, *Large time behavior of the vorticity of two-dimensional viscous flow and its application to vortex formation*, Comm. Math. Phys. **117** (1988), 549–568.

[GM] Y. Giga and T. Miyakawa, *Navier-Stokes flow in* $\mathbb{R}^3$ *with measures as initial vorticity and Morrey spaces*, Comm. in Partial Differential Equations, **14** (1989), 577–618.

[GMO] Y. Giga, T. Miyakawa and H. Osada, *Two-dimensional Navier-Stokes flow with measures as initial vorticity*, Arch. Rat. Mech. Anal. **104** (1988), 223–250.

[G] G. Grubb, *Initial value problems for the Navier-Stokes equations with Neumann conditions*, The Navier-Stokes Equations II (J. G. Heywood, K. Masuda, R. Rautmann and S. A. Solonnikov, eds.), Proc. Conf. Oberwolfach 1991, Lecture Notes in Math. **1530**, Springer, Berlin, 1992, pp. 262–283.

[HW] A. Haraux and F. B. Weissler, *Non-uniqueness for a semilinear initial value problem*, Indiana Univ. Math. J. **31** (1982), 167–189.

[K1] T. Kato, *Strong* L^p*-solutions of the Navier-Stokes equation in* $\mathbf{R}^m$, *with applications to weak solutions*, Math. Z. **187** (1984), 471–480.

[K2] T. Kato, *Strong solutions of the Navier-Stokes equations in Morrey spaces*, Boll. Soc. Brasil. Mat. (N. S.) **22-2** (1992), 127–155.

[K3] T. Kato, *The Navier-Stokes equation for an incompressible fluid in* $\mathbb{R}^2$ *with a measure as the initial vorticity*, Differential and Integral

Equations **7** (1994), 949–966.

[KP] T. Kato and G. Ponce, *The Navier-Stokes equation with weak initial data*, Int. Math. Res. Notices **10** (1994), 435–444.

[KM] T. Kobayashi and T. Muramatu, *Abstract Besov space approach to the non-stationary Navier-Stokes equations*, Math. Methods in the Appl. Sci. **15** (1992), 599–620.

[KS] H. Kozono and H. Sohr, *On stationary Navier-Stokes equations in unbounded domains*, Ricerche Mat. **42** (1993), 69–86.

[KY1] H. Kozono and M. Yamazaki, *Semilinear heat equations and the Navier-Stokes equation with distributions in new function spaces as initial data*, Comm. in Partial Differential Equations **19** (1994), 959–1014.

[KY2] H. Kozono and M. Yamazaki, *The stability of small stationary solutions in Morrey spaces of the Navier-Stokes equation*, Indiana Univ. Math. J. (to appear).

[KY3] H. Kozono and M. Yamazaki, *The Navier-Stokes exterior problem*, (preprint).

[MR] M. Michaux and J. M. Rakotoson, *Remarks on Navier-Stokes equations with measures as data*, Appl. Math. Lett. **6-6** (1993), 75–77.

[MY] T. Miyakawa and M. Yamada, *Planar Navier-Stokes flows in a bounded domain with measures as initial vorticities*, Hiroshima Math. J. **22** (1992), 401–420.

[N] Y. Niwa, *Semilinear heat equations with measures as initial data*, Thesis, Univ. of Tokyo, 1986.

[P] J. Peetre, *On the theory of $L_{p,\lambda}$ spaces*, J. Funct. Anal. **4** (1969), 71–87.

[Ta] M. E. Taylor, *Analysis on Morrey spaces and applications to Navier-Stokes and other evolution equations*, Comm. in Partial Differential Equations **17** (1992), 1407–1456.

[Tr] H. Triebel, *Theory of Function Spaces*, Birkhäuser, Basel, 1983.

[W] F. B. Weissler, *Local existence and nonexistence for semilinear parabolic equations in L^p*, Indiana Univ. Math. J. **29** (1980), 79–102.

Bloch function in an external electric field and Berry-Buslaev phase

Shinichi Tajima

Department of Information Engineering, Faculty of Engineering, Niigata University, Niigata 950-21, Japan

The study of the behavior of Bloch electrons in an uniform external electric field is as old as the quantum theory of solids. Analysis of the motion of electrons in such external fields turned out, perhaps rather surprisingly, to be quit complicated and even at present the subject is very much alive. The source of the difficulties of this problem is that no matter how small the electron field strength is, for sufficiently large distances, the perturbed potential becomes arbitraly strong. In fact the perturbation created by electric field is singular from the spectral theoretic point of view. In consequence, strightforward application of the naive perturbation method is dangerous and rigorous results are hard to come by. For a better understanding of the difficulties, let us remined here the case of the Stark effect in atomic physics : although the Stark effects were the first example of quantum mechanical perturbation theory, it needed half a centry to develop a satisfactory mathematical description. Actually the existence of the Stark-Wannier resonance states, a quantum mechanical concept proposed by Wannier in solid state physics about 40 years ago, was strongly debated until recently.

The proposal and demonstration of superlattice effects in semiconductors by Esaki-Tsu [17] at the beginning of the 1970s and subsequent advance of technologies have revived interest in this problem.

In 1985, Agler and Froese [2] proved mathematically the existence of the resonances for the one dimensional Schrödinger equations with large external electric fields for certain cases. In 1988, Voisin et al [32] confirmed experimentally its existence in semiconductor superlattice devices. Since then, several experiments have observed various aspects of the Stark-Wannier resonances. More recently Bentosela-Grecchi [9], Combes-Hislop [15] and Buslaev-Dmitrieva [14] independently investigated the Stark-Wannier resonance states. They not only proved the existence of the resonances but also obtained more detailed results.

The basic ideas and the methods developed by Buslaev and Dmitrieva are very natural. They applied the standard multiple-scales method succesfully to reconcile the effects of the perturbation created by external electric field and the band concept for Bloch electrons. The leading terms of the asymptotic solution they constructed contains a geometric phase factor, which can be interpreted as Berry phase [10] associated with some Laglangian surface. The geometric phase derived by Buslaev [11] turned out to be important for the investigation of the

Stark-Wannier resonance states [14].

In this paper we examine the most typical case: the perturbed Lamé equation with a one-gap potential. We use ideas and methods borrowed from the papers of Buslaev [11] and Buslaev-Dmitrieva [14]. We derive an adiabatic connection on a complex torus and obtain an explicite formula of the geometric phase in this case.

1 Physical backgroung

In this section we breafly recall the Bloch theory for one dimensional crystal and its implications, which will be helpfull to understand naively what a Stark-Wannier resonance would be.

Let us consider the stationary Schrödinger equation for a single electron in a perfect one-dimensional crystal

$$\left(-\frac{\hbar^2}{2m}\frac{d^2}{du^2}+V(u)\right)\psi(u)=E\psi(u), \qquad -\infty<u<+\infty,$$

where $V(u)$ is an effective periodic potential with period d.

The classical Bloch's theorem says that the eigenstates ψ of the Schrödinger equation above can be choosen to have the form of a plane wave e^{iku} times a function with the crystalline periodicity d:

$$\psi_n(u)=e^{iku}b_{n,k}(u), \quad n=0,1,2,\cdots,$$

where the wave vector k is real. The corresponding electron energies $E_n(k)$ vary continuously as k varies. The index n is called the band index. The solution of the form above is called a Bloch function or Bloch electron.

Note that a Bloch wave function ψ_n extends without attenuation through a crystal. When k is imaginary, the function $\psi_n(u)$ defined by the formula above diverges for one direction in real space. Hence the ranges of electron energy E for which k must be imaginary are the energy gaps.

In the following, let us recall the classical concept of Bloch-Zerner oscillation, which was suggested in 1927 by Bloch. In order to discuss what an electron ought to do in an external electric field, let us imagine a band which is empty except for single electron. We shall imagine that there is no scattering for this electron. In a semiclassical picture, the time evolution of the position and crystal momentum $\hbar k$ of the wave packet of an electron with band index n are determined by the following Newton like equations of motion :

$$\frac{du}{dt}=\frac{1}{\hbar}\frac{dE_n}{dk}(k), \quad \frac{dk}{dt}=\frac{-eF}{\hbar}.$$

The motion of an electron is governed by the dispersion law describing the relation between the wave vector k and the energy of the electron. It follows easily that $k(t)=k(0)-eF/\hbar t$. Since $E_n(k)$ is a periodic funtion with the period $2\pi/d$, the electron would move back and forth, becoming localized in a finite region of the crystal. This is a striking contrast to the free electron case! We arrived at the concept of Bloch-Zener oscillation.

Let us continue the calculation. The period of this oscillation would be $T = \frac{2\pi}{d}\frac{\hbar}{eF}$. The amplitude of this oscillation would be :

$$\int_0^{\frac{T}{2}} \frac{du}{dt}dt = \int_0^{\frac{2\pi}{d}\frac{\hbar}{eF}} \frac{1}{\hbar}\frac{dE_n}{dk}\frac{dt}{dk}dk$$
$$= \int_0^{\frac{\pi}{d}} \frac{1}{\hbar}(-\frac{\hbar}{eF})dE_n(k) = \frac{\Delta E_n}{eF},$$

where ΔE_n is the bandwidth. We conclude that the electron would be localized within a region of length $\Delta E_n/eF$.

Example (a bulk crystal) Let a lattice periodicity $d = 8\AA$, a bandwidth $\Delta E = 3eV$ and an electric field strength $F = 10^5 V/cm$, then we have

$$\frac{\Delta E}{eF} = \frac{3 \times 1.6 \times 10^{-19} J}{1.6 \times 10^{-19} C \times 10^7 V/m} = 3 \times 10^{-7} m = 3 \times 10^3 \AA.$$

Example (a superlattice semiconductor) Let $d = 60\AA, \Delta E = 70meV$ and $F = 10^5 V/cm$, then we have

$$\frac{\Delta E}{eF} = 70\AA.$$

Note that one of the motivation by Esaki-Tsu [17] for the growth of semiconductor superlattice was the hope of achieving a negative differential resistance by realizing a Bloch-Zener oscillation. But there is no experimental confermation of such oscillation yet.

The Stark-Wannier resonances could be considered as the counterpart in the energy domain of the Bloch-Zener oscillations in the time domain. We refer to Bastard-Brum-Ferreira [6] for recent rsults on Stark-Wannier resonances in semiconductor supperlattices.

2 Band structure of the Lamé equation

We briefly recall basic properties of a Lamé equation and fix our notation. Let $\wp(u)$ be the Weierstrass elliptic function with primitive periods $2\omega_1$ and $2\omega_3$, where ω_1 and ω_3 are real and purely imaginary respectively. Let consider the Lamé equation of the form

$$-\frac{d^2}{du^2}\psi + 2\wp(u + \omega_3)\psi = E\psi. \tag{1}$$

Note that $2\wp(u+\omega_3)$ is a real-valued periodic function.

Let $\zeta(u)$ be the Weierstrass zeta function and let $\sigma(u)$ be the Weierstrass sigma function:

$$\zeta(u) = \frac{1}{u} + \sum_{\omega\in\Omega'} \left\{ \frac{1}{u-\omega} + \frac{1}{\omega} + \frac{u}{\omega^2} \right\},$$
$$\sigma(u) = u \prod_{\omega\in\Omega'} \left(1 - \frac{u}{\omega}\right) e^{\frac{u}{\omega}+\frac{u^2}{2\omega^2}}.$$

Let us introduce a complex variable z and we set:

$$E = E(z) = -\wp(z),$$
$$\chi = \zeta(u+\omega_3+z) - \zeta(u+\omega_3) - \zeta(z).$$

The additions formulae

$$\zeta(u_1+u_2) - \zeta(u_1) - \zeta(u_2) = \frac{1}{2}\left(\frac{\wp'(u_1)-\wp'(u_2)}{\wp(u_1)-\wp(u_2)}\right),$$
$$\wp(u_1+u_2) + \wp(u_1) + \wp(u_2) = \frac{1}{4}\left(\frac{\wp'(u_1)-\wp'(u_2)}{\wp(u_1)-\wp(u_2)}\right)^2$$

imply that the function χ satisfies the Riccati equation :

$$\chi^2 + \frac{d\chi}{du} + 2\wp(u+\omega_3) - E = 0.$$

This yields the factorization of the linear differential operator of the form

$$\frac{d^2}{du^2} - 2\wp(u+\omega_3) + E = \left(\frac{d}{du}+\chi\right)\left(\frac{d}{du}-\chi\right).$$

It is easy to see that the function ψ defined by

$$\psi(u,z) = e^{-\zeta(z)u} \cdot \frac{\sigma(u+\omega_3+z)}{\sigma(u+\omega_3)}$$

satisfies the following equation :

$$\begin{aligned}\frac{d}{du}\log\psi(u,z) &= \chi \\ &= \zeta(u+\omega_3+z) - \zeta(u+\omega_3) - \zeta(z).\end{aligned}$$

This yields that the function ψ is a solution of the Lamé equation.

Now we set

$$k = k(z) = \frac{i}{\omega_1}\left(\zeta(z)\omega_1 - \zeta(\omega_1)z\right), \tag{2}$$

$$b(u,z) = e^{-\frac{\zeta(\omega_1)}{\omega_1}zu} \cdot \frac{\sigma(u+\omega_3+z)}{\sigma(u+\omega_3)}. \tag{3}$$

Then we have the following:

$$\psi(u,z) = e^{ik(z)u} b(u,z). \tag{4}$$

Further more we set:

$$e_1 = \wp(\omega_1),\ e_3 = \wp(\omega_3),\ e_2 = \wp(\omega_1+\omega_3). \tag{5}$$

It is easy to verify the following results.

Lemma 1. *(1) Functions $\psi(u, \pm z)$ are Bloch solutions of the Lamé equation. (2) If $E(z) \neq -e_1, -e_2, -e_3$, then $\psi(u,z)$ and $\psi(u,-z)$ are lineary independent.*

Note that the energy E, the crystal-momentum k and the Bloch solutions are explicitely parametrized by the complex variable z.

Let

$$\begin{aligned}
I_1 &= \{z = x+iy \mid 0 \le x \le \omega_1, y = 0\}, \\
I_2 &= \{z = x+iy \mid 0 \le x \le \omega_1, y = \Im m\omega_3\}, \\
J_1 &= \{z = x+iy \mid x = \omega_1, 0 \le y \le \Im m\omega_3\}, \\
J_2 &= \{z = x+iy \mid x = 0, 0 \le y \le \Im m\omega_3\}.
\end{aligned}$$

Since $E(z) = -\wp(z)$ we have

$$\begin{aligned}
E(I_1) &= (-\infty, -e_1], \quad E(I_2) = [-e_2, -e_3], \\
E(J_1) &= (-e_1, -e_2), \quad E(J_2) = (-e_3, \infty).
\end{aligned}$$

Let us recall the following classical result (cf.[3]) .

Lemma 2. *The spectrum of the Lamé equation is continuous and consists of the intervals $[-e_1, -e_2]$, $[-e_3, \infty)$ separated by the forbidden band $(-e_2, -e_3)$.*

Proof. Let $z \in I_1$, then $2i\omega_1 k(z) = -2\zeta(z)\omega_1 + 2\zeta(\omega_1)z$ is real. This implies that $E(I_1) = (-\infty, -e_1]$ is a forbidden band. Let $z \in I_2$, then

$$-2\zeta(z)\omega_1 + 2\zeta(\omega_1)(z) = 2(\zeta(\omega_3)\omega_1 - \zeta(\omega_1)\omega_3) + 2\zeta(x)\omega_1 - 2\zeta(\omega_1)x + \frac{\wp'(x)}{\wp(x)-\wp(\omega_3)}\omega_1,$$

where $z = x + \omega_3$. Since

$$2\zeta(x)\omega_1 - 2\zeta(\omega_1)x + \frac{\wp'(x)}{\wp(x)-\wp(\omega_3)}\omega_1$$

is real, Legendre relation

$$e^{2(\zeta(\omega_3)\omega_1-\zeta(\omega_1)\omega_3)} = e^{\pi i} = -1$$

implies that $E(I_2)$ is a forbidden band. One can verify in a same way that $E(J_1)$ and $E(J_2)$ are stable bands. Q.E.D.

Note that Hochstadt [21] proved the converse of this statement.

3 Adiabatic connection on a complex torus

We consider, as a model equation, the perturbed Lamé equation of the form

$$\left(-\frac{d^2}{du^2} + 2\wp(u+\omega_3) + \varepsilon u - E_0\right)\psi(u,\varepsilon) = 0 \tag{6}$$

where the parameter $\varepsilon > 0$ is sufficiently small. Following the argument of Buslaev [11], we construct an asymptotic solution of this equation. We derive, in particular, the connection formula on a complex torus for the phase factor which is contained in the leading term of the asymptotic solution. We first introduce a new scaled variable $r = \varepsilon u$ and we apply the method of multiple-scales to this equation. We replace the unkown function $\psi(u,\varepsilon)$ by a new unkown function $f(u,r,\varepsilon)$ of two independent variables, which should satisfy the following condition:

$$f(u,r,\varepsilon)|_{r=\varepsilon u} = \psi(u,\varepsilon).$$

The function $f(u,r,\varepsilon)$ must satisfy the following linear partial differential equation:

$$\left(-\left(\frac{\partial}{\partial u} + \varepsilon\frac{\partial}{\partial r}\right)^2 + 2\wp(u+\omega_3) + r - E_0\right) f(u,r,\varepsilon) = 0.$$

Let us look for a formal solution of the form

$$f(u,r,\varepsilon) = e^{\frac{i}{\varepsilon}S(r)}a(u,r,\varepsilon), \tag{7}$$

where $a(u,r,\varepsilon) = a_0(u,r)+\varepsilon a_1(u,r)+\varepsilon^2 a_2(u,r)+\cdots$ and the amplitude $a(u,r,\varepsilon)$ is assumed to have the period $2\omega_1$ in u.

It follows that the amplitude $a(u,r,\varepsilon)$ satisfies the equation

$$\left(L_0 + \varepsilon L_1 + \varepsilon^2 L_2\right) a(u,r,\varepsilon) = 0, \tag{8}$$

where

$$L_0 = -\frac{\partial^2}{\partial u^2} - 2i\frac{\partial S}{\partial r}\frac{\partial}{\partial u} + \left(\frac{\partial S}{\partial r}\right)^2 + 2\wp(u+\omega_3) + r - E_0,$$

$$L_1 = -2\frac{\partial^2}{\partial u \partial r} - 2i\frac{\partial S}{\partial r}\frac{\partial}{\partial r} - i\frac{\partial^2 S}{\partial r^2}$$

and

$$L_2 = -\frac{\partial^2}{\partial r^2}.$$

We equate correspondint powers of ε in the usual way to get the following partial differential equations :

$$\begin{aligned} L_0 a_0(u,r) &= 0, \\ L_0 a_1(u,r) &= -L_1 a_0(u,r), \\ L_0 a_{j+2}(u,r) &= -L_1 a_{j+1}(u,r) - L_2 a_j(u,r) \qquad j \geq 0. \end{aligned}$$

In particular we have

$$\left(-\frac{\partial^2}{\partial u^2} - 2i\frac{\partial S}{\partial r}\frac{\partial}{\partial u} + \left(\frac{\partial S}{\partial r}\right)^2 + 2\wp(u+\omega_3) + r - E_0 \right) a_0(u,r) = 0.$$

Recall that the periodic function $b(u,z)$ satisfies the following differential equation:

$$\left(-\frac{d^2}{du^2} - 2ik(z)\frac{d}{du} + k^2(z) + 2\wp(u+\omega_3) - E(z) \right) b(u,z) = 0.$$

In the following we use the complex variable z instead of the variable r and rewrite everything we need in terms of z. We set:

$$k(z) = \frac{\partial S}{\partial r}(r) \quad \text{and} \quad r = E_0 - E(z).$$

Then we have

$$L_0 = -\frac{\partial^2}{\partial u^2} - 2ik(z)\frac{\partial}{\partial u} + k^2(z) + 2\wp(u+\omega_3) - E(z).$$

We introduce a formal series $b(u,z,\varepsilon) = b_0(u,z) + \varepsilon b_1(u,z) + \varepsilon^2 b_2(u,z) + \cdots$ by

$$b(u,z,\varepsilon) = a(u, E_0 - E(z), \varepsilon).$$

We have the following partial differential equations :

$$M_0\left(b_0(u,z)\right) = 0 \quad \text{and} \quad M_0\left(b_1(u,z)\right) = -M_1\left(b_0(u,z)\right), \tag{9}$$

where

$$M_0 = -\frac{\partial^2}{\partial u^2} - 2ik(z)^2\frac{\partial}{\partial u} + k(z)^2 + p(u) - E(z),$$

$$M_1 = \left(\frac{\partial r}{\partial z}\right)^{-1} \left(-2\frac{\partial^2}{\partial u \partial z} - 2ik(z)\frac{\partial}{\partial z} - i\frac{\partial k}{\partial z}(z) \right).$$

Since the function $b(u,z)$ satisfies the first equation of the system (9) we set

$$b_0(u,z) = N(z)b(u,z).$$

Note that the $N(z)$ is undetermined function of z at this stage. Let us recall that the functions $b_0(u,z)$ and $b_1(u,z)$ are assumed to be periodic functions of u with the period $2\omega_1$. It follows by the Fredholm alternatives that the second equation of the system (9) has a periodic solution $b_1(u,z)$ if and only if

$$\int_0^{2\omega_1} b(u,-z)M_1\left(N(z)b(u,z)\right)du = 0.$$

This yields the following first order linear differential equation for N :

$$\frac{\partial N}{\partial z}(z) + \left(\frac{\langle b(u,-z), \frac{\partial b}{\partial z}(u,z)\rangle}{\langle b(u,-z), b(u,z)\rangle} + \frac{1}{2}\left(\frac{\frac{\partial^2 E}{\partial z^2}}{\frac{\partial E}{\partial z}} - \frac{\frac{\partial^2 k}{\partial z^2}}{\frac{\partial k}{\partial z}} \right) \right) N(z) = 0, \tag{10}$$

where

$$< f, g >= \int_0^{2\omega_1} f(u)g(u)du.$$

We set

$$b_0(u,z) = U(z)\left(\frac{\frac{\partial E}{\partial z}}{\frac{\partial k}{\partial z}}\right)^{-\frac{1}{2}} b(u,z), \tag{11}$$

$$\theta(z) = -i\frac{\langle b(u,-z), \frac{\partial b}{\partial z}(u,z)\rangle}{\langle b(u,-z), b(u,z)\rangle}. \tag{12}$$

We thus arrive at the following Buslaev's result:

Theorem 1 (Buslaev [11]). *The function $U(z)$ satisfies the following equation:*

$$\frac{dU}{dz}(z) + i\theta(z)U(z) = 0.$$

Note that if $z \in J_1 \cup J_2$, i.e. the energy $E = -\wp(z)$ lies in the energy bands, then the phase function $\theta(z)$ is real.

4 Geometric Phase

In the previous section we have verified that the leading term of the asymptotic solution contains a gemetric phase factor of the form :

$$U(z) = e^{i\int \theta(z)dz},$$

where

$$\theta(z) = -i\frac{\langle b(u,-z), \frac{\partial b}{\partial z}(u,z)\rangle}{\langle b(u,-z), b(u,z)\rangle}.$$

Note that such kind of phase factor was already obtained by Adames II [1] in the early 50-th, but there were no interpretation for this phase. In 1984, Buslaev [11] derived this geometric phase factor by using the multiple-scales method and clarified its true nature. The geometric phase derived by Buslaev can be interpreted as an adiabatic phase found by Berry [10] in his study of the quantum adiabatic theorem. In this section we calculate explicitly the geometric phase associated to the perturbed Lamé equation. For this aim it is sufficient to calculate the following integrals:

$$< b(u,-z), b(u,z) >= \int_0^{2\omega_1} b(u,-z)b(u,z)du,$$

$$< b(u,-z), \frac{\partial b}{\partial z}(u,z) >= \int_0^{2\omega_1} b(u,-z)\frac{\partial b}{\partial z}(u,z)du.$$

Let us start the calculation. Since

$$b(u,z) = e^{-\frac{\zeta(\omega_1)}{\omega_1}zu} \cdot \frac{\sigma(u+\omega_3+z)}{\sigma(u+\omega_3)}.$$

and

$$\wp(u_1) - \wp(u_2) = -\frac{\sigma(u_1-u_2)\sigma(u_1+u_2)}{\sigma^2(u_1)\sigma^2(u_2)}$$

we have

$$b(u,-z)b(u,z) = -\sigma^2(z)\left(\wp(u+\omega_3) - \wp(z)\right).$$

By integrating the formula above, we have

$$\begin{aligned}
< b(u,-z), b(u,z) > &= \int_0^{2\omega_1} b(u,-z)b(u,z)du \\
&= -\sigma^2(z)\int_0^{2\omega_1} \left(\wp(u+\omega_3) - \wp(z)\right) du \\
&= -\sigma^2(z)\left[\zeta(u+\omega_3)\right]_{u=0}^{u=2\omega_1} + \sigma^2(z)2\omega_1\wp(z) \\
&= \sigma^2(z)\left\{\zeta(\omega_3+2\omega_1) - \zeta(\omega_3) + 2\omega_1\wp(z)\right\} \\
&= 2\sigma^2(z)\left(\zeta(\omega_1) + \omega_1\wp(z)\right).
\end{aligned}$$

Lemma 3. *The following equality holds:*

$$\begin{aligned}
&b(u,-z)\frac{\partial b}{\partial z}(u,z) \\
&= \sigma^2(z)\left(\frac{\zeta(\omega_1)}{\omega_1}u - \zeta(u+\omega_3) - \zeta(z)\right)\left(\wp(u+\omega_3) - \wp(z)\right) + \frac{1}{2}\sigma^2(z)\left(\wp'(z) - \wp'(u+\omega_3)\right).
\end{aligned}$$

Proof. Since

$$\frac{\partial b}{\partial z}(u,z) = \frac{\zeta(\omega_1)}{\omega_1}ub(u,z) + e^{\frac{\zeta(\omega_1)}{\omega_1}zu} \cdot \frac{\sigma'(u+\omega_3+z)}{\sigma(u+\omega_3)}$$
$$= \left(-\frac{\zeta(\omega_1)}{\omega_1}u + \zeta(u+\omega_3+z)\right) b(u,z),$$

We have

$$b(u,-z)\frac{\partial b}{\partial z}(u,z) = \left(-\frac{\zeta(\omega_1)}{\omega_1}u + \zeta(u+\omega_3+z)\right) b(u,-z)b(u,z)$$
$$= \sigma^2(z)\left(\frac{\zeta(\omega_1)}{\omega_1}u - \zeta(u+\omega_3+z)\right)(\wp(u+\omega_3)-\wp(z)).$$

We thus have:

$$b(u,-z)\frac{\partial b}{\partial z}(u,z)$$
$$= \sigma^2(z)\left(\frac{\zeta(\omega_1)}{\omega_1}u - \zeta(u+\omega_3) - \zeta(z) - \frac{1}{2}\frac{\wp'(u+\omega_3)-\wp'(z)}{\wp(u+\omega_3)-\wp(z)}\right)(\wp(u+\omega_3)-\wp(z))$$

This conclude the result. Q.E.D.

Let us continue our calculation. We have

$$\int_0^{2\omega_1} b(u,-z)\frac{\partial b}{\partial z}(u,z)du = \sigma^2(z)\left(\frac{\zeta(\omega_1)}{\omega_1}I_1 - I_2 + I_3 + R\right).$$

where

$$I_1 = \int_0^{2\omega_1} u\wp(u+\omega_3)du,$$
$$I_2 = \int_0^{2\omega_1} \zeta(u+\omega_3)\wp(u+\omega_3)du,$$
$$I_3 = \int_0^{2\omega_1} \zeta(u+\omega_3)du$$

and

$$R = -\frac{\zeta(\omega_1)}{\omega_1}\wp(z)\int_0^{2\omega_1} udu - \zeta(z)\int_0^{2\omega_1}\wp(u+\omega_3)du + \omega_1\left(2\zeta(z)\wp(z) - \wp'(z)\right)$$
$$= 2\zeta(z)(\zeta(\omega_1)+\omega_1\wp(z)) - 2\omega_1\zeta(\omega_1)\wp(z) + \omega_1\wp'(z).$$

By direct calculation we have

Lemma 4. *The following equalities hold:*

$$\begin{aligned} I_1 &= I_3 - 2\omega_1\zeta(2\omega_1+\omega_3), \\ I_2 &= -\zeta(\omega_1)\left(\zeta(2\omega_1+\omega_3)+\zeta(\omega_3)\right), \\ I_3 &= 2\omega_1\left(\zeta(\omega_1)+\zeta(\omega_3)\right)+2\pi i. \end{aligned}$$

We thus have

$$\begin{aligned} &\int_0^{2\omega_1} b(u,-z)\frac{\partial b}{\partial z}(u,z)du \\ &= \sigma^2(z)\left\{(2\zeta(\omega_3)+\frac{2\pi i}{\omega_1})(\zeta(\omega_1)+\omega_1\wp(z))+2\omega_1\zeta(\omega_1)\wp(z)\right\} \\ &\quad +\sigma^2(z)\left\{2\zeta(z)(\zeta(\omega_1)+\omega_1\wp(z))-2\omega_1\zeta(\omega_1)\wp(z)+\omega_1\wp'(z)\right\} \\ &= 2\sigma^2(z)\left\{(\zeta(\omega_3)+\zeta(z)+\frac{2\pi i}{\omega_1})(\zeta(\omega_1)+\omega_1\wp(z))+\frac{1}{2}\omega_1\wp'(z)\right\}. \end{aligned}$$

We arrive at the following explicite formula:

$$i\theta(z) = \zeta(\omega_3)+\zeta(z)+\frac{\pi i}{\omega_1}+\frac{1}{2}\cdot\frac{\omega_1\wp'(z)}{\zeta(\omega_1)+\omega_1\wp(z)}. \tag{13}$$

We summarize our results in the following theorem

Theorem 2. *Let*

$$\begin{aligned} E &= E(z) = -\wp(z), \\ k &= k(z) = \frac{i}{\omega_1}\left(\zeta(z)\omega_1-\zeta(\omega_1)z\right), \end{aligned}$$

and

$$b(u,z) = e^{-\frac{\zeta(\omega_1)}{\omega_1}zu}\cdot\frac{\sigma(u+\omega_3+z)}{\sigma(u+\omega_3)}.$$

Let $r \neq E_0+e_1,\ E_0+e_2,\ E_0+e_3$ and let $z(r)$ denote the inverse function of $r = E_0 - E(z)$. Then the leading term of the asymptotic solution of the perturbed Lamé equation

$$\left(-\frac{d^2}{du^2}+2\wp(u+\omega_3)+\varepsilon u-E_0\right)\psi(u,\varepsilon)=0,$$

corresponding to the Bloch solution $e^{iku}b(u,z)$, has the following form :

$$e^{\frac{i}{\epsilon}S(r)}b_0\left(u, z(r)\right),$$

where

$$b_0(u,z) = e^{i\int \theta(z)dz}\left(\frac{\frac{\partial E}{\partial z}}{\frac{\partial k}{\partial z}}\right)^{-\frac{1}{2}} b(u,z),$$

$$i\theta(z) = \zeta(\omega_3) + \zeta(z) + \frac{\pi i}{\omega_1} + \frac{1}{2}\cdot\frac{\omega_1\wp'(z)}{\zeta(\omega_1)+\omega_1\wp(z)},$$

$$\left(\frac{\frac{\partial E}{\partial z}}{\frac{\partial k}{\partial z}}\right) = -i\frac{\omega_1\wp'(z)}{\zeta(\omega_1)+\omega_1\wp(z)}.$$

Recently Maruyama [27] extended the result above for the one-gap potential case to the case of Lame equation with a two-gaps potential.

Note that, by using a general formula in [12] and the explicit formula for the geometric phase, one can calculate "quantization condition for the Stark-Wannier localized state" for the perturbed Lamé equation. This line of attack should be interesting for the further investigation of the Stark-Wannier resonance states.

References

1. E. N. Adams, II, *Motion of electron in a perturbed periodic potential*, Phys. Rev. **85** (1952), 41-50.
2. J. Agler and R. Froese, *Existence of Stark ladder resonances*, Comm. Math. Phys. **100** (1985), 161-171.
3. N. I. Akhiezer, *On the spectral theory of Lamé's equation*, Istor. Mat. Issled. **23** (1978), 77-86 (in Russian).
4. W. Ashcroft and N. D. Mermin, *Solid State Physics*, Saunders Campany, 1976.
5. J. E. Avron, *On the lifetime of Wannier ladder states*, Annals of Phys. **143** (1982), 33-53.
6. G. Bastard, J. A. Brum and R. Ferreira, *Electronic states in semiconductor heterostructures*, Solid State Phys. **44** (1991), 229-415.
7. F. Bentosela, *Bloch electrons in constant electric field*, Comm. Math. Phys. **68** (1979), 173-182.
8. F. Bentosela, V. Grecchi and F. Zironi, *Approximate ladder of resonances in a semi-infinite crystal*, J. Phys. C. Solid State Phys. **15** (1982), 7119-7131.
9. F. Bentosela and V. Grecchi, *Stark-Wannier ladders*, Comm. Math. Phys. **142** (1991), 169-192.
10. M. V. Berry, *Quantum phase factors accompanying adiabatic changes*, Proc. R. Soc. London A **392** (1984), 45-57.
11. V. S. Buslaev, *Adiabatic perturbation of a periodic potential*, Theoret. and Math. Phys. **58** (1984), 153-159.

12. V. S. Buslaev, *Semiclassical approximation for equations with periodic coefficients*, Russian Math. Survey **42** (1987), 97-125.
13. V. S. Buslaev and L. A. Dmitrieva, *Geometric aspects of the Bloch electrons theory in external field*, Topological Phases in Quantum Theory (B. Markovski and S. I. Vinitsky, eds.), World Scientific, 1989, pp. 218-250.
14. V. S. Buslaev and L. A. Dmitrieva, *A Bloch electron in an external field*, Leninglad Math. **1** (1990), 287-320.
15. J. M. Combes and P. D. Hislop, *Stark ladder resonances for small electric fields*, Comm. Math. Phys. **140** (1991), 291-320.
16. B. A. Dubrovin, V. B. Matveev and S. P. Novikov, *Non-linear equations of Korteweg-de Vries type*, finite-zone linear operators, and Abelian varieties, Russian Math. Surveys **31** (1976), 59-146.
17. L. Esaki and R. Tsu, *Superlattice and negative differential conductivity in semiconductors*, IBM J. Res. Develop. **14** (1970), 61-65.
18. L. Fritsche, *Representation of a lattice electron in a uniform electric field*, Phys. Stat. Sol **13** (1966), 487-497.
19. J. C. Guillot, J. Ralston and E. Trubowitz, *Semi-classical asymptotics in solid state physics*, Comm. Math. Phys. **116** (1988), 401-415.
20. I. W. Herbst and J. S. Howland, *The Stark ladder and other one-dimensional external field problems*, Comm. Math. Phys. **80** (1981), 23-42.
21. H. Hochstadt, *On the determination of a Hill's equation from its spectrum*, Arch. Rat. Mech. Anal. **19** (1965), 353-362.
22. W. V. Houston, *Acceleration of electrons in a crystal lattice*, Phys. Review **57** (1940), 184-186.
23. W. Hunziker, *Schrödinger operators with electric or magnetic fields*, Mathematical Problems in Theoretical Physics (K. Osterwalder, ed.), Lecture Notes in Physics, **116** (1979), pp. 25-44.
24. A. R. Its and V. B. Matveev, *On a class of solutions of the KdV equation*, Problems of mathematical physics **8** Izdat Leningrad Univ. (1976), 70-92 (in Russian).
25. W. Kohn, *Analytic properties of Bloch waves and Wannier functions*, Phys. Review **115** (1959), 809-821.
26. W. Magnus and S. Winkler, *Hill's Equation*, 1979, Dover, 1979.
27. F. Maruyama, *Sur la phase géometrique de l'équation de Lamé perturbée*, (preprint).
28. E. E. Mendez, F. Agulló-Rueda and J. M. Hong, *Stark localization in GaAs-GaAlAs superlattices under an electric field*, Phys. Rev. Lett. **60** (1988), 2426-2429.
29. G. Nenciu, Dynamics of band electrons in electric and magnetic fields, Rev. Mod. Phys. **63** (1991), 91-127.
30. S. Tajima, *Geometric phase in a perturbed Lamé equation*, First Korean-Japanese Colloquium on Finite or Infinite Dimensional Complex Analysis (J. Kajiwara, H. Kazama and K. H. Shon, eds.), 1993, pp. 55-61.
31. E. C. Titchmarsh, *Some theorems on perturbation theory, III, IV, V*, Proc. Roy. Soc. **A207** (1951), 321-328; **A210** (1951), 30-47; J. Analyse Math. **4** (1954/56), 187-208.

32. P. Voisin, J. Bleuse, C. Bouch, S. Gaillard, C. Alibert and A. Regreny, *Observation of the Wannier-Stark quantization in a semiconductor superlattice*, Phys. Rev. Lett. **61** (1988), 1639-1642.
33. G. H. Wannier, *Wave functions and effective Hamiltonian for Bloch electrons in an electric field*, Phys. Rev. **117** (1960), 432-439.
34. J. Zak, *Localized states and effective Hamiltonians in perturbed crystals*, Comm. in Physics **1** (1976), 73-79.
35. C. Zener, *A theory of the electrical breakdown of solid dielectrics*, Proc. Roy. Soc. London, ser A **145** (1934), 523-529.

Part III
Algebraic Analysis
$\mathcal{D}$-modules and Sheave Theory

An application of symbol calculus

Emmanuel Andronikof[†]

1 Introduction

Let $X = \mathbb{C}^{1+n}_{(t,x)}$, $t \in \mathbb{C}$, $x = (x_1, \ldots, x_n) \in \mathbb{C}^n$, and let $(t, x; \tau, \xi)$ be the associated symplectic coordinates in T^*X. In Kashiwara and Oshima's study of regular systems (cf [5]), the following definition occurs (with a slightly different vocabulary): a matrix of microdifferential operators $A(x, D_x)$ is essentially of order ≤ 0 if there exists $\nu > 0$ such that the coefficients of any power of A are microdifferential operators of order at most ν. It is shown in [5] that any regular system of microdifferential equations with regular singularities along $V = \{t = \xi_1 = \cdots = \xi_r = 0, \tau \neq 0\}$, is a quotient of a system of the form $(tD_t - A(x, D_x))u = D_{x_1}u = \cdots = D_{x_r}u = 0$, with A essentially of order ≤ 0.

When investigating e.g. distribution solutions of regular systems from a microlocal point of view, that is, solutions with values in the sheaf $\mathcal{C}^f_{\mathbb{R}^n}$ of tempered microfunctions on $\mathbb{R}^n$, one is led to look for the simplest normal form over some extension of $\mathcal{E}_X$ to a ring of operators acting on $\mathcal{C}^f_{\mathbb{R}^n}$ (see Remark 4 (i) below).

In particular, one has on $\mathcal{C}^f_{\mathbb{R}^n}$ an action of $\mathcal{E}^{\mathbb{R},f}_X$, the ring of tempered microlocal operators (cf [1]), and the purpose of this note is to prove the

Theorem 1. *Let $A = A(x, D_t, D_x) \in \mathbf{M}_N(\mathcal{E}_X)_{(0;1,\xi_0)}$ be essentially of order ≤ 0. Then*

(i) The operator $(D_tI_N)^A$ is well defined in $\mathbf{M}_N(\mathcal{E}^{\mathbb{R},f}_X)_{(0;1,\xi_0)}$ and invertible, its inverse being $(D_tI_N)^{-A}$,

(ii) one has:

$$(D_tI_N)^A(tD_tI_N - A)(D_tI_N)^{-A} = tD_tI_N.$$

It is reasonable to think that this result could also be deduced from [5], but, to our knowledge, this has never been achieved. We proved a partial version of the above theorem in [2].

2 Symbol calculus

Let us recall here the following facts about $\mathcal{E}_X^{\mathbf{R},f}$. Let X be an n-dimensional complex manifold. The sheaf of rings $\mathcal{E}_X^{\mathbf{R},f}$ is the tempered version of [1] of the sheaf of rings $\mathcal{E}_X^{\mathbf{R}}$ on T^*X of holomorphic microlocal operators of [6], of which it is a subsheaf.

One has $\gamma^{-1}\gamma_*\mathcal{E}_X^{\mathbf{R},f} = \mathcal{E}_X$. Also $\mathcal{E}_X^{\mathbf{R},f}$ is faithfully flat over $\mathcal{E}_X$.

We are going to make use of a few topics of the theory of symbols of holomorphic microlocal operators as developped by Kataoka and Aoki (see [4] and the literature quoted there), which adapt to the framework of $\mathcal{E}_X^{\mathbf{R},f}$.

Let $x_0^* = (x_0;\xi_0) \in T^*X$ and U a $\mathbb{R}_{>0}$-conical open neighborhood of x_0^*. Denote by

$$S(U) \qquad (\text{resp. } S^f(U), \text{ resp. } R(U)),$$

the space of holomorphic functions $p(x,\xi)$ on U such that for any compactly generated cone $U' \subset U$ one has:

$$\begin{cases} \text{for any } \varepsilon > 0 \quad p(x,\xi) = O(e^{\varepsilon|\xi|}) \text{ on } U' \\ (\text{resp. there exists } m > 0 \text{ such that } p(x,\xi) = O(|\xi|^m) \text{ on } U'), \\ (\text{resp. there exists } \delta > 0 \text{ such that } p(x,\xi) = O(e^{-\delta|\xi|}) \text{ on } U'). \end{cases}$$

The notations $S(U)$, $R(U)$ are borrowed from [4].

Proposition 2. *(i) (cf [4]) There is an isomorphism of vector spaces*

$$\varinjlim_{U \ni x_0^*} S(U)/R(U) \xrightarrow{\sim} \mathcal{E}^{\mathbf{R}}_{X,x_0^*}.$$

(ii) The isomorphism of (i) induces an isomorphism

$$\varinjlim_{U \ni x_0^*} S^f(U)/R(U) \xrightarrow{\sim} \mathcal{E}^{\mathbf{R},f}_{X,x_0^*}.$$

In a local coordinate system $(x_1,\dots,x_n)$, the above morphisms take x_i to x_i, ξ_i to D_{ξ_i}, and $x_i\xi_i$ to $x_iD_{\xi_i}$. In fact (ii) is easily deduced from the calculation in [4].

A representative $p(x,\xi) \in S(U)$ of an operator $P \in \mathcal{E}^{\mathbf{R}}_{X,x_0^*}$ for a suitable neighborhood U of x_0^* is called a symbol of P, and the lower bound of $m \in \mathbb{R}$ such that $p(x,\xi) = O(|\xi|^m)$ in a conical neighborhood of x_0^* is called the order of P at x_0^*, e.g. $m < \infty$ iff $P \in \mathcal{E}^{\mathbf{R},f}_{X,x_0^*}$. As noted in [4], (i) of Proposition 2 entails that if $p(x,\xi) \in S(U)$ satisfies $p(x,\xi) = o(|\xi|)$ in a conical neighborhood of x_0^*, then $\exp(p(x,\xi))$ is a symbol of an operator of $\mathcal{E}^{\mathbf{R}}_{X,x_0^*}$. By (ii) we get also:

$$\begin{cases} \text{if } p(x,\xi) \in S(U) \text{ satisfies } p(x,\xi) = O(\log|\xi|) \\ \text{in a conical neighborhood of } x_0^*, \\ \text{then } \exp(p(x,\xi)) \text{ is a symbol of an operator of } \mathcal{E}^{\mathbf{R},f}_{X,x_0^*}. \end{cases}$$

For example, if $X = \mathbb{C}^n_x$ with coordinates $x = (x_1,\dots,x_n)$, one has $\exp(x_1 \log D_{x_1}) \in \mathcal{E}^{\mathbf{R},f}_{X,dx_1}$, whereas $\exp(x_1\sqrt{D_{x_1}}) \in \mathcal{E}^{\mathbf{R}}_{X,dx_1} \setminus \mathcal{E}^{\mathbf{R},f}_{X,dx_1}$.

3 Some norm estimates

Lemma 3. *Let $A \in \mathbf{M}_N(\mathcal{E}_X)_p$ be essentially of order zero. Then the series*

$$\sum_{j\geq 0} \frac{A^j}{\lambda^j}$$

converges in $\mathbf{M}_N(\mathcal{E}_X)_p$ for $|\lambda| \gg 1$.

Proof. Since A is essentially of order zero, by definition there exists ν such that $A^j \in \mathbf{M}_N(\mathcal{E}_X(\nu))$ for every $j > 0$. What we need is to get some estimates of the Boutet de Monvel-Krée norms:

$$N_\nu(A^j, s) = \sum_{k\geq 0} C_{\alpha,\beta,k} |\partial_x^\alpha \partial_\xi^\beta (A^j_{\nu-k})| s^{2k+|\alpha+\beta|},$$

where we recall that

$$C_{\alpha,\beta,k} = \frac{2}{(2n)^k} \frac{k!}{(|\alpha|+k)!(|\beta|+k)!}. \tag{1}$$

Since it is clear that

$$N_{j\nu}(A^j, s) \ll N_\nu(A, s)^j,$$

if we had an estimate of the form

$$N_\nu(B, s) \ll M c^j N_{j\nu}(B, s), \tag{2}$$

taking $B = A^j$ we would indeed deduce that the series $\sum A^j/\lambda^j$ converges for $|\lambda| \gg 1$.

To prove (2), let us restrict ourselves to the simpler case of a single operator $P \in \mathcal{E}_X(\nu)$, the case of a matrix B being obtained similarly.

One has for $\nu' > \nu$:

$$\begin{aligned} N_{\nu'}(P, s) &= \sum_{k\geq 0} C_{\alpha,\beta,k} |\partial_x^\alpha \partial_\xi^\beta (P_{\nu'-k})| s^{2k+|\alpha+\beta|} \\ &= \sum_{k\geq \nu'-\nu} C_{\alpha,\beta,k} |\partial_x^\alpha \partial_\xi^\beta (P_{\nu-(k-(\nu'-\nu))})| s^{2k+|\alpha+\beta|} \\ &= \sum_{k'\geq 0} C_{\alpha,\beta,k'+(\nu'-\nu)} |\partial_x^\alpha \partial_\xi^\beta (P_{\nu-k'})| s^{2k'+|\alpha+\beta|} s^{2(\nu'-\nu)}, \end{aligned}$$

which we may rewrite as

$$N_{\nu+\mu}(P, s) = \left(\sum_{k\geq 0} C_{\alpha,\beta,k+\mu} |\partial_x^\alpha \partial_\xi^\beta (P_{\nu-k})| s^{2k+|\alpha+\beta|} \right) s^{2\mu}.$$

One has:

$$\begin{aligned} C_{\alpha,\beta,k+\mu} &= \frac{2}{(2n)^{k+\mu}} \frac{(k+\mu)!}{(|\alpha|+k+\mu)!(|\beta|+k+\mu)!} \\ &= \frac{1}{(2n)^{\mu}} C_{\alpha,\beta,k} \frac{(k+\mu)!}{k!} \frac{(|\alpha|+k+\mu)!(|\beta|+k+\mu)!}{(|\alpha|+k)!(|\beta|+k)!} \\ &\geq \frac{1}{(2n)^{\mu}} C_{\alpha,\beta,k} (\mu!)^3, \end{aligned}$$

where, in the last inequality, we used the estimate

$$\frac{(m+p)!}{m!} \geq p!.$$

It follows that

$$N_\nu(P,s) \ll \frac{(2n)^\mu}{(\mu!)^3} N_{\nu+\mu}(P,s) s^{-2\mu},$$

and, finally, by induction

$$N_\nu(P,s) \ll \left(\frac{(2n)^\nu}{(\nu!)^3} s^{-2\nu} \right)^j N_{j\nu}(P,s).$$

This proves (2).

Example 1. If A is essentially of order zero, then $\exp(A) \in \mathbf{M}_N(\mathcal{E}_X)$.

We will need a sharper control of the symbol.

Lemma 4. *$A \in \mathbf{M}_N(\mathcal{E}_X)_p$ is essentially of order zero if and only if there exist $\nu > 0$, $M > 0$, and a conic neighborhood V of p such that for every $j \in \mathbb{N}$ the estimate $\|A^j(x;\xi)\| \leq M^j(1+|\xi|)^\nu$ holds uniformly in $(x;\xi) \in V$.*

Proof. Let us prove that the condition is necessary (the sufficiency being obvious). Let $\lambda \in \mathbb{C}$ be an independent variable. By Kashiwara-Oshima [5], we know that the Sato-Kashiwara determinant of $\lambda - A(x, \partial_x)$ (considered as a matrix operator in $\mathbf{M}_N(\mathcal{E}_{\mathbb{C}\times X})$) has the form

$$\det(\lambda - A(x,\partial_x)) = \lambda^N + p_1(x;\xi)\lambda^{N-1} + \cdots + p_N(x;\xi),$$

where the p_js are holomorphic functions homogeneous of degree zero in ξ, defined in a conic neighborhood of p. Let $\lambda_j = \lambda_j(x,\xi)$, $1 \leq j \leq N$ be the roots of $\det(\lambda - A(x,\partial_x)) = 0$. Fixing a conic neighborhood V of p, we may find a loop γ in $\mathbb{C}$ surrounding every λ_j clockwise.

Recall that if P is a microdifferential operator in $\mathcal{E}_{X\times\mathbb{C}}$ of order $\leq \nu$, of the form

$$P(\lambda, x, \partial_x) = \sum_{j\geq 0} P_{\nu-j}(\lambda, x; \xi),$$

we denote by $\int_\gamma P(\lambda, x; \partial_x)\, d\lambda$ the operator defined by $\sum_{j\geq 0} \int_\gamma P_{\nu-j}(\lambda, x; \partial_x)\, d\lambda$. The same notation is used for matrices, integrating entrywise.

Since A is essentially of order zero, by definition there exists ν such that $A^j \in \mathbf{M}_N(\mathcal{E}_X(\nu))$ for every $j > 0$. For $|\lambda| \gg 1$, one has

$$(\lambda - A)^{-1} = \frac{1}{\lambda}(I + \frac{1}{\lambda}A + \cdots + \frac{1}{\lambda^k}A^k + \cdots) \tag{3}$$

where the series converges in $\mathbf{M}_N(\mathcal{E}_X(\nu))$, due to Lemma 3. It follows that for every integer $j \geq 1$

$$A^j = \frac{1}{2\pi j}\int_\gamma \lambda^j(\lambda - A)^{-1}\, d\lambda.$$

Hence, for $M = |\lambda| \gg 1$,

$$\|A^j(x,\xi)\| \leq cM^k \int_\gamma \|(\lambda - A)^{-1}(x,\xi)\|\, |d\lambda|.$$

The conclusion follows, since, in view of (3), $(\lambda - A)^{-1}$ is an operator of order $\leq \nu$.

4 Proof of the main theorem

Let now $X = \mathbb{C}^{1+n}_{(t,x)}$, $t \in \mathbb{C}$, $x = (x_1, \ldots, x_n) \in \mathbb{C}^n$, and let $(t, x; \tau, \xi)$ be the associated symplectic coordinates in T^*X.

According to Lemma 4, we may find a conic neighborhood V of $(0; 1, \xi_0)$ and costants c, M so that

$$\|A^j(x;\tau,\xi)\| \leq cM^j(1 + |\tau| + |\xi|)^\nu$$

for every $j \geq 1$. Set

$$p(x;\tau,\xi) = \sum_{j\geq 0} \frac{A^j(x;\tau,\xi)(\log\tau)^j}{j!},$$

where we took the determination of $\log\tau$ on Re$\tau >$Imτ such that $\log 1 = 0$. For $|\tau| > e^\pi$, we have $|\log\tau| \leq \sqrt{2}\log|\tau|$. In view of the above, we can then write:

$$\begin{aligned}\|p(x;\tau,\xi)\| &\leq c(1 + |\tau| + |\xi|)^\nu \sum_{j\geq 0} \frac{M^j|\log\tau|^j}{j!} \\ &\leq c(1 + |\tau| + |\xi|)^\nu \sum_{j\geq 0} \frac{M^j(\sqrt{2}\log|\tau|)^j}{j!} \\ &= c(1 + |\tau| + |\xi|)^\nu |\tau|^{M\sqrt{2}},\end{aligned}$$

for $(x;\tau,\xi) \in V$ and $|\tau| \gg 1$. Hence $p(x;\tau,\xi) \in \mathbf{M}_N(S^f_X(V))$.

Writing D_t instead of $D_t I_N$ for sake of brevity, we define D_t^A as the $N \times N$ matrix of operators in $\mathcal{E}^{\mathbb{R},f}_{X,(0;1,\xi_0)}$ associated to the symbol p. We have a similar definition replacing A by $-A$. Since $[A, D_t] = 0$, we also have $D_t^A D_t^{-A} = D_t^{-A} D_t^A = I_N$.

To prove (ii), from $[A, D_t] = 0$, we deduce also that $[A, D_t^A] = 0$ and $[D_t^A, t] = AD_t^{A-I_N}$. Hence,

$$\begin{aligned} D_t^A(tD_t - A) &= tD_t^{A+I_N} + AD_t^{A-I_N} - D_t^A A \\ &= tD_t^{A+I_N} + AD_t^A - D_t^A A \\ &= tD_t D_t^A. \end{aligned}$$

This achieves the proof of Theorem 1

Remark. (i) Distribution solutions of regular operators are investigated in [3]. If we assume in Theorem 1 that $A = A(x, D_x)$ is a differential operator (i.e. $\xi_0 = 0$), then we recover Proposition 3.1 of loc. cit.

(ii) Assuming $n = 0$, and A of the form

$$A = A(D_t) = A_1 D_t + A_0 + A_{-1} D_t^{-1} + \cdots$$

(a microdifferential operator matrix with constant coefficients), then

$$D_t^A(tD_t - A)D_t^{-A} = tD_t.$$

References

1. E. Andronikof, *Microlocalisation temperée*, Mém. Soc. Math. France **57** (1994), Supl. au Bull. de la Soc. Math. France **122 (2)**.
2. ———, *A conjugacy class of regular operators*, Microlocal geometry (Kyoto 1992), RIMS kôkyûroku, Kyoto Univ. **845**, 1993, pp. 8–12.
3. E. Andronikof and T. Monteiro Fernandes, *On the tempered solutions of regular systems*, Proc. Intern. Conf. "$\mathcal{D}$-modules and microlocal geometry" Lisbon (1990), de Gruyter, Berlin, New-York.
4. T. Aoki, *Calcul exponentiel des opérateurs microdifférentiels d'ordre infini*, Ann. Inst. Fourier, Grenoble, **33 (4)** (1983), 227–250.
5. M. Kashiwara and T. Oshima, *Systems of differential equations with regular singularities and their boundary value problems*, Ann. of Math. **106** (1977), 145–200.
6. M. Sato, T. Kawai, and M. Kashiwara, *Microfunctions and pseudo-differential equations*, Hyperfunctions and pseudo-differential equations (H. Komatsu, ed.), Lecture Notes in Math. **287**, Springer, 1973, Proceedings Katata 1971, pp. 265–529.

Elliptic boundary value problems in the space of distributions

EmmanuelAndronikof † *and NobuyukiTose*[1]

[1] Mathematics, Hiyoshi Campus, Keio University, Hiyoshi, Yokohama, Kanagawa 223, Japan (e-mail: `tose@math.hc.keio.ac.jp`)

Introduction

Elliptic boundary value problems have their own long history. For the general system they were, however, first clearly fomulated microlocally by M. Kashiwara and T. Kawai [K-K]. Their theorem has enjoyed many applications, for example, to solvability of operators of simple characteristics, hypoelliptic operators, and tangential Cauchy-Riemann systems. The theorem does not give, however, much information if we restrict ourselves in the space of distributions. This note aims at giving an analogous theorem of Kashiwara-Kawai type in case function spaces are tempered. See Theorem 3 in Section 1 for the main theorem. By this theorem, we can obtain many application to distribution boundary values of holomorphic functions (e.g. M. Uchida[U]).

1 Main theorem

Let M be a real analytic manifolfd of dimension n with a complex neighborhood X. Let $\mathcal{M}$ be a coherent $\mathcal{D}_X$ module on X and assume that $\mathcal{M}$ is elliptic on M, *i.e.*

$$\text{(1)} \qquad \operatorname{char}(\mathcal{M}) \cap T^*_M X \subset T^*_X X.$$

Let N be a real analytic submanifold of M of codimension $d \geq 1$ in M, and Y be a complexification of N in X. We assume that Y is non-characteristic for $\mathcal{M}$, *i.e.*

$$\text{(2)} \qquad \operatorname{char}(\mathcal{M}) \cap T^*_Y X \subset T^*_X X.$$

In this situation, we have the canonical morphisms

$$T^*_N M \underset{\rho}{\longleftarrow} T^*_N X \underset{\varpi}{\xrightarrow{\simeq}} T^*_N X.$$

Under the above notation we have

Theorem 1. *The natural morphism*

$$\mathbf{R}\rho_*\mathbf{R}\underline{\mathrm{Hom}}_{\mathcal{D}_X}(\mathcal{M}, \mathcal{C}^f_{N|X}) \leftarrow \mathbf{R}\underline{\mathrm{Hom}}_{\mathcal{D}_X}(\mathcal{M}, T\text{–}\mu_N(\mathcal{D}b_M)) \otimes or_{N/M}$$

is an isomorphism.

In the above theorem $or_{N/M}$ denotes the relativs orientation sheaf of N in M. The sheaf $\mathcal{C}^f_{N|X}$ on $T^*_N X$ is the tempered version of $\mathcal{C}_{N|X}$ and is given, with the tempered microlocalization due to E. Andronikof[An], by

$$\mathcal{C}^f_{N|X} := T\text{–}\mu_N(\mathcal{O}_X) \otimes or_M[n].$$

We remark that the above object in the derived category is concentrated in degree 0. For a point $\mathring{x} \in T^*_N X$, the stalk of $\mathcal{C}^f_{N|X}$ at $\mathring{x}$ is given, with the aid of local cohomology with bounds, by

$$\mathcal{C}^f_{N|X,\mathring{x}} \simeq \varinjlim \underline{\mathrm{H}}^n_{[Z]}(\mathcal{O}_X)_{\pi_X(\mathring{x})}.$$

Here π_X denotes the projection $\pi_X : T^*X \longrightarrow X$ and the inductive limit is taken for all closed subanalytic sets Z in X satisfying the property

$$C_N(Z)_{\pi_X(\mathring{x})} \subset \{v \in T_N X;\ <\mathring{x}, v>< 0\} \cup \{0\}.$$

Refer here to Kashiwara-Schapira[K-S2] for the notion of normal cones $C_N(\cdot)$. The sheaf $T\text{–}\mu_N(\mathcal{D}b_M)$ on $T^*_N M$ is also constructed by E. Andronikof[An1,2]. We just explain that its stalk at $\mathring{x} \in T^*_N M$ is given by the isomorphism

$$T\text{–}\mu_N(\mathcal{D}b_M)_{\mathring{x}} \simeq \varinjlim_Z \Gamma_Z(\mathcal{D}b_M)_{\pi_M(\mathring{x})}.$$

Here the inductive limit is taken for any closed subanalytic set Z in M with the property

$$C_N(Z)_{\pi_M(\mathring{x})} \subset \{v \in T_N M;\ <\mathring{x}, v>< 0\} \cup \{0\}$$

$(\pi_M : T^*M \longrightarrow M)$.

Next we give another theorem, which is analogous to Theorem 6.3.1 of Kashiwara-Schapira [K-S1] (refer also to Kashiwara-Kawai[K-K] where we find the theorem of [K-S1] in its original form).

Theorem 2. *Let* $\tilde{\mathcal{M}} = \mathcal{E}_X \otimes_{\pi_X^{-1}\mathcal{D}_X} \pi_X^{-1}\mathcal{M}$. *Then the natural morphism*

$$\begin{aligned}&\mathbf{R}\underline{\mathrm{Hom}}_{\mathcal{E}_X}(\tilde{\mathcal{M}}, \mathcal{C}^f_{N|X})\\ &\longleftarrow \mathbf{R}\underline{\mathrm{Hom}}_{\mathcal{E}_X}(\tilde{\mathcal{M}}, \mathcal{E}_{X\leftarrow Y}) \overset{\mathbf{L}}{\underset{End(\mathcal{E}_{X\leftarrow Y})}{\otimes}} \mathbf{R}\underline{\mathrm{Hom}}_{\mathcal{E}_X}(\mathcal{E}_{X\leftarrow Y}, \mathcal{C}^f_{N|X})\end{aligned}$$

is an isomorphism outside of $T^*_N X \cap T^*_Y X$. *This entails an isomorphism*

$$\mathbf{R}\underline{\mathrm{Hom}}_{\mathcal{E}_X}(\mathcal{M}, \mathcal{C}^f_{N|X}) \simeq \mathbf{R}\underline{\mathrm{Hom}}_{\mathcal{E}_X}(\mathcal{M}, \mathcal{E}_{X\leftarrow Y}) \otimes^{\mathbf{L}}_{p^{-1}\mathcal{E}_Y} p^{-1}\mathcal{C}^f_N$$

on $T^*_N X \setminus T^*_Y X$ *where* p *is the canonical morphism*

$$p : \ T^*_N X \setminus T^*_Y X \longrightarrow T^*_N Y.$$

In the above theorem, the object $\mathcal{C}^f_N$ on $T^*_N Y$ is the sheaf of temperate microfunctions. This is a subsheaf of $\mathcal{C}_N$ and describes microlocal analytic singularities of distributions on N. By the notation of E. Andronikof[An1,2], this sheaf is defined as

$$\mathcal{C}^f_N := T\text{–}\mu_N(\mathcal{O}_Y)[n-d] \otimes or_{N/Y}.$$

The proof of this theorem is essentially the same as in Theorem 6.3.1 of [K-S1] and relies on the division theorem of temperate microfunctions with holomorphic parameters with respect to microdifferential operators. We also remark that only the non-charactericity of Y is utilized in its proof.

By combining the above theorems into one, we get the main theorem of this note. Let q denote the restriction of ρ to $\mathring{T}^*_N X \setminus T^*_M X$;

$$q : \ \mathring{T}^*_N X \setminus T^*_M X \longrightarrow T^*_N M$$

and p the projection $\mathring{T}^*_N X \setminus T^*_Y X \longrightarrow \mathring{T}^*_N Y$. Then we have

Theorem 3. *We have a canonical isomorphism on* $\mathring{T}^*_N Y$

$$\begin{aligned}&\mathbf{R}q_*\left(\mathbf{R}\underline{\mathrm{Hom}}_{\mathcal{E}_X}(\tilde{\mathcal{M}}, \mathcal{E}_{X\leftarrow Y}|_{\mathring{T}^*_N X}) \overset{\mathbf{L}}{\underset{p^{-1}\mathcal{E}_Y}{\otimes}} p^{-1}\mathcal{C}^f_N\right)\\ &\simeq \mathbf{R}\underline{\mathrm{Hom}}_{\mathcal{D}_X}(\mathcal{M}, T\text{–}\mu_N(\mathcal{D}b_M)) \otimes or_{N/M}.\end{aligned}$$

2 Idea of Proof

What is left to us is now to construct the morphism in Theorem 1 and to show it an isomorphism.

First we construct a commutative diagram

$$\text{(A)}\qquad \begin{array}{ccc} \mathbf{R}\rho_! \mathcal{C}^f_{N|X} \otimes or_{N/X} & \longrightarrow & T\text{–}\mu_N(\mathcal{A}_M) \\ \downarrow & & \downarrow \\ \mathbf{R}\rho_* \mathcal{C}^f_{N|X} \otimes or_{N/X} & \longleftarrow & T\text{–}\mu_N(\mathcal{D}b_M) \end{array}$$

where $T\text{–}\mu_N(\mathcal{A}_M)$ is the tempered microlocalization of the sheaf $\mathcal{A}_M$ along N and is constructed by E. Andronikof[An1,2]. This object is the Fourier transform of the tempered specialization $T\text{–}\nu_N(\mathcal{A}_M)$ whose stalk at $\mathring{v} \in T_N M$ is given by

$$T\text{–}\nu_N(\mathcal{A}_M)_{\mathring{v}} \simeq \varinjlim_U \{u \in \mathcal{A}(U);\ u \text{ is tempered on } M \text{ as a distribution}\}.$$

Here U in the inductive limit ranges through any open subanalytic set in M with the property

$$\mathring{v} \notin C_N(M \setminus U).$$

To construct (A), it is sufficient to construct its image by the inverse Fourier transformation

$$\text{(A')}\qquad \begin{array}{ccc} \iota^{-1} T\text{–}\nu_N(\mathcal{O}_X) \otimes or_{N/X} & \longrightarrow & T\text{–}\nu_N(\mathcal{A}_M) \\ \downarrow & & \downarrow \\ \iota^! T\text{–}\nu_N(\mathcal{O}_X) \otimes or_{N/X} & \longleftarrow & T\text{–}\nu_N(\mathcal{D}b_M). \end{array}$$

Here ι is the canonical embedding

$$\iota : T_N M \longrightarrow T_N X,$$

and $T\text{–}\nu_N(\mathcal{O}_X)$ is the tempered specialization of the sheaf $\mathcal{O}_X$ along N, which is concentrated in degree 0. The stalk of $T\text{–}\nu_N(\mathcal{O}_X)$ at $\mathring{v} \in T_N X$ is given by

$$T\text{–}\nu_N(\mathcal{O}_X)_{\mathring{v}} \simeq \varinjlim_U \{u \in \mathcal{O}(U);\ u \text{ can be extended to } X \text{ as a distribution}\}$$

where U runs through all open subanalytic sets in X with $\mathring{v} \notin C_N(M \setminus U)$. The diagram (A') can be constructed easily if we scrutinize the construction by E. Andronikof[An1,2].

Next we apply $\mathbf{R}\underline{\mathrm{Hom}}_{\mathcal{D}_X}(\mathcal{M}, \cdot)$ to the diagram (A') and obtain the commutative diagram

$$\begin{array}{ccc} \mathbf{R}\underline{\mathrm{Hom}}_{\mathcal{D}_X}(\mathcal{M}, \iota^{-1} T\text{–}\nu_N(\mathcal{O}_X)) \otimes or_{N/X} & \xrightarrow{\Phi_1} & \mathbf{R}\underline{\mathrm{Hom}}_{\mathcal{D}_X}(\mathcal{M}, T\text{–}\nu_N(\mathcal{A}_M)) \\ {\scriptstyle \Phi_4}\downarrow & & \downarrow{\scriptstyle \Phi_2} \\ \mathbf{R}\underline{\mathrm{Hom}}_{\mathcal{D}_X}(\mathcal{M}, \iota^! T\text{–}\nu_N(\mathcal{O}_X)) \otimes or_{N/X} & \xleftarrow[\Phi_3]{} & \mathbf{R}\underline{\mathrm{Hom}}_{\mathcal{D}_X}(\mathcal{M}, T\text{–}\nu_N(\mathcal{D}b_M)). \end{array}$$

It is easy to see from the ellipticity of $\mathcal{M}$ that Φ_4 and Φ_2 are isomorphisms. (To show Φ_4 is an isomorphism, it is easier to consider its image by Fourier transformation). Thus to prove that Φ_3 and thus its image by Fourier transformation are isomorphisms, it suffices to show that Φ_1 is an isomorphism. The problem for Φ_1 can be reduced to the case where $\mathcal{M}$ is a single equation; i.e. $\mathcal{M} = \mathcal{D}_X/\mathcal{D}_X P$. Moreover it is sufficient to show that

$$\underline{\mathrm{Hom}}_{\mathcal{D}_X}(\mathcal{D}_X/\mathcal{D}_X P, \iota^{-1}T\text{-}\nu_M(\mathcal{O}_X))\otimes or_{N/X} \longrightarrow \underline{\mathrm{Hom}}_{\mathcal{D}_X}(\mathcal{D}_X/\mathcal{D}_X P, T\text{-}\nu_N(\mathcal{A}_M))$$

is surjective. This problem can be solved by using the construction of the elementary solution of P by means of Radon transformation and microdifferential operators.

References

[An1] Andronikof, E., *Microlocalisation tempérée des distributions et des fonctions holomorphes I*, C.R. Acad. Sci. **t.303** (1986), 347-350; II **t.304** (1987), 511-514; See also Thèse d'Etat, Paris-Nord (juin 1987).

[An2] ———, *Microlocalisation tempérée*, Mémoire **57** (1994), Supl. au Bull. de la Soc. Math. France **122 (2)**.

[K-K] Kashiwara M. and T. Kawai, *On the Boundary Value Problems for Elliptic Systems of Linear Differential Equations I*, Proc. Japan Academy **48** (1972), 712-715; II **49** (1973), 164-168.

[K-S1] Kashiwara, M. and P. Schapira, *Microhyperbolic Systems*, Acta Math. **142** (1974), 1-55.

[K-S2] ———, *Microlocal Study of Sheaves*, Astérisque **128** (1985); *Sheaves on Manifolds*, Grndlehren der Math. **292**, Springer-Verlag, 1994.

[U] Uchida, M., *A Generalization of Bochner's Tube Theorem for Elliptic Boundary Value Problems*, RIMS Kôkyûroku, Kyoto Univ. **845**, 1993, pp. 129-138.

On the Holonomic Character of the Elementary Solution of a Partial Differential Operator

En l'honneur de H. Komatsu, pour son soixantième anniversaire

Louis Boutet de Monvel

Institut de Mathématiques, Analyse Algébrique, Université Pierre et Marie Curie, Case 247, 4, place Jussieu, F-75252 Paris Cedex 05, France

Abstract. We describe an elementary regular holonomic system of partial differential equations which should be satisfied by an elementary solution of a differential operator P(d) with constant coefficients and simple characteristics. This is heuristic, but it is exact for strictly hyperbolic operators, and exact mod. holomorphic functions for elliptic operators or operators with real principal part. For these this explains again why the elementary solution extends holomorphically, with the expected ramification

1 Introduction

Let $P(x, d)$ be a differential operator with analytic coefficients on $\mathbf{R}^n$. It is known, particularily since the work of J. Leray, that under suitable conditions, e.g. if P is strictly hyperbolic, or elliptic, with simple characteristics (real and complex), that P has an elementary solution E of Nilsson class, i.e. E extends holomorphically, with ramification (and moderate growth) along the complex bicharacteristic cone.

We will denote by $\mathcal{O}$, resp. $\mathcal{D}$ the sheaf of holomorphic functions, resp. of holomorphic differential operators, on $X = \mathbf{C}^n$ or $\mathbf{R}^n$. We recall that a system of analytic partial differential equations $P_j(x, d)f = 0$ on X is holonomic if its characteristic manifold is lagrangian, i.e. if the submanifold of the cotangent bundle T^*X with equations $p_j(x, \xi) = 0$ is of dimension n ($p_j(x, \xi)$ denotes the symbol, or principal part, of P_j; the characteristic manifold is always involutive, hence of dimension $\geq n$).

Equivalently we say that the $\mathcal{D}$-module $M = \mathcal{D}/\mathcal{I}$, with $\mathcal{I}$ the ideal generated by the P_j in $\mathcal{D}$, is holonomic if its characteristic set is lagrangian. [1]

We say furthermore that M is "regular holonomic" if it has a good filtration $M = \cup M_k$ [2] such that $PM_k \subset M_{k+p-1}$ if $P \in \mathcal{D}_p$ and $\sigma(P)$ vanishes on $char M$.

---

[1] the characteristic set $char M$ is the set of common zeros of the symbols $\sigma(P_j)$, at least if the generators p_j are in sufficient number so that any $Q \in \mathcal{I}$ is a linear combination $Q = \sum A_j P_j$ with $deg(Aj) \leq deg(Q) - deg(Pj)$.

[2] i.e. the M_k are subsheaves of M and are coherent $\mathcal{O}$-modules, $M_k = 0$ for $k \ll 0$, and we have $\mathcal{D}_p M_k \subset M_{p+k}$, with equality if $k \gg 0$.

As was shown by M. Kashiwara or Z. Mebkhout there is a close relation to regular holonomic $\mathcal{D}$-modules functions of Nilsson class; in particular functions of Nilsson class are solutions of regular holonomic system of differential equations.

So it is a natural idea to try to construct a priori a holonomic sysptem of differential equations which should be satisfied by the elementary solution.

If $P = P(d)$ is a differential operator on $\mathbf{R}^n$ with constant coefficients, it has an elementary solution which is the convolution operator by a temperate distribution E, whose Fourier transform $\hat{E}$ is essentially $\frac{1}{P(\xi)}$. It follows easily from the work of Bernstein and Gel'fand on the distributions f^λ (f a polynomial) that we may choose E so that $\hat{E}$ (hence also E) satisfies a holonomic system of differential equations with polynomial coefficients. However this is described in terms of the Fourier transform $\hat{E}$ and the information for E may remain complicated to deduce (e.g. the system for $\hat{E}$ is always regular holonomic whereas that for E is usually not if P has multiple characteristics).

Here we will produce a very simple system of differential equations which should be satisfied by an elementary solution E of a differential operator P with constant coefficients. In general our argument heuristic and incorrect or insufficient, and does not yield a holonomic system; but it does give a holonomic system if P has simple characteristics, and it is exact for strictly hyperbolic operators, or exact for elliptic operators with simple (complex) characteristics or more generally for operators with real principal part (it is then possible to extract it from the Bernstein-Gel'fand system). The knowledge of such a holonomic system enables one to predict the ramification locus of the holomorphic extension of E, i.e. the complex bicharacteristic flow of P issued from the origin, whose projection is the complex bicharacteristic conoid of P), so as the monodromy, inasmuchas this is known for solutions of a regular holonomic system.

2 Heuristic Argument

Let $P = P(d)$ be a differential operator with constant coefficients on $X = R^n$. The elementary solution E of P satisties and its Fourier transform should satisfy $P(\xi)\widehat{E} = 1$.

Formally the convolution operator $f \to E * f$ is the inverse of the operator $P(d)$, so it should commute with all operators commuting with P. The Fourier transform *widehat*$E(\xi)$ should be constant on level surfaces of $P(\xi)$. Thus (still formally)

$$\partial\widehat{E} \wedge \partial P = 0, \qquad \text{i.e. } P_i(\xi)\hat{E}_j - P_j(\xi)\widehat{E}_i \text{ pour } i,j = 1\ldots n$$

where ∂ denotes the exterior differentiation with respect to ξ, and we have set $P_i = \partial P/\partial\xi_i \quad , \hat{E}_i = \partial\hat{E}/\partial\xi_i$.

Since $\partial\widehat{E}$ is the Fourier transform of xE (up to a constant factor), this can also be written

$$Q_{ij}E = 0 \quad with\ Q_{ij} = x_i P_j(d) - x_j P_i(d)$$

which we may also write (with a slight abuse)

$$(x \wedge P'(d))E = 0$$

Note that the Q_{ij} commute with P (because the second derivatives P_{ij} and P_{ji} are equal). They also form an involutive system, i.e. the commutator $[Q_{ij}, Q_{kl}]$ is a linear combination of the Q_{pq} with coefficients differential operators of the right degree $m-2$, as one can check easily.

Thus we get heuristically the following system of differential equations for the elementary solution E:

$$P(d)E = \delta \qquad or \quad xP(d)E = 0 \tag{1}$$

$$Q_{ij}E = 0 \qquad or \quad x \wedge P'(d)E = 0 \tag{2}$$

We will call this system of partial differential equations the basic system for the elementary solution E.

3 The basic system for E

Let us examine more precisely the system (1), (2).

As noticed above the Q_{ij} commute with $P(d)$ and form an involutive system, i.e. the commutators $[Q_{ij}, Q_{kl}]$ are linear combinations of the Q_{ij} (in fact the Fourier transforms $P_i(\xi)\frac{\partial}{\partial \xi_j} - P_j(\xi)\frac{\partial}{\partial \xi_i}$ generate all tangent vectors to the level sets $P = constant$, where $P'(\xi) \neq 0$)

Let us determine the characteristic manifold of system (1)(2): this is in any case contained in the zero variety of the symbols (principal parts) of the operators which appear in (1)(2). This consists of T_0^*X, of the zero section $\xi = 0$ (if the degree d of P is > 1) and of the closure Λ of the variety defined by

$$P_m(\xi) = 0 \quad (\xi \neq 0) \quad x \wedge P_m'(\xi) = 0 \tag{3}$$

If P has simple characteristics (real and complex), Λ is exactly the bichar-acteristic flow of P from the origin, i.e. the set of all complex covectors (x, ξ) such that $P_m(\xi) = 0$, $\xi \neq 0$ and x is parallel to $P_m'(\xi)$; it is obviously Lagrangian (of dimension n) smooth outside of the zero section ($\xi = 0$); its projection is the complex characteristic conoid of P.

In this case the system (1), (2) is regular holonomic, of multiplicity 1 outside of the zero section, and its solutions are holomorphic functions of Nilsson class, ramified (and of moderate growth) along the characteristic conoid of P, as one would optimally expect of the elementary solution.

4 Justification of the Canonical System

The argumentation above is of course heuristic. First (and least importantly) because the elementary solution E may also satify further equations which do not follow from (1)(2). For instance if E behaves as the inverse of P we must also have

$$AoE_* = E_*oB \quad \text{for any differential operators } A,\ B \text{ such that} \quad BoP = PoA, \tag{4}$$

where we denote E_* the convolution operator $f \to E * f$. These imply further differential equations for E which do not necessarily exactly follow from (1)(2): e.g. if $P(\xi)$ is homogeneous of degree k, i.e. $[x.\partial, P] = -kP$, E_* is homogeneous of degree $-k$, and the resulting equation for E is $(x.\partial + n - k)E = 0$ (E is homogeneous of degree -n+k as a generalized function), ans this does not follow from (1)(2) if $k \geq 2$).

Secondly, because the elementary solution of P is not unique, and system (1)(2) is not expected to be satisfied by all elementary solutions - only a suitably chosen one. But, most importantly because our heuristic argumentation supposes that we can construct the inverse of P in a suitable associative algebra of operators, which is of course usually not true. So in general one does not really expect P to have any elementary solution satisfying (1)(2).

However there are important cases where this works exacly, or essentially exactly:

1. If P is hyperbolic, it is invertible in the algebra of operators A with kernel $A(x, y)$ supported in the forward cone. This algebra contains the algebra $\mathcal{D}$ of all analytic differential operators, so in this case our argumentation is rigorous.

2. If P is elliptic it is invertible in the algebra $\hat{E}$ of pseudodifferential operators, so here again our argumentation is correct, mod. analytic (or even entire holomorphic) functions. More precisely in that case there exists an elementary solution E such that $P(d)E = \delta$, $x \wedge P'(d)E$ is an entire holomorhic function, and of course any other elementary solution of P differs from this by an operator whose Schwartz-kernel is analytic near the reals.

3. If P is homogeneous, real and has simple characteristics, then $\hat{E} = \frac{1}{P(\xi) \pm i0}$ is well defined and satifies the Fourier transforms of equations (1)(2) ($P(\xi)\hat{E} = 1, \partial P(\xi)\partial\hat{E}) = 0$) for large real ξ real (in fact for $\xi \neq 0$). An elementary perturbation argument shows that this is still true if P is simply characteristic and has a real principal part (more generally if the set of real charactristic points is of real codimension 1). Since we can always extend $\hat{E}$ for small ξ so that $P(\xi)\hat{E} = 1$, we get again in that cade an elementary solution E such that $x \wedge P'(d)E$ is analytic (although there is no pertinent algebra there).

Note that in the last two examples, system (1)(2) mod. analytic functions still gives the right ramification locus, but only gives the monodromy mod. unipotent matrices.

Remark 1. The "basic" system we described is no longer correct for complex differential operators. In general one still expects P to posess a holonomic elementary solution E, but if the set of real points of *charP* is of real codimension ≥ 2, the system (1)(2) may be to restrictive (fail to be "positive") and have no distribution solution; in fact one no longer really expects that the microsupport of E is reduced to the one described above (i.e. the covectors above the origin, the hamiltonian flow of P from the origin and the zero covectors). For instance if $P = \frac{1}{2}(\partial_1 + i\partial_2)$ is the partial Cauchy-Riemann operator on $\mathbf{R}^n$, the elementary solution $E = \frac{-1}{2i\pi(x_1+ix_2)}\,\delta(x_3, \ldots x_n)$ is of course holonomic, but does not satisfy (1)(2); in fact if $n > 2$ the characteristic manifold is not the one predicted by (1)(2) : it also contains the hamiltonian bicharacteristic flow of the system $(P, \overline{P})$ from 0; (we see this equivalently on the Fourier transform: if $\widehat{P}(\xi) = \frac{1}{2}(\xi_1 + i\xi_2)$ there is no distribution $\widehat{E}$ such that $\widehat{P}\widehat{E}$, $\partial\widehat{P}\partial\widehat{E} = 0$, near any real ξ where $\widehat{P}$ vanishes.

Remark 2. Our construction fails completely, even heuristally, for operators P with multiple charatacteristics: on one hand the manifold Λ above by 3 is no longer smooth of dimension n outside of the zero section (since if $P(\xi)$ and $P'(\xi)$ both vanish at a covector $\xi \neq 0$, Λ contains the subvariety $\mathbf{C}^n \times \mathbf{C}^*\xi$ which is of dimension $n + 1$). Moreover in this case the principal parts of equations (1)(2) no longer determine the characteristic set of (1)(2) , and this could be strictly smaller than $\Lambda \cup T_0^*X \cup T_X^*X$. So in this case system (1)(2) is not necessarily holonomic (at least it is not for obvious reasons), and if it is, it is expected to be regular.

For instance, let $P = \partial_1^2 + c\,\partial_2$, avec $c \neq 0$ be the heat operator: in this case system (1)(2) reduces to

$$(\partial_1^2 + c\,\partial_2)E = \delta \tag{5}$$

$$(c\,x_1 - 2x_2\partial_1)E = 0 \tag{6}$$

these imply

$$\frac{1}{c}(\partial_1\,(6) + 2x_2\,(5)\,)E = (\partial_1 x_1 + 2x_2\partial_2)E = 0$$

which could not be seen on the principal parts. The relsulting system is still holonomic, but not regular; it admits as solution the standard elementary solution of the heat equation (or of the Schrödinger equation if c is chosen pure imaginary).

5 Final Remark

We end with a remark on the elementary solution $E(x, y)$ of an analytic differential operator P with simple characteristics on a manifold X (E is a distribution on $X \times X$. For the existence of a global holonomic system controlling E, one obviously needs some nice geometric properties properties of the bicharacteristic flow of P, ensuring that the bicharacteristic flow Λ of P from the diagonal be a closed manifold. If P has simple characteristics, this bicharacteristic flow at least defines a closed Lagrangian manifold in a neighborhood of the diagonal.

The remarks above incite to characterize the elementary solution E using the commutator of P. Although there may not be many differential operators which commute with P, there are many pseudodifferential operators (when P has simple characteristics); these form a sheaf, microlocally isomorphic to the commutator of ∂_1. We may then form the system of pseudo-differential equations:

$$P(x, d_x)\, E(x, y) = \delta(x - y) \tag{7}$$

$$(Q(x, d_x) -{}^t Q(y, d_y))\, E(x, y) = 0 \quad \text{if } Q \text{ commutes with } P \tag{8}$$

Near the diagonal of $X \times X$, this is again a holonomic system of pseudo-differential equations, whose characteristic variety is the union of $T^*_{diag}(X \times X)$ and Λ, the bicharacteristic hamiltonian flowout of P out of the conormal bundle of the diagonal. Hopefully this should correspond to a differential system, controlling the elementary solution in good cases.

References

[G1] Garding L., *Linear hyperbolic partial differential equations with constant coefficients*, Acta Math. **85** (1951), 1-62.

[G1] Garding L., *Solution directe du problème de Cauchy pour les équations hyperboliques*, Coll. Int. CNRS, Nancy 1956, 71-90.

[GKK] Guillemin V., Kashiwara M., Kawai T., Seminar on microlocal analysis, Ann. of Math. studies **93**, Princeton University Press, 1979.

[J] John F., Plane waves qnd spherical means applied to partial fdifferential equations, Interscience, New York, 1955.

[K1] Kashiwara M., *On the maximally overdetermined systems of linear differential equations I*, Publ. RIMS, Kyoto University **10** (1975), 563-5.

[K2] Kashiwara M., *Faisceaux constructibles et systèmes holonômes d'équations aux dérivées partielles linéaires à points singuliers réguliers*, Sém. Goulaouic-Schwartz 1979-80, exposé n°19.

[K3] Kashiwara M., *Systems of microdifferential equations*, Cours à l'Université Paris Nord, Progress in Math. **34**, Birkhaüser, 1983.

[K4] Kashiwara M., Kawai T., Kiwura T., *Foundations of algebraic analysis*, Princeton Math. Series n° **37**, Princeton University Press, Princeton N.J., 1986.

[KW] Kashiwara M., Kawai T., *On holonomic systems of microdifferential equations III - systems with regular singularitie*, Publ. RIMS, Kyoto University **17** (1981) 813-979.

[KKS] Kashiwara M., Kawai T., Sato M., *Microfunctions and pseudodifferential equations*, Lecture Notes in Math. **287**, Springer-Verlag, 1973, pp. 265-524.

[L3] Leray J., *Le problème de Cauchy pour une équation linéaire à coefficients polynomiaux*, C.R. Acad Sc. Paris **242** (1956), 1483-1488.

[L4] Leray J., *Uniformisation de la solution du problème linéaire analytique de Cauchy près de la variété qui porte les données de Cauchy (problème de Cauchy I)*, Bull. Soc. Math. France **85** (1957), 389-429.

[L5] Leray J., *La solution élémentaire d'un opérateur différentiel linéaire*, Bull. Soc. Math. France **86** (1958), 389-429.

[L6] Leray J., *Le calcul différentiel et intégral sur une variété complexe (problème de Cauchy III)*, Bull. Soc. Math. France **87** (1959), 81-180.

[L7] Leray J., *Un prolongement de la transformée de Laplace qui transforme la solution unitaire d'un opérateur hyperbolique en sa solution élémentaire (problème de Cauchy IV)*, Bull. Soc. Math. France **90** (1962), 39-156.

[L8] Garding L., Kotake T., Leray J., *Uniformisation et développement asymptotique de la solution du problème de Cauchy linéaire, à données holomorphes; analogie avec la théorie des ondes asymptotiques et approchées (problème de Caucy Ibis et VI)*, Bull. Soc. Math. France **92** (1964), 263-361.

[L9] Leray J., *Un complement au theoreme de N. Nilsson sur les integrales de formes differentielles a support singulier algebrique*, Bull.-Soc.-Math.-France **95** (1967), 313–374.

[L10] Leray J., *Solutions asymptotiques et groupe métaplectique*, Séminaire sur les équations aux dérivées partielles 1973-74, Collège de France, Paris.

[L11] Leray J., *Analyse Lagrangienne et mécanique quantique*, Séminaire sur les équations aux dérivées partielles 1976-77, Collège de France, Paris.

[M1] Malgrange B., *L'involutivité des caractéristiques des systèmes différentiels et micro-différentiels*, Séminaire Bourbaki 1977-78, n° **552**.

[M2] Malgrange B., *Equations différentielles à coefficients polynomiaux*, Progress in Math. **96**, Birkhaüser, 1991.

[N] Nilsson N., *Some growth and ramification properties of certain integrals on algebraic manifolds*, Ark.-Mat. **5** 1965 (1965), 463-476.

[P1] Petrowsky I.G.,*Über das Cauchysche Problem für eis System linearer partieller Differentialgleichungen*, Mat. Sb. **2 (44)** (1937), 815-870.

[P2] Petrowsky I.G., *Über das Cauchysche Problem für Systeme von partiellen Differentialgleichungen im Gebiete der nichtanalytischen Funktionen*, Bull. Univ. Moscow Sér. Int. **1**, n° **7** (1938), 1-74.

[Ph] Pham F., *Singularités des systèmes différentiels de Gauss-Manin*, Progress in Math. **2**, Birkhäuser, 1980.

[Sch] Schapira P., *Microdifferential systems in the complex domain*, Grundlehren der math. Wiss. **269**, Springer, 1985.

[Sj] Sjöstrand J., *Singularités analytiques microlocales*, Astérisque **95** (1982), 1-166.

Kernel calculus and extension of contact transformations to $\mathcal{D}$-modules

Dedicated to Professor Hikosaburo Komatsu

Andrea D'Agnolo and Pierre Schapira

Institut de Mathématiques; Analyse Algébrique; Université Pierre et Marie Curie; Case 247; 4, place Jussieu; F-75252 Paris Cedex 05

1 Introduction

There is an important literature dealing with integral transformations. In our papers [6], [7] we proposed a general framework to the study of such transforms in the language of sheaves and $\mathcal{D}$-modules. In particular, we showed that there are two natural adjunction formulas which split many difficulties into two totally different kind of problems: one of analytical nature, the calculation of the transform of a $\mathcal{D}$-module, the other one topological, the calculation of the transform of a constructible sheaf. Similar adjunction formulas for temperate and formal cohomology are obtained by M. Kashiwara and P. S. in [15], and allow one to treat C^∞-functions and distributions in this framework.

Here, we shall first recall the four above mentioned adjunction formulas, and then concentrate our study on the $\mathcal{D}$-module theoretical transform. Given two complex manifolds X and Y and a $\mathcal{D}$-module kernel which defines a quantized contact transformation on an open subset of $\dot{T}^*(X \times Y)$, our main result (Theorem 15 below) gives a geometrical condition to extend it as an isomorphism of locally free $\mathcal{D}$-modules of rank one. This improves our previous result of [7].

These results apply to classical problems of integral geometry, in the line of Leray [16], Martineau [18], Gelfand-Gindikin-Graev [9], Helgason [10] or Penrose [8]. In particular, they allowed us to treat projective duality and the twistor correspondence. (Refer to [6], [7], [5]. See also [17] for other flag correspondences.)

2 Review on the calculus of kernels

In this section we will develop the formalism of kernels in the framework of sheaves and $\mathcal{D}$-modules. The results below concerning kernels for sheaves and $\mathcal{D}$-modules are well-known from the specialists: let us mention in particular M. Kashiwara and also J.-P. Schneiders, with whom we had many discussions on this subject.

2.1 A review on sheaves, $\mathcal{D}$-modules and temperate cohomology

References are made to [14] for the theory of sheaves, and to [19] and [11] for the theory of $\mathcal{D}$-modules (see [22] for a detailed exposition).

Let X be a real analytic manifold, and denote by a_X the map from X to the set consisting of a single element. We denote by $\mathbf{D}^{\mathrm{b}}(\mathbb{C}_X)$ the derived category of the category of bounded complexes of sheaves of $\mathbb{C}$-vector spaces on a topological space X. If $A \subset X$ is a locally closed subset, we denote by $\mathbb{C}_A$ the sheaf on X which is the constant sheaf on A with stalk $\mathbb{C}$, and zero on $X \setminus A$. We consider the "six operations" of sheaf theory $R\mathcal{H}om(\cdot,\cdot)$, $\cdot\otimes\cdot$, $Rf_!$, Rf_*, f^{-1}, $f^!$, and we denote by $\boxtimes$ the exterior tensor product. Recall that $\mathrm{RHom}(\cdot,\cdot) = Ra_{X*}R\mathcal{H}om(\cdot,\cdot)$. For $F \in \mathbf{D}^{\mathrm{b}}(\mathbb{C}_X)$ we set $D'F = R\mathcal{H}om(F, \mathbb{C}_X)$, $DF = R\mathcal{H}om(F, \omega_M)$, where $\omega_X \simeq or_X[\dim^{\mathbb{R}} X]$ denotes the dualizing complex, and or_X the orientation sheaf.

We denote by $SS(F)$ the micro-support of F, a closed conic involutive subset of T^*X. We denote by $\mathbf{D}^{\mathrm{b}}_{\mathbb{R}-c}(\mathbb{C}_X)$ the full triangulated subcategory of $\mathbf{D}^{\mathrm{b}}(\mathbb{C}_X)$ of objects with $\mathbb{R}$-constructible cohomology. If X is a complex manifold, one defines similarly the category $\mathbf{D}^{\mathrm{b}}_{\mathbb{C}-c}(\mathbb{C}_X)$ of $\mathbb{C}$-constructible objects.

Let X be a complex manifold of dimension d_X. We denote by $\mathcal{O}_X$ the structural sheaf, by Ω_X the sheaf of holomorphic forms of maximal degree, and by $\mathcal{D}_X$ the sheaf of rings of linear differential operators. We denote by $\mathrm{Mod}(\mathcal{D}_X)$ the category of left $\mathcal{D}_X$-modules, and by $\mathrm{Mod}_{\mathrm{good}}(\mathcal{D}_X)$ the full subcategory of $\mathrm{Mod}(\mathcal{D}_X)$ consisting of good $\mathcal{D}_X$-modules. This is the smallest thick subcategory of $\mathrm{Mod}(\mathcal{D}_X)$ containing the coherent modules which can be endowed with good filtrations on a neighborhood of any compact subset of X. Note that in the algebraic case, coherent $\mathcal{D}$-modules are good. We denote by $\mathbf{D}^{\mathrm{b}}(\mathcal{D}_X)$ the derived category of the category of bounded complexes of left $\mathcal{D}_X$-modules, and by $\mathbf{D}^{\mathrm{b}}_{\mathrm{good}}(\mathcal{D}_X)$ its full triangulated subcategory whose objects have cohomology groups belonging to $\mathrm{Mod}_{\mathrm{good}}(\mathcal{D}_X)$. We consider the operations in the derived category of (left or right) $\mathcal{D}$-modules: $R\mathcal{H}om_{\mathcal{D}_X}(\cdot,\cdot)$, $\cdot \otimes^L_{\mathcal{O}_X} \cdot$, $\underline{f}^{-1}$, $\underline{f}_!$, $\underline{f}_*$. In particular, if $\mathcal{M} \in \mathbf{D}^{\mathrm{b}}(\mathcal{D}_X)$, $\mathcal{N} \in \mathbf{D}^{\mathrm{b}}(\mathcal{D}_Y)$, and $f : Y \longrightarrow X$:

$$\underline{f}^{-1}\mathcal{M} = \mathcal{D}_{Y\to X} \otimes^L_{f^{-1}\mathcal{D}_X} f^{-1}\mathcal{M}, \quad \underline{f}_!\mathcal{N} = Rf_!(\mathcal{D}_{X\leftarrow Y} \otimes^L_{\mathcal{D}_Y} \mathcal{N}),$$

where $\mathcal{D}_{Y\to X}$ and $\mathcal{D}_{X\leftarrow Y}$ are the transfer bimodules associated to f. We denote by $\underline{\boxtimes}$ the exterior tensor product, and we also use the notation:

$$\underline{D}_X\mathcal{M} = R\mathcal{H}om_{\mathcal{D}_X}(\mathcal{M}, \mathcal{K}_X),$$

where $\mathcal{K}_X$ denotes the dualizing complex for left $\mathcal{D}_X$-modules, defined by $\mathcal{K}_X = \mathcal{D}_X \otimes_{\mathcal{O}_X} \Omega_X^{\otimes -1}[d_X]$. If $\mathcal{F}$ is a holomorphic vector bundle on X, we set:

$$\mathcal{F}^* = \mathcal{H}om_{\mathcal{O}_X}(\mathcal{F}, \mathcal{O}_X), \quad \mathcal{D}\mathcal{F} = \mathcal{D}_X \otimes_{\mathcal{O}_X} \mathcal{F}.$$

Let us briefly recall some constructions of [12] and [15].

First, assume X is a real analytic manifold. Denote by $\mathcal{D}b_X$ the sheaf of Schwartz's distributions on X, and by $\mathcal{C}^\infty_X$ the sheaf of functions of class $\mathcal{C}^\infty$.

There exist unique contravariant functors, exact for the natural t-structures:

$$\mathcal{T}hom(\cdot, \mathcal{D}b_X) : \mathbf{D}^{\mathrm{b}}_{\mathbb{R}-c}(\mathbb{C}_X)^{\mathrm{op}} \longrightarrow \mathbf{D}^{\mathrm{b}}(\mathcal{D}_X),$$
$$\cdot \overset{\mathrm{w}}{\otimes} \mathcal{C}^\infty_X : \mathbf{D}^{\mathrm{b}}_{\mathbb{R}-c}(\mathbb{C}_X) \longrightarrow \mathbf{D}^{\mathrm{b}}(\mathcal{D}_X),$$

such that if Z is a closed subanalytic subset of X, then

$$\mathcal{T}hom(\mathbb{C}_Z, \mathcal{D}b_X) = \Gamma_Z \mathcal{D}b_X,$$
$$\mathbb{C}_{X\setminus Z} \overset{\mathrm{w}}{\otimes} \mathcal{C}^\infty_X = \mathcal{I}^\infty_{Z,X},$$

where $\mathcal{I}^\infty_{Z,X}$ denotes the ideal of $\mathcal{C}^\infty_X$ of functions vanishing to infinite order on Z.

Now, assume that X is a complex manifold. Denote by $\overline{X}$ the associated anti-holomorphic manifold, by $X^{\mathbb{R}}$ the underlying real analytic manifold, and identify $X^{\mathbb{R}}$ to the diagonal of $X \times \overline{X}$. For $F \in \mathbf{D}^{\mathrm{b}}_{\mathbb{R}-c}(\mathbb{C}_X)$ one sets:

$$\mathcal{T}hom(F, \mathcal{O}_X) = R\mathcal{H}om_{\mathcal{D}_{\overline{X}}}(\mathcal{O}_{\overline{X}}, \mathcal{T}hom(F, \mathcal{D}b_{X^{\mathbb{R}}})),$$
$$F \overset{\mathrm{w}}{\otimes} \mathcal{O}_X = R\mathcal{H}om_{\mathcal{D}_{\overline{X}}}(\mathcal{O}_{\overline{X}}, F \overset{\mathrm{w}}{\otimes} \mathcal{C}^\infty_{X^{\mathbb{R}}}).$$

In other words, one defines $\mathcal{T}hom(F, \mathcal{O}_X)$ and $F \overset{\mathrm{w}}{\otimes} \mathcal{O}_X$ as the Dolbeault complexes with coefficients in $\mathcal{T}hom(F, \mathcal{D}b_{X^{\mathbb{R}}})$, and $F \overset{\mathrm{w}}{\otimes} \mathcal{C}^\infty_{X^{\mathbb{R}}}$ respectively.

If $F \in \mathbf{D}^{\mathrm{b}}_{\mathbb{C}-c}(\mathbb{C}_X)$, then $\mathcal{T}hom(F, \mathcal{O}_X)$ has regular holonomic cohomology groups (this is the way Kashiwara proves the Riemann-Hilbert equivalence of categories). In such a case, one has:

$$\mathcal{T}hom(D'F, \mathcal{O}_X) \simeq \underline{D}\,\mathcal{T}hom(F, \mathcal{O}_X).$$

If Z is a closed complex submanifold of codimension d of X, we shall consider the holonomic left $\mathcal{D}_X$-module $\mathcal{B}_{Z|X} = \mathcal{T}hom(\mathbb{C}_Z[-d], \mathcal{O}_X)$ of [19]. Recall that $\mathcal{B}_{Z|X} \simeq H^d_{[Z]}(\mathcal{O}_X)$ (algebraic cohomology) is a subsheaf of $\mathcal{B}^\infty_{Z|X} = H^d_Z(\mathcal{O}_X)$.

2.2 Kernels for sheaves

Here, all manifolds and morphisms of manifolds will be complex analytic.

Let X and Y be complex manifolds of dimension d_X and d_Y respectively. Denote by $r : X \times Y \longrightarrow Y \times X$ the map $r(x,y) = (y,x)$, and by q_1, q_2 the first and second projection from $X \times Y$ to the corresponding factor. If Z is another manifold, for $i,j = 1,2,3$ we denote by q_{ij} the projections from $X \times Y \times Z$ to the corresponding factor (e.g., $q_{23} : X \times Y \times Z \longrightarrow Y \times Z$).

Definition 1. For $K \in \mathbf{D}^{\mathrm{b}}(\mathbb{C}_{X\times Y})$ and $L \in \mathbf{D}^{\mathrm{b}}(\mathbb{C}_{Y\times Z})$, we set:

$$K \circ L = Rq_{13!}(q_{12}^{-1}K \otimes q_{23}^{-1}L),$$
$${}^tK = r_* D'K.$$

Note that the operation $\circ$ is associative.

For $K \in \mathbf{D}^{\mathrm{b}}(\mathbb{C}_{X\times Y})$, $L \in \mathbf{D}^{\mathrm{b}}(\mathbb{C}_{Y\times Z})$, consider the hypotheses:

$$(\operatorname{supp}(K) \times Z) \cap (X \times \operatorname{supp}(L)) \text{ is proper over } X \times Z, \tag{1}$$

$$(SS(K) \times T^*_Z Z) \cap (T^*_X X \times SS(L)) \subset T^*_{X\times Y\times Z}(X \times Y \times Z). \tag{2}$$

Proposition 2. *Let $K \in \mathbf{D}^{\mathrm{b}}_{\mathbb{R}-c}(\mathbb{C}_{X\times Y})$, $L \in \mathbf{D}^{\mathrm{b}}(\mathbb{C}_{Y\times Z})$, and $H \in \mathbf{D}^{\mathrm{b}}(\mathbb{C}_{X\times Z})$. Assume (1), (2). Then:*

$$\mathrm{RHom}(H, K \circ L) \simeq \mathrm{RHom}(({}^tK) \circ H, L)[-2d_X].$$

Proof. One has the chain of isomorphisms:

$$\begin{aligned}
\mathrm{RHom}(H, K \circ L) &= \mathrm{RHom}(H, Rq_{13!}(q_{12}^{-1}K \otimes q_{23}^{-1}L)) \\
&\simeq \mathrm{RHom}(H, Rq_{13*}(q_{12}^{-1}K \otimes q_{23}^{-1}L)) \\
&\simeq \mathrm{RHom}(q_{13}^{-1}H, q_{12}^{-1}K \otimes q_{23}^{-1}L) \\
&\simeq \mathrm{RHom}(q_{13}^{-1}H, R\mathcal{H}om(q_{12}^{-1}D'K, q_{23}^{-1}L)) \\
&\simeq \mathrm{RHom}(q_{13}^{-1}H \otimes q_{12}^{-1}D'K, q_{23}^{-1}L) \\
&\simeq \mathrm{RHom}(q_{13}^{-1}H \otimes q_{12}^{-1}D'K, q_{23}^{!}L)[-2d_X] \\
&\simeq \mathrm{RHom}(Rq_{23!}(q_{13}^{-1}H \otimes q_{12}^{-1}D'K[2d_X]), L) \\
&= \mathrm{RHom}(({}^tK) \circ H, L)[-2d_X].
\end{aligned}$$

Here, we used hypothesis (1) in the first isomorphism, and hypothesis (2) in the third.

Corollary 3. *Let $K \in \mathbf{D}^{\mathrm{b}}_{\mathbb{R}-c}(\mathbb{C}_{X\times Y})$, and assume*

$$\begin{gathered}
\operatorname{supp}(K) \textit{ is proper over } X, \\
SS(K) \cap (T^*_X X \times T^*Y) \subset T^*_{X\times Y}(X \times Y).
\end{gathered}$$

Then there are natural morphisms:

$$\begin{aligned}
\mathbb{C}_{\Delta_X} &\longrightarrow K \circ {}^tK[2d_X], \\
K \circ {}^tK[2d_Y] &\longrightarrow \mathbb{C}_{\Delta_X}.
\end{aligned}$$

Proof. Applying Proposition 2 for $Z = X$, $H = \mathbb{C}_{\Delta_X}$, $L = {}^tK$ we obtain the first morphism. Choosing instead $Z = Y$, $L = \mathbb{C}_{\Delta_Y}$, $H = K$, we get a morphism ${}^tK \circ K[2d_X] \longrightarrow \mathbb{C}_{\Delta_Y}$, from which the second morphism in the statement is easily deduced.

Assuming (1), (2), and assuming that K or L is $\mathbb{R}$-constructible, one proves similarly that

$$D'(K \circ L) \simeq D'K \circ D'L\,[2d_Y].$$

2.3 Kernels for $\mathcal{D}$-modules

Definition 4. For $\mathcal{K} \in \mathbf{D}^{\mathrm{b}}(\mathcal{D}_{X\times Y})$ and $\mathcal{L} \in \mathbf{D}^{\mathrm{b}}(\mathcal{D}_{Y\times Z})$, we set:

$$\mathcal{K} \underline{\circ} \mathcal{L} = \underline{q_{13}}_!(\underline{q_{12}}^{-1}\mathcal{K} \otimes^{L}_{\mathcal{O}_{X\times Y\times Z}} \underline{q_{23}}^{-1}\mathcal{L})$$

Note that

$$\mathcal{K} \underline{\circ} \mathcal{L} \simeq \underline{q_{13}}_!\underline{\delta}^{-1}(\mathcal{K} \underline{\boxtimes} \mathcal{L}),$$

where δ denotes the diagonal embedding $X \times Y \times Z \longrightarrow X \times Y \times Y \times Z$.

Proposition 5. [1] *Let $\mathcal{K} \in \mathbf{D}^{\mathrm{b}}(\mathcal{D}_{X\times Y})$ and $\mathcal{N} \in \mathbf{D}^{\mathrm{b}}(\mathcal{D}_Y)$. Then, there is a natural isomorphism in $\mathbf{D}^{\mathrm{b}}(\mathcal{D}_X)$:*

$$\mathcal{K} \underline{\circ} \mathcal{N} \simeq Rq_{1!}(\mathcal{K}^{(0,d_Y)} \otimes^{L}_{q_2^{-1}\mathcal{D}_Y} q_2^{-1}\mathcal{N}),$$

where $\mathcal{K}^{(0,d_Y)} = \mathcal{K} \otimes_{q_2^{-1}\mathcal{O}_Y} q_2^{-1}\Omega_Y$ is endowed with its natural $(q_1^{-1}\mathcal{D}_X, q_2^{-1}\mathcal{D}_Y)$-bimodule structure.

Proof. By definition,

$$\mathcal{K} \underline{\circ} \mathcal{N} \simeq Rq_{1!}((\mathcal{D}_X \underline{\boxtimes} \Omega_Y) \otimes^{L}_{\mathcal{D}_{X\times Y}} (\mathcal{K} \otimes^{L}_{\mathcal{O}_{X\times Y}} (\mathcal{O}_X \underline{\boxtimes} \mathcal{N})))$$

Moreover, one has the following chain of isomorphisms:

$$\begin{aligned}
&(\mathcal{D}_X \underline{\boxtimes} \Omega_Y) \otimes^{L}_{\mathcal{D}_{X\times Y}} (\mathcal{K} \otimes^{L}_{\mathcal{O}_{X\times Y}} (\mathcal{O}_X \underline{\boxtimes} \mathcal{N})) \\
&\simeq q_2^{-1}\Omega_Y \otimes^{L}_{q_2^{-1}\mathcal{D}_Y} (\mathcal{K} \otimes^{L}_{q_2^{-1}\mathcal{O}_Y} q_2^{-1}\mathcal{N}) \\
&\simeq (q_2^{-1}\Omega_Y \otimes^{L}_{q_2^{-1}\mathcal{O}_Y} \mathcal{K}) \otimes^{L}_{q_2^{-1}\mathcal{D}_Y} q_2^{-1}\mathcal{N} \\
&\simeq \mathcal{K}^{(0,d_Y)} \otimes^{L}_{q_2^{-1}\mathcal{D}_Y} q_2^{-1}\mathcal{N}.
\end{aligned}$$

Proposition 6. *Let $K \in \mathbf{D}^{\mathrm{b}}_{\mathbb{C}-c}(\mathbb{C}_{X\times Y})$, $L \in \mathbf{D}^{\mathrm{b}}_{\mathbb{C}-c}(\mathbb{C}_{Y\times Z})$, and assume (1). Then there is a natural isomorphism:*

$$\mathcal{T}hom(K, \mathcal{O}_{X\times Y}) \underline{\circ} \mathcal{T}hom(L, \mathcal{O}_{Y\times Z}) \xrightarrow{\sim} \mathcal{T}hom(K \circ L, \mathcal{O}_{X\times Z})[-d_Y].$$

Proof. Consider the chain of isomorphisms:

$$\begin{aligned}
&\mathcal{T}hom(K, \mathcal{O}_{X\times Y}) \underline{\circ} \mathcal{T}hom(L, \mathcal{O}_{Y\times Z}) \\
&= \underline{q_{13}}_!(\underline{q_{12}}^{-1}\mathcal{T}hom(K, \mathcal{O}_{X\times Y}) \otimes^{L}_{\mathcal{O}_{X\times Y\times Z}} \underline{q_{23}}^{-1}\mathcal{T}hom(L, \mathcal{O}_{Y\times Z})) \\
&\simeq \underline{q_{13}}_!(\mathcal{T}hom(q_{12}^{-1}K, \mathcal{O}_{X\times Y\times Z}) \otimes^{L}_{\mathcal{O}_{X\times Y\times Z}} \mathcal{T}hom(q_{23}^{-1}L, \mathcal{O}_{X\times Y\times Z})) \\
&\simeq \underline{q_{13}}_!\mathcal{T}hom(q_{12}^{-1}K \otimes q_{23}^{-1}L, \mathcal{O}_{X\times Y\times Z}) \\
&\simeq \mathcal{T}hom(Rq_{13!}(q_{12}^{-1}K \otimes q_{23}^{-1}L), \mathcal{O}_{X\times Z})[-d_Y] \\
&= \mathcal{T}hom(K \circ L, \mathcal{O}_{X\times Z})[-d_Y].
\end{aligned}$$

For the proof of the above isomorphisms, see [12], [1], [15], and [3]. Note that we used hypothesis (1) in the third isomorphism.

[1] As pointed out to us by Andrei Baran, Proposition B.5 of [7] holds only in the algebraic case. In the analytic case, it should be replaced by Proposition 5 above.

Proposition 7. *For $\mathcal{K} \in \mathbf{D}^{\mathrm{b}}_{\mathrm{good}}(\mathcal{D}_{X\times Y})$ and $\mathcal{L} \in \mathbf{D}^{\mathrm{b}}_{\mathrm{good}}(\mathcal{D}_{Y\times Z})$, consider the analogous hypotheses to (1) (2):*

$$(\mathrm{supp}(\mathcal{K}) \times Z) \cap (X \times \mathrm{supp}(\mathcal{L})) \text{ is proper over } X \times Z, \tag{3}$$

$$(\mathrm{char}(\mathcal{K}) \times T^*_Z Z) \cap (T^*_X X \times \mathrm{char}(\mathcal{L})) \subset T^*_{X\times Y\times Z}(X \times Y \times Z). \tag{4}$$

Then there is a natural isomorphism:

$$\underline{D}(\mathcal{K} \underline{\circ} \mathcal{L}) \simeq \underline{D}\mathcal{K} \underline{\circ} \underline{D}\mathcal{L}. \tag{5}$$

Proof. Under the above assumptions, $\underline{D}$ commutes to the operations appearing in the definition of $\underline{\circ}$.

Proposition 8. *Let $\mathcal{K}$ be a regular holonomic $\mathcal{D}_{X\times Y}$-module, and let K be the associated perverse sheaf $K = R\mathcal{H}om_{\mathcal{D}_{X\times Y}}(\mathcal{K}, \mathcal{O}_{X\times Y})$. Set*

$${}^t\mathcal{K} = \underline{r}_*\underline{D}\mathcal{K} \simeq \mathcal{T}hom({}^tK, \mathcal{O}_{Y\times X}).$$

Assume (1) and (2) (or equivalently (3) and (4)) with $Z = X$ and $L = {}^tK$ (or equivalently $\mathcal{L} = {}^t\mathcal{K}$). Then there are natural morphisms:

$$\mathcal{K} \underline{\circ} {}^t\mathcal{K}[d_Y - d_X] \longrightarrow \mathcal{B}_{\Delta_X|X\times X},$$
$$\mathcal{B}_{\Delta_X|X\times X} \longrightarrow \mathcal{K} \underline{\circ} {}^t\mathcal{K}[d_X - d_Y].$$

Proof. Applying Corollary 3, we get the morphism:

$$\mathcal{T}hom(\mathbb{C}_{\Delta_X}[-d_X], \mathcal{O}_{X\times X}) \longleftarrow \mathcal{T}hom(K \circ {}^tK[d_Y], \mathcal{O}_{X\times X})[d_Y - d_X].$$

The first morphism follows by applying Proposition 6. The second morphism is similarly obtained.

Remark. Consider a correspondence:

$$X \xleftarrow{f} S \xrightarrow{g} Y, \tag{6}$$

and denote by $h : S \longrightarrow X \times Y$ the morphism $h = (f, g)$. It is then immediate to check that for $F \in \mathbf{D}^{\mathrm{b}}(\mathbb{C}_X)$, $K \in \mathbf{D}^{\mathrm{b}}(\mathbb{C}_S)$, one has:

$$F \circ (Rh_!K) \simeq Rg_!(K \otimes f^{-1}F).$$

Moreover, assuming h is proper it is easy to check that for $\mathcal{M} \in \mathbf{D}^{\mathrm{b}}(\mathcal{D}_X)$, $\mathcal{K} \in \mathbf{D}^{\mathrm{b}}(\mathcal{D}_S)$:

$$\mathcal{M} \underline{\circ} (\underline{h}_!\mathcal{K}) \simeq \underline{g}_!(\mathcal{K} \otimes^L_{\mathcal{O}_S} \underline{f}^{-1}\mathcal{M}).$$

Recall that, in the particular case when h is a closed embedding, one has $\underline{h}_!\mathcal{O}_S \simeq \mathcal{B}_{S|X\times Y}$.

2.4 Adjunction formulas

Formulas (10), (11) below appeared in a slightly more particular situation in [6]. Formulas (12), (13) are due to [15].

For $\mathcal{K}$ a regular holonomic $\mathcal{D}_{X\times Y}$-module, set $K = R\mathcal{H}om_{\mathcal{D}_{X\times Y}}(\mathcal{K}, \mathcal{O}_{X\times Y})$. Consider the hypotheses:

$$(\mathrm{supp}(\mathcal{M}) \times Y) \cap \mathrm{supp}(\mathcal{K}) \text{ is proper over } Y, \tag{7}$$

$$(\mathrm{char}(\mathcal{M}) \times T^*_Y Y) \cap \mathrm{char}(\mathcal{K}) \subset T^*_{X\times Y}(X \times Y), \tag{8}$$

$$(X \times \mathrm{supp}(G)) \cap \mathrm{supp}(K) \text{ is proper over } X. \tag{9}$$

Theorem 9. [2] *(i) Assuming hypotheses (7) and (8) above, we have isomorphisms:*

$$\begin{aligned} &\mathrm{R}\Gamma_c(X; R\mathcal{H}om_{\mathcal{D}_X}(\mathcal{M}, (K \circ G) \otimes \mathcal{O}_X))[d_X] \\ &\quad \simeq \mathrm{R}\Gamma_c(Y; R\mathcal{H}om_{\mathcal{D}_Y}(\mathcal{M} \underline{\circ} \mathcal{K}, G \otimes \mathcal{O}_Y)), \end{aligned} \tag{10}$$

$$\begin{aligned} &\mathrm{R}\Gamma(X; R\mathcal{H}om_{\mathcal{D}_X}(\mathcal{M}, R\mathcal{H}om(G \circ {}^tK, \mathcal{O}_X)))[d_X] \\ &\quad \simeq \mathrm{R}\Gamma(Y; R\mathcal{H}om_{\mathcal{D}_Y}(\mathcal{M} \underline{\circ} \mathcal{K}, R\mathcal{H}om(G, \mathcal{O}_Y)))[2d_Y]. \end{aligned} \tag{11}$$

If moreover (9) is satisfied, the formulas above hold interchanging Γ and Γ_c.

(ii) Assuming hypotheses (7) and (9) above, we have an isomorphism:

$$\begin{aligned} &\mathrm{R}\Gamma(X; R\mathcal{H}om_{\mathcal{D}_X}(\mathcal{M}, (K \circ G) \overset{\mathrm{w}}{\otimes} \mathcal{O}_X))[d_X] \\ &\quad \simeq \mathrm{R}\Gamma(Y; R\mathcal{H}om_{\mathcal{D}_Y}(\mathcal{M} \underline{\circ} \mathcal{K}, G \overset{\mathrm{w}}{\otimes} \mathcal{O}_Y)). \end{aligned} \tag{12}$$

If moreover (8) holds, then we have an isomorphism:

$$\begin{aligned} &\mathrm{R}\Gamma_c(X; R\mathcal{H}om_{\mathcal{D}_X}(\mathcal{M}, \mathcal{T}hom(G \circ {}^tK, \mathcal{O}_X)))[d_X] \\ &\quad \simeq \mathrm{R}\Gamma_c(Y; R\mathcal{H}om_{\mathcal{D}_Y}(\mathcal{M} \underline{\circ} \mathcal{K}, \mathcal{T}hom(G, \mathcal{O}_Y)))[2d_Y]. \end{aligned} \tag{13}$$

Under the same hypotheses, formulas (12) and (13) hold interchanging Γ and Γ_c.

3 Generalized QCTs

3.1 Kernels for $\mathcal{E}$-modules

We recall here some definitions from the theory of $\mathcal{E}$-modules. We refer to [19] and to [20] for an exposition.

We denote by $\pi_X : T^*X \longrightarrow X$ the cotangent bundle to X, by $\dot{\pi}_X : \dot{T}^*X \longrightarrow X$ the cotangent bundle with the zero-section removed, and by $T^*_M X$ the conormal bundle to a submanifold M of X. For a subset V of T^*X, we set $\dot{V} = V \cap \dot{T}^*X$. We denote by p_1 and p_2 the first and second projection from

[2] In formulas (C.4) and (C.7) of [7], $K \circ G$ should be replaced by $G \circ {}^tK$ as in formulas (11) and (13) above.

$T^*(X \times Y) \simeq T^*X \times T^*Y$ to the corresponding factor, and by p_2^a the composite of p_2 with the antipodal map of T^*Y.

Let $\mathcal{E}_X$ denote the sheaf of microdifferential operators of finite order on T^*X. We denote by $\mathbf{D}^{\mathrm{b}}(\mathcal{E}_X)$ the derived category of the category of bounded complexes of left $\mathcal{E}_X$-modules.

To $f : Y \longrightarrow X$ one associates the natural maps:

$$T^*Y \underset{{}^tf'}{\longleftarrow} Y \times_X T^*X \underset{f_\pi}{\longrightarrow} T^*X.$$

We will denote by $\underline{f}^\mu$ and $\underline{f}_\mu$ the inverse and direct images in the sense of $\mathcal{E}$-modules. Hence, for $\mathcal{M} \in \mathbf{D}^{\mathrm{b}}(\mathcal{E}_X)$ and $\mathcal{N} \in \mathbf{D}^{\mathrm{b}}(\mathcal{E}_Y)$:

$$\underline{f}^\mu \mathcal{M} = R^tf'_*(\mathcal{E}_{Y\to X} \otimes^L_{f_\pi^{-1}\mathcal{E}_X} f_\pi^{-1}\mathcal{M}), \quad \underline{f}_\mu \mathcal{N} = Rf_{\pi *}(\mathcal{E}_{X\leftarrow Y} \otimes^L_{{}^tf'^{-1}\mathcal{E}_Y} {}^tf'^{-1}\mathcal{N}),$$

where $\mathcal{E}_{Y\to X}$ and $\mathcal{E}_{X\leftarrow Y}$ are the transfer bimodules.

If $\mathcal{M} \in \mathbf{D}^{\mathrm{b}}(\mathcal{D}_X)$, we set:

$$\mathcal{E}\mathcal{M} = \mathcal{E}_X \otimes_{\pi_X^{-1}\mathcal{D}_X} \pi_X^{-1}\mathcal{M},$$

an object of $\mathbf{D}^{\mathrm{b}}(\mathcal{E}_X)$, and if $\mathcal{F}$ is a holomorphic vector bundle on X, we set:

$$\mathcal{E}\mathcal{F} = \mathcal{E}(\mathcal{D}\mathcal{F}).$$

If Z is a closed complex submanifold of codimension d of X, we shall consider the holonomic left $\mathcal{E}_X$-module $\mathcal{C}_{Z|X} = \mathcal{E}\mathcal{B}_{Z|X}$.

Definition 10. Let $\mathcal{K} \in \mathbf{D}^{\mathrm{b}}(\mathcal{D}_{X\times Y})$ and $\mathcal{L} \in \mathbf{D}^{\mathrm{b}}(\mathcal{D}_{Y\times Z})$. Denoting by $\delta : X \times Y \times Z \longrightarrow X \times Y \times Y \times Z$ the diagonal embedding, we set:

$$\mathcal{E}\mathcal{K} \underline{\circ}^\mu \mathcal{E}\mathcal{L} = \underline{q_{13}}_\mu \underline{\delta}^\mu(\mathcal{E}\mathcal{K} \underline{\boxtimes} \mathcal{E}\mathcal{L}).$$

Proposition 11. *(i) Let $\mathcal{M} \in \mathbf{D}^{\mathrm{b}}_{\mathrm{good}}(\mathcal{D}_X)$, and assume that $f : Y \longrightarrow X$ is non-characteristic for $\mathcal{M}$. Then*

$$\mathcal{E}\underline{f}^{-1}\mathcal{M} \simeq \underline{f}^\mu\mathcal{E}\mathcal{M}.$$

(ii) Let $\mathcal{N} \in \mathbf{D}^{\mathrm{b}}_{\mathrm{good}}(\mathcal{D}_Y)$. Assume that f is proper on $\operatorname{supp}\mathcal{N}$. *Then*

$$\mathcal{E}\underline{f}_*\mathcal{N} \simeq \underline{f}_\mu\mathcal{E}\mathcal{N}.$$

(iii) Let $\mathcal{M} \in \mathbf{D}^{\mathrm{b}}_{\mathrm{good}}(\mathcal{D}_X)$, $\mathcal{N} \in \mathbf{D}^{\mathrm{b}}_{\mathrm{good}}(\mathcal{D}_Y)$. Then

$$\mathcal{E}(\mathcal{M} \underline{\boxtimes} \mathcal{N}) \simeq (\mathcal{E}\mathcal{M}) \underline{\boxtimes} (\mathcal{E}\mathcal{N}).$$

Proof. Assertion (iii) is obvious, and (ii) is proved in [21]. Assertion (i) follows from the division theorem of [19], using the techniques of that paper.

¿From Proposition 11 it immediately follows:

Corollary 12. *Assuming (3), (4), there is a natural isomorphism:*

$$\mathcal{E}(\mathcal{K} \underline{\circ} \mathcal{L}) \simeq \mathcal{E}\mathcal{K} \underline{\circ}^{\mu} \mathcal{E}\mathcal{L}.$$

Theorem 13. *Let* $\mathcal{K} \in \mathbf{D}^{\mathrm{b}}_{\mathrm{good}}(\mathcal{D}_{X\times Y})$, $\mathcal{N} \in \mathbf{D}^{\mathrm{b}}_{\mathrm{good}}(\mathcal{D}_Y)$. *Assume:*

$$\begin{aligned}&\mathrm{supp}(\mathcal{K}) \text{ is proper over } X,\\ &\mathrm{char}(\mathcal{K}) \cap (T^*_X X \times T^*Y) \subset T^*_{X\times Y}(X\times Y).\end{aligned}$$

Then there is a natural isomorphism:

$$\mathcal{E}\mathcal{K} \underline{\circ}^{\mu} \mathcal{E}\mathcal{N} \xrightarrow{\sim} Rp_{1*}(\mathcal{E}\mathcal{K}^{(0,d_Y)} \otimes^L_{p_2^{a-1}\mathcal{E}_Y} p_2^{a-1}\mathcal{E}\mathcal{N}).$$

Proof. By Corollary 12, the left hand side is isomorphic to $\mathcal{E}(\mathcal{K} \underline{\circ} \mathcal{N})$. Hence, by Proposition 5, it is enough to prove the natural isomorphism:

$$\mathcal{E}_X Rq_{1!}(\mathcal{K} \otimes^L_{\mathcal{D}_Y} \mathcal{N}) \xrightarrow{\sim} Rp_{1!}(\mathcal{E}_{X\times Y}\mathcal{K} \otimes^L_{\mathcal{D}_Y} \mathcal{N}).$$

Let $\mathcal{E}_{X\times Y/Y}$ denote the subsheaf of $\mathcal{E}_{X\times Y}$ of sections which commute with $\mathcal{O}_Y$. The second hypothesis, and the division theorem of [19] gives:

$$\mathcal{E}_{X\times Y/Y}\mathcal{K} \simeq \mathcal{E}_{X\times Y}\mathcal{K}$$

Let $\mathcal{K}_0$ be a coherent $\mathcal{O}_{X\times Y}$-module which generates $\mathcal{K}$, and $\mathcal{N}_0$ be a coherent $\mathcal{O}_Y$-module which generates $\mathcal{N}$. Let $\mathcal{L}_0 = \mathcal{K}_0 \otimes_{\mathcal{O}_Y} \mathcal{N}_0$. It is enough to check:

$$\mathcal{E}_X(0) \otimes_{\mathcal{O}_X} Rp_{1!}\mathcal{L}_0 \xrightarrow{\sim} Rp_{1!}(\mathcal{E}_{X\times Y/Y}(0) \otimes_{\mathcal{O}_{X\times Y}} \mathcal{L}_0). \tag{14}$$

$\mathcal{E}_X(0)$ is an $\mathcal{O}_X$-module of type DFN and $\mathcal{E}_{X\times Y/Y}(0) \simeq \mathcal{E}_X(0)\widehat{\otimes}\mathcal{O}_Y$ (see [21, I §7]). Then the proof goes as that of Theorem 7.3 of [15], using Proposition 3.13 of [21] (or Theorem 8.1 of [15]).

3.2 Extending QCTs to $\mathcal{D}$-modules

In this section, we assume X and Y have the same dimension n. Let $\mathcal{L}$ be a regular holonomic $\mathcal{D}_{X\times Y}$-module, and set:

$$\begin{aligned}L &= R\mathcal{H}om_{\mathcal{D}_{X\times Y}}(\mathcal{L}, \mathcal{O}_{X\times Y}),\\ \Lambda &= \mathrm{char}(\mathcal{L}) \subset T^*(X\times Y).\end{aligned}$$

Let $\mathcal{F}$ and $\mathcal{G}$ be holomorphic line bundles on X and Y respectively, and set:

$$\mathcal{L}^{(n,0)}(\mathcal{F},\mathcal{G}) = q_1^{-1}(\mathcal{F} \otimes_{\mathcal{O}_X} \Omega_X) \otimes_{q_1^{-1}\mathcal{O}_X} \mathcal{L} \otimes_{q_2^{-1}\mathcal{O}_Y} q_2^{-1}\mathcal{G}.$$

Lemma 14. *Assuming* q_2 *is proper on* $\mathrm{supp}\,\mathcal{L}$, *there is a natural isomorphism:*

$$\alpha : \Gamma(X\times Y; \mathcal{L}^{(n,0)}(\mathcal{F},\mathcal{G}^*)) \simeq \mathrm{Hom}_{\mathbf{D}^{\mathrm{b}}(\mathcal{D}_Y)}(\mathcal{D}\mathcal{G}, \mathcal{D}\mathcal{F} \underline{\circ} \mathcal{L}). \tag{15}$$

Proof. It is enough to apply $H^0(\cdot)$ to the following chain of isomorphisms:

$$\begin{aligned} Ra_{X\times Y*}\mathcal{L}^{(n,0)}(\mathcal{F},\mathcal{G}^*) &\simeq \\ &\simeq Ra_{Y*}Rq_{2*}R\mathcal{H}om_{q_2^{-1}\mathcal{D}_Y}(q_2^{-1}\mathcal{D}\mathcal{G},\mathcal{L}^{(n,0)}\otimes_{q_1^{-1}\mathcal{O}_X} q_1^{-1}\mathcal{F}) \\ &\simeq Ra_{Y*}R\mathcal{H}om_{\mathcal{D}_Y}(\mathcal{D}\mathcal{G}, Rq_{2*}(\mathcal{L}^{(n,0)}\otimes_{q_1^{-1}\mathcal{O}_X} q_1^{-1}\mathcal{F})) \\ &\simeq Ra_{Y*}R\mathcal{H}om_{\mathcal{D}_Y}(\mathcal{D}\mathcal{G},\mathcal{D}\mathcal{F}\underline{\circ}\mathcal{L}). \end{aligned}$$

Here, in the first isomorphism we used the fact that $\mathcal{D}\mathcal{G}$ is $\mathcal{D}_Y$-coherent, and in the last one we used the fact that q_2 is proper on $\operatorname{supp}\mathcal{L}$.

Assuming q_2 is proper on $\operatorname{supp}\mathcal{L}$, by Lemma 14 we associate to $s\in\Gamma(X\times Y;\mathcal{L}^{(n,0)}(\mathcal{F},\mathcal{G}^*))$ the $\mathcal{D}_Y$-linear morphism:

$$\alpha(s):\mathcal{D}\mathcal{G}\longrightarrow\mathcal{D}\mathcal{F}\underline{\circ}\mathcal{L}.$$

Let U_X and U_Y be two open conic subsets of $\dot{T}^*X$ and $\dot{T}^*Y$ respectively. Set

$$\Lambda_0=\Lambda\cap(U_X\times U_Y^a),\quad \Sigma=\Lambda\setminus\Lambda_0,\quad W=T^*Y\setminus U_Y.$$

We assume:

(a) $\dot{W}$ is a $\mathbb{C}$-analytic closed conic subset of $\dot{T}^*Y$ of codimension ≥ 2,
(b) Λ_0 is a smooth Lagrangian manifold, and $p_1:\Lambda_0\longrightarrow U_X$, $p_2^a:\Lambda_0\longrightarrow U_Y$ are isomorphisms (in other words, Λ_0 defines a contact transformation),
(c) $p_2^{a-1}\dot{W}\cap\dot{\Lambda}=\dot{\Sigma}$, and this analytic set has dimension $<n$,
(d) $\mathcal{L}$ has no submodules isomorphic to $\mathcal{O}_{X\times Y}$,
(e) there exists $s\in\Gamma(X\times Y;\mathcal{L}^{(n,0)}(\mathcal{F},\mathcal{G}^*))$, which is non-degenerate on Λ_0.

We refer to [19] for the notion of non-degenerate section of a holonomic module.

Theorem 15. *Assume hypotheses* (a)–(e) *above. Then:*

$$H^0(\alpha(s)):\mathcal{D}\mathcal{G}\longrightarrow H^0(\mathcal{D}\mathcal{F}\underline{\circ}\mathcal{L})$$

is an isomorphism.

Proof. Applying Corollary 12, it is enough to prove the isomorphism $\mathcal{E}\mathcal{G}\xrightarrow{\sim} H^0(\mathcal{E}\mathcal{F}\underline{\circ}^\mu\mathcal{E}\mathcal{L})$ all over T^*Y. Consider the morphism of distinguished triangles:

$$\begin{array}{ccccccc} \mathrm{R}\Gamma_W(\mathcal{E}\mathcal{G}) & \longrightarrow & R\pi_{Y*}(\mathcal{E}\mathcal{G}) & \longrightarrow & \mathrm{R}\Gamma_{U_Y}(\mathcal{E}\mathcal{G}) & \xrightarrow{+1} \\ \downarrow{\scriptstyle \alpha_W(s)} & & \downarrow{\scriptstyle \alpha(s)} & & \downarrow{\scriptstyle \alpha_{U_Y}(s)} & \\ \mathrm{R}\Gamma_W(\mathcal{E}\mathcal{F}\underline{\circ}^\mu\mathcal{E}\mathcal{L}) & \longrightarrow & R\pi_{Y*}(\mathcal{E}\mathcal{F}\underline{\circ}^\mu\mathcal{E}\mathcal{L}) & \longrightarrow & \mathrm{R}\Gamma_{U_Y}(\mathcal{E}\mathcal{F}\underline{\circ}^\mu\mathcal{E}\mathcal{L}) & \xrightarrow{+1} \end{array}\tag{16}$$

By (b) and (c), $\mathrm{R}\Gamma_{U_Y}(\mathcal{E}\mathcal{F}\underline{\circ}^\mu\mathcal{E}\mathcal{L})\simeq\mathrm{R}\Gamma_{U_Y}(\mathcal{E}\mathcal{F}|_{U_X}\underline{\circ}^\mu\mathcal{E}\mathcal{L}|_{\Lambda_0})$. Since s is non-degenerate on Λ_0, $\alpha_{U_Y}(s)$ is an isomorphism by [19]. Applying $H^0(\cdot)$, we get the commutative diagram in which the horizontal lines are exact:

$$\begin{array}{ccccccc} H^0_W\mathcal{E}\mathcal{G} & \to & \mathcal{D}\mathcal{G} & \to & H^0_{U_Y}\mathcal{E}\mathcal{G} & \longrightarrow & H^1_W\mathcal{E}\mathcal{G} \\ \downarrow & & \downarrow{\scriptstyle H^0(\alpha(s))} & & \downarrow\wr & & \downarrow \\ H^0_W(\mathcal{E}\mathcal{F}\underline{\circ}^\mu\mathcal{E}\mathcal{L}) & \to & H^0(\mathcal{D}\mathcal{F}\underline{\circ}\mathcal{L}) & \to & H^0_{U_Y}(\mathcal{E}\mathcal{F}\underline{\circ}^\mu\mathcal{E}\mathcal{L}) & \xrightarrow{\beta} & H^1_W(\mathcal{E}\mathcal{F}\underline{\circ}^\mu\mathcal{E}\mathcal{L}). \end{array}$$

We shall prove

(i) $H^0_W \mathcal{E}\mathcal{G} = 0$,
(ii) $H^0_W(\mathcal{E}\mathcal{F} \underline{\circ}^\mu \mathcal{E}\mathcal{L}) = 0$,
(iii) $H^1_W \mathcal{E}\mathcal{G} = 0$.

This will imply the result, for then β will be the zero morphism.

Since the problem is local on T^*Y, in (i) and (iii) we may assume $\mathcal{G} = \mathcal{O}_Y$. Then (i) and (iii) follow from [13, Theorem 1.2.2]. To prove (ii), consider the isomorphisms:

$$\begin{aligned} \mathrm{R}\Gamma_W(\mathcal{E}\mathcal{F} \underline{\circ}^\mu \mathcal{E}\mathcal{L}) &\simeq \mathrm{R}\Gamma_W Rp^a_{2*}(\mathcal{E}\mathcal{L}^{(n,0)} \otimes_{p_1^{-1}\mathcal{O}_X} p_1^{-1}\mathcal{F}) \\ &\simeq Rp^a_{2*}\mathrm{R}\Gamma_\Sigma(\mathcal{E}\mathcal{L}^{(n,0)} \otimes_{p_1^{-1}\mathcal{O}_X} p_1^{-1}\mathcal{F}). \end{aligned}$$

We thus reduce to a local problem on Λ. Applying $H^0(\cdot)$, it is enough to prove that

$$\Gamma_\Sigma \mathcal{E}\mathcal{L} = 0.$$

Since Σ contains the zero-section, using the exact sequence:

$$0 \longrightarrow \Gamma_{X\times Y}\mathcal{E}\mathcal{L} \longrightarrow \Gamma_\Sigma \mathcal{E}\mathcal{L} \longrightarrow \Gamma_{\dot{\Sigma}}\mathcal{E}\mathcal{L},$$

and recalling hypothesis (d), it remains to prove that $\Gamma_{\dot{\Sigma}}\mathcal{E}\mathcal{L} = 0$. Since $\dot{\Sigma}$ is an analytic subset of dimension smaller than that of $X \times Y$, it follows by the involutivity theorem that $\mathcal{L}$ has no germs of sections supported by $\dot{\Sigma}$.

Remark. (i) In [7] we obtained a similar result in the case where $\Lambda_0 = \dot{\Lambda}$, a situation which applies to projective duality.
(ii) Other applications of Theorem 15 will be found in [17].

References

1. E. Andronikof, *Microlocalisation temperée*, Mém. Soc. Math. France **57** (1994), Supl. au Bull. de la Soc. Math. France **122 (2)**.
2. R. J. Baston and M. G. Eastwood, *The Penrose transform: its interaction with representation theory*, Oxford Univ. Press, 1989.
3. J-E. Björk, Analytic $\mathcal{D}$-modules and Applications. Kluwer Academic Publisher, Dordrecht-Boston-London, 1993.
4. J. L. Brylinski, *Transformations canoniques, dualité projective, théorie de Lefschetz, transformations de Fourier et sommes trigonométriques*, Astérisque **140-141** (1986), 3–134.
5. A. D'Agnolo, *Nonlocal Differentials, Radon Transform, and Cavalieri Condition: a Cohomological Approach*, Prépublication Univ. Paris VI and Paris VII (1996), and article to appear.
6. A. D'Agnolo and P. Schapira, *Radon-Penrose transform for $\mathcal{D}$-modules*, to appear in J. of Functional Analysis; see also: *La transformée de Radon-Penrose des $\mathcal{D}$-modules*, C. R. Acad. Sci. Paris Sér. I Math. **319** (1994), 461–466.
7. ———, *Leray's quantization of projective duality*, Duke Math. J. **84** (1996), 453–496.

8. M. G. Eastwood, R. Penrose, and R. O. Jr. Wells. *Cohomology and massless fields*. Comm. Math. Phys. **78** (1981), 305–351.
9. I. M. Gelfand, S. G. Gindikin and M. I. Graev, *Integral geometry in affine and projective spaces*, Journal of Soviet Math. **18** (1982), 39–167.
10. S. Helgason, The Radon Transform, Progress in Math. **5**, Birkhäuser, 1980.
11. M. Kashiwara, *Systems of microdifferential equations*, Progress in Math. **34**, Birkhäuser, 1983.
12. ———, *The Riemann-Hilbert problem for holonomic systems*, Publ. RIMS, Kyoto Univ. **20(20)** (1984), 319–365.
13. M. Kashiwara and T. Kawai, *On holonomic systems of micro-differential equations III – systems with regular singularities*, Publ. RIMS, Kyoto Univ. **17** (1981), 813–979.
14. M. Kashiwara and P. Schapira, *Sheaves on manifolds*, Grundlehren der Math. Wiss. **292**, Springer, 1990.
15. ———, *Moderate and formal cohomology associated with constructible sheaves*, Mém. Soc. Math. France (N.S.) **64** (1996).
16. J. Leray, *Le calcul différentiel et intégral sur une variété analytique complexe*, Bull. Soc. Math. France **87** (1959), 81–180.
17. C. Marastoni, *Grassmann duality and $\mathcal{D}$-modules*, (in preparation).
18. A. Martineau, *Indicatrice des fonctions analytiques et inversion de la transformation de Fourier-Borel par la transormation de Laplace*, C. R. Acad. Sci. Paris Sér. I Math. **255** (1962), 2888–2890.
19. M. Sato, T. Kawai, and M. Kashiwara, *Microfunctions and pseudo-differential equations*, Hyperfunctions and pseudo-differential equations (Komatsu, ed.), Lecture Notes in Math. **287**, Springer, 1973, Proceedings Katata 1971, pp. 265–529.
20. P. Schapira, *Microdifferential systems in the complex domain*, Grundlehren der Math. Wiss. **269**, Springer, 1985.
21. P. Schapira and J.-P. Schneiders, *Index theorems for elliptic pairs*, Astérisque **224**, 1994.
22. J.-P. Schneiders, *Introduction to $\mathcal{D}$-modules*, Bull. Soc. Roy. Sci. Liège, **63(3-4)** (1994).

Microfunction solutions of holonomic systems with irregular singularities

Dedicated to Professor Hikosaburo Komatsu on the occasion of his sixtieth birthday

Naofumi Honda

Department of Mathematics, Graduate School of Science, Hokkaido University, Sapporo, Hokkaido 060, Japan

0 Introduction

One important feature of a holonomic system with regular singularities is that, if we consider the problem in the "complex holomorphic" category, its solution complex is cohomologically $\mathbb{C}$ constructible. For example, for a pair of complex manifolds $Y \hookrightarrow X$ and a holonomic $\mathcal{E}_X$ module $\mathcal{M}$ with regular singularities, the solution complex

$$\mathbb{R}\mathcal{H}om_{\mathcal{E}_X}(\mathcal{M}, \mathcal{C}^{\mathbb{R},f}_{Y|X}) \tag{0.0}$$

has $\mathbb{C}$ constructible cohomologies where $\mathcal{C}^{\mathbb{R},f}_{Y|X}$ denotes the sheaf of tempered holomorphic microfunctions. However, for a holonomic module with irregular singularities, the solution complex (0.0) is no longer $\mathbb{C}$ constructible in general. Such a breakdown of complex structure is deeply connected with Stokes lines of ordinary differential equations with irregular singularities. The purpose of this article is to investigate the relations between solutions of a system and the classical Stokes lines.

In section 1, we quickly review the definitions of tempered and Gevrey class holomorphic microfunctions and related objects to be used in this article. In section 2, we introduce the notion of a Stokes set for a holonomic module, which is used in section 3 to estimate the micro-support of a solution. An explicit example related to the Bessel equation is also discussed in section 3.

1 Holomorphic microfunctions

We review briefly the definitions of several kinds of holomorphic microfunctions and microdifferential operators which play an important role in this article.

Let X be a complex manifold of dimension n and $\pi_X : T^*X \to X$ its cotangent bundle with a coordinate $(z_1, \cdots, z_n; \zeta_1, \cdots, \zeta_n)$. Set $\overset{\circ}{T}{}^*X = T^*X \setminus T^*_X X$ and denote by $\overset{\circ}{\pi}_X$ the restriction of π_X to $\overset{\circ}{T}{}^*X$. The sheaf ${\mathcal{E}_X}^\infty$ of microdifferential operators of infinite order is defined on T^*X (Sato-Kawai-Kashiwara [23]). We denote by $\mathcal{E}_X$ (resp. $\mathcal{E}_X(m)$) the subsheaf of ${\mathcal{E}_X}^\infty$ consisting of microdifferential operators of finite order (resp. microdifferential operators of order at most m).

Let Y be a complex submanifold of X with a complex codimension d and $T^*_Y X$ its conormal bundle.

The sheaf of holomorphic microfunctions in $T^*_Y X$ is defined by

$$\mathcal{C}^{\mathbf{R}}_{Y|X} = \mu_Y(\mathcal{O}_X)[d], \tag{1.0}$$

where $\mu_Y(\bullet)$ is Sato's microlocalization functor . We also define

$$\mathcal{C}^{\infty}_{Y|X}|_{\overset{\circ}{T}{}^*X} = \tau\tau^{-1}\mathcal{C}^{\mathbf{R}}_{Y|X}|_{\overset{\circ}{T}{}^*X}, \quad \mathcal{C}^{\infty}_{Y|X}|_{T^*_X X} = \mathbb{R}\Gamma_{[Y]}(\mathcal{O}_X)[d], \tag{1.1}$$

where $\tau : \overset{\circ}{T}{}^*X \to P^*X$ is the canonical projection. A stalk of $\mathcal{C}^{\mathbf{R}}_{Y|X}$ is represented by boundary values of holomorphic functions (see Example 1.1 below).

We now introduce the sheaf $\mathcal{C}^{\mathbf{R},f}_{Y|X}$ of tempered holomorphic microfunctions and $\mathcal{C}^{\mathbf{R},(s)}_{Y|X}$ of holomorphic microfunctions of Gevrey class (s) using Sato's microlocalization functor.

Let $\overline{X}$ be a complex conjugate space of X, $X_{\mathbf{R}}$ an underlying real manifold of X and $s \in (1, \infty)$. As usual, we denote by $\mathcal{O}_{\overline{X}}$ (resp. $\mathcal{D}b_{X_{\mathbf{R}}}$, $\mathcal{D}b^{(s)}_{X_{\mathbf{R}}}$) the sheaf of holomorphic functions in $\overline{X}$, which is regarded as a $\mathcal{D}_{\overline{X}}$ module (resp. distributions, ultradistributions of Gevrey class (s) in $X_{\mathbf{R}}$). The $\mathcal{D}_{\overline{X}}$ module $\mathcal{O}_{\overline{X}}$ is the Cauchy-Riemann system in $X_{\mathbf{R}}$.

Definition 1.0. *The sheaf $\mathcal{C}^{\mathbf{R},(s)}_{Y|X}$ of holomorphic microfunctions of Gevrey class (s) is defined by*

$$\mathcal{C}^{\mathbf{R},(s)}_{Y|X} = \mathbb{R}\mathcal{H}om_{\mathcal{D}_{\overline{X}}}(\mathcal{O}_{\overline{X}}, \tau_{\leq 0} \circ \mu_Y(\mathcal{D}b^{(s)}_{X_{\mathbf{R}}}))[d] \tag{1.2}$$

where $\tau_{\leq 0}$ is the usual way out functor.

The sheaf $\mathcal{C}^{\mathbf{R},f}_{Y|X}$ of tempered holomorphic microfunctions is also defined by (1.2) replacing $\mathcal{D}b^{(s)}_{X_{\mathbf{R}}}$ with $\mathcal{D}b_{X_{\mathbf{R}}}$.

Sheaves $\mathcal{C}_{Y|X}$ and $\mathcal{C}^{(s)}_{Y|X}$ are defined in a similar manner.

Remark. The sheaf of tempered holomorhic microfunctions was first introduced by Andronikof[1,2]. He constructed the tempered microlocalizaion functor $T\text{-}\mu_\bullet(\bullet)$ and defined $\mathcal{C}^{\mathbb{R},f}_{Y|X}$ as $T\text{-}\mu_Y(\mathcal{O}_X)[d]$.

Although the definitions of holomorphic microfunctions seem to be a little complicated, the explicit expression of its stalk as boundary values of holomorphic functions has a familiar form, in particular, when Y is a hypersurface.

Example 1.1. Let Y be a smooth hypersurface of X with $z_1 = 0$, $T^*_Y X$ its conormal bundle with a coordinate $(\zeta_1, z_2, \ldots.., z_n)$ and $p = (0; -dz_1)$. Set for any $\epsilon > 0$

$$S_\epsilon = \{(z_i); 0 < |z_1| < \epsilon,\ |z_i| < \epsilon\,(i \neq 1),\ |\arg(z_1)| < \frac{\pi}{2} + \epsilon\} \subset X.$$

Then the stalk of $\mathcal{C}^{\mathbb{R}}_{Y|X}$ at p is written as

$$\mathcal{C}^{\mathbb{R}}_{Y|X\,,p} = \frac{\lim\limits_{\epsilon \to 0} \mathcal{O}_X(S_\epsilon)}{\mathcal{O}_{X,\pi_X(p)}}. \tag{1.3}$$

Denote by $\mathcal{O}^f_X(S_\epsilon)$ (resp. $\mathcal{O}^{(s)}_X(S_\epsilon)$) a subset of $\mathcal{O}_X(S_\epsilon)$ consisting of holomorphic functions $f \in \mathcal{O}_X(S_\epsilon)$ with the following estimate (f) (resp. (s)).

(f) There exist $C, N > 0$ for which f satisfies $|f(z)| < C|z_1|^{-N}$.

(s) There exist $C, l > 0$ for which f satisfies $|f(z)| < C\exp(l|z_1|^{\frac{1}{1-s}})$.

Then $\mathcal{C}^{\mathbb{R},f}_{Y|X\,,p}$ (resp. $\mathcal{C}^{\mathbb{R},(s)}_{Y|X\,,p}$) is also written as (1.3) replacing $\mathcal{O}_X(S_\epsilon)$ with $\mathcal{O}^f_X(S_\epsilon)$ (resp. $\mathcal{O}^{(s)}_X(S_\epsilon)$).

Finally we introduce several sheaves of operators. Let Δ_X be the diagonal set in $X \times X$ and p_1, p_2 the first and the second projection from $T^*(X \times X)$ to T^*X respectively. We identify $T^*_{\Delta_X}(X \times X)$ with T^*X. The sheaf $\mathcal{E}^{\mathbb{R},f}_X$ of tempered microdifferential operators and the sheaf $\mathcal{E}^{\mathbb{R},(s)}_X$ of microdifferential operators of Gevrey class (s) are defined by

$$\mathcal{E}^{\mathbb{R},*}_X = \mathcal{C}^{\mathbb{R},*}_{\Delta_X|X\times X} \underset{p_2^{-1}\mathcal{O}_X}{\otimes} p_2^{-1}\Omega^n_X,$$

$$\mathcal{E}^*_X = \mathcal{C}^*_{\Delta_X|X\times X} \underset{p_2^{-1}\mathcal{O}_X}{\otimes} p_2^{-1}\Omega^n_X,$$

where * stands for f or (s) and Ω^n_X is the sheaf of holomorphic n forms. A section of these sheaves of operators is expressed by formal symbols. For example, an alternative definition of $\mathcal{E}^{(s)}_X$ is:

Definition 1.2. *For an open set $U \subset T^*X$, a formal sum $\sum_{i \in \mathbf{Z}} P_i(z,\zeta) \in \mathcal{E}_X^\infty(U)$ belongs to $\mathcal{E}_X^{(s)}(U)$ if and only if $\sum_{i \in \mathbf{Z}} P_i(z,\zeta)$ satisfies the following estimate:*

For any compact subset K of U, there exists a positive constant C_K satisfying

$$\sup_K \mid P_i(z,\zeta) \mid \leq \frac{C_K^i}{i!^s} \quad (i \geq 0).$$

Let $p = (0; \zeta_0) \in \overset{\circ}{T}{}^*X$ and set

$$V_\epsilon = \{(z;\zeta); |\zeta| > \frac{1}{\epsilon},\ |z| < \epsilon,\ |\frac{\zeta}{|\zeta|} - \frac{\zeta_0}{|\zeta_0|}| < \epsilon\} \subset \overset{\circ}{T}{}^*X.$$

We define spaces of holomorphic functions with Gevrey growth as

$$S^{(s)}(V_\epsilon) = \{f \in \mathcal{O}_{T^*X}(V_\epsilon); \text{there exist}\, C, l > 0\, \text{such that}\, |f| < C\exp(l|\zeta|^{\frac{1}{s}})\},$$
$$S^f(V_\epsilon) = \{f \in \mathcal{O}_{T^*X}(V_\epsilon); \text{there exist}\, C, N > 0\, \text{such that}\, |f| < C|\zeta|^N\},$$
$$R(V_\epsilon) = \{f \in \mathcal{O}_{T^*X}(V_\epsilon); \text{there exist}\, C, \delta > 0\, \text{such that}\, |f| < C\exp(-\delta|\zeta|)\}.$$

The symbol theory for the sheaf $\mathcal{E}_X^{\mathbf{R},*}$ was deeply studied by Aoki [5,6]. Employing the Radon transformation (see Kataoka [18]), we can write a stalk of $\mathcal{E}_X^{\mathbf{R},*}$ at p also as

$$\mathcal{E}_{X\ ,p}^{\mathbf{R},*} = \lim_{\epsilon \to 0} \frac{S^*(V_\epsilon)}{R(V_\epsilon)}$$

where $* = f$ or (s).

For convenience, we set $\mathcal{C}_{Y|X}^{\mathbf{R},(1)} = \mathcal{C}_{Y|X}^{\mathbf{R}}$, $\mathcal{C}_{Y|X}^{\mathbf{R},(\infty)} = \mathcal{C}_{Y|X}^{\mathbf{R},f}$, $\mathcal{E}_X^{\mathbf{R},(1)} = \mathcal{E}_X^{\mathbf{R}}$, $\mathcal{E}_X^{\mathbf{R},(\infty)} = \mathcal{E}_X^{\mathbf{R},f}$ and so on.

2 Irregularity of a holonomic module and a Stokes set

2.0 Irregularity of a holonomic module.

In this subsection, we quickly review the definition of the irregularity of a holonomic module. There are several equivalent definitions of the irregularities. We follow Kashiwara-Kawai[13] here.

Let Λ be a regular or maximally degenerate involutive submanifold of codimension $d \geq 1$ in $\overset{\circ}{T}{}^*X$ and $\sigma = \frac{q}{p} \in \mathbf{Q}$ ($\sigma \geq 1$, $p, q \in \mathbf{N}$ prime to each other). A subsheaf $I_\Lambda \subset \mathcal{E}_X(1)$ is defined by

$$I_\Lambda = \{P \in \mathcal{E}_X(1); \sigma_1(P) \mid_\Lambda \equiv 0\}.$$

Here $\sigma_1(\bullet)$ denotes a symbol map of degree 1. Using the sheaf I_Λ, we can define a filtration $Fr_\Lambda^{(\sigma)}(\mathcal{E}_X)$ of $\mathcal{E}_X$ as follows:

$$Fr_\Lambda^{(\sigma)}(\mathcal{E}_X)(k) = \sum_{qi+(q-p)j\leq k} \mathcal{E}_X(i) I_\Lambda^j.$$

We set $\mathcal{E}_\Lambda^{(\sigma)} = Fr_\Lambda^{(\sigma)}(\mathcal{E}_X)(0)$. Remark that $\mathcal{E}_\Lambda^{(1)}$ coincides with the sheaf $\mathcal{E}_\Lambda$ in Kashiwara-Oshima [15] and [13].

We list up some main properties of $\mathcal{E}_\Lambda^{(\sigma)}$:

(1) $\mathcal{E}_X(0) \subset \mathcal{E}_\Lambda^{(\sigma)}$, and $\mathcal{E}_\Lambda^{(\sigma)}$ is a left and right $\mathcal{E}_X(0)$ module.

(2) $\mathcal{E}_\Lambda^{(\sigma)}$ is a sheaf of Noetherian ring and any coherent $\mathcal{E}_X$ module is pseudo-coherent over $\mathcal{E}_\Lambda^{(\sigma)}$.

(3) $\mathcal{E}_\Lambda^{(\sigma)}$ is stable under a quantized contact transformation.

Let $\mathcal{M}$ be a holonomic $\mathcal{E}_X$ module in a neighborhood of $p \in \overset{\circ}{T}{}^*X$. We first define the irregularity of $\mathcal{M}$ when its support is smooth.

Definition 2.0.0. *The irregularity of $\mathcal{M}$ along Λ at p is at most σ if and only if $\mathcal{M}$ satisfies one of the following equivalent conditions.*

(1) *There exist an open neighborhood U of p, a maximally degenerate involutive submanifold V with its singular locus Λ, and a coherent $\mathcal{E}_V^{(\sigma)}$ module $\mathcal{M}_0$ on U which generates $\mathcal{M}$ over $\mathcal{E}_X$ and is finitely generated over $\mathcal{E}_X(0)$ at any point of a dense subset in $\Lambda \cap U$.*

(2) *There exist an open neighborhood U of p and an $\mathcal{E}_\Lambda^{(\sigma)}$ module $\mathcal{M}_0$ on U which generates $\mathcal{M}$ over $\mathcal{E}_X$ and is $\mathcal{E}_X(0)$ coherent.*

It is easy to see there exists a unique number σ such that the above conditions (1) or (2) is satisfied for σ and that for any $\sigma' < \sigma$ (1) and (2) does not hold. We call such a number σ the irregularity of $\mathcal{M}$ along Λ.

One important difference between the condition (1) and (2) is that it is enough to estimate the irregularity along a larger submanifold than its support. Their equivalence is shown with the aid of a b-function.

Definition 2.0.1. *(1) A holonomic $\mathcal{E}_X$ module $\mathcal{M}$ has the irregularity at most σ at p if and only if there exists an open neighborhood $U \subset \overset{\circ}{T}{}^*X$ of p such that for any smooth point $p' \in \mathrm{supp}(\mathcal{M}) \cap U$ $\mathcal{M}$ has the irregularity at most σ along its support at p'.*
(2) A holonomic $\mathcal{D}_X$ module $\mathcal{N}$ has the irregularity at most σ if and only if $\mathcal{E}_X \underset{\pi_X^{-1}\mathcal{D}_X}{\otimes} \pi_X^{-1}\mathcal{N}$ has the irregularity at most σ at any point in $\overset{\circ}{T}{}^*X$.

As we see below, Definition 2.0.1 coincides with the classical definition for ordinary differential operators.

Example 2.0.2. Let $X = \mathbb{C}$, $V = \overset{\circ}{T}{}^*_{\{0\}}X = \{(x;\xi); x = 0, \xi \neq 0\}$ and $p = (0; dx)$. Let $P(x, D) = x^d D^m + a_{m-1}(x) D^{m-1} + \cdots + a_0(x)$ be a differential operator of order m. We define a rational number $\mathrm{Irr}_{V,p}(P)$ by

$$\mathrm{Irr}_{V,p}(P) = \max_{0\le k<m}\{1, \frac{d-\mathrm{Van}_{\{0\}}(a_k(x))}{m-k}\},$$

where $\mathrm{Van}_{\{0\}}(f)$ is the vanishing order of f at the origin. Then a coherent $\mathcal{E}_X$ module $\mathcal{E}_X/\mathcal{E}_XP$ has the irregularity $\mathrm{Irr}_{V,p}(P)$ along V. Conversely we have

Lemma 2.0.3. *Let $\mathcal{M}$ be a coherent $\mathcal{E}_X$ module which has the irregularity at most σ along V at p. Then for any $u \in \mathcal{M}$, there exists a differential operator P such that $Pu = 0$ and $\mathrm{Irr}_{V,p}(P) \le \sigma$. (Remark that, although $\mathcal{M}$ is an $\mathcal{E}_X$ module, we can find such an operator in the category of differential operators.)*

2.1 Stokes set.

Let $\Lambda \subset \overset{\circ}{T}{}^*X$ be a $\mathbb{C}^*$ conic Lagrangian submanifold, $E \subset \Lambda$ a $\mathbb{C}^*$ conic submanifold in Λ, and p a point in E.

Since E has a $\mathbb{C}^*$ action on its fiber, we have a canonical map induced by an Euler vector field

$$\theta_E : T^*E \to \mathbb{C}.$$

If we take a coordinate $(z', z''; \zeta', \zeta'')$ of T^*X with $E = \{z' = z'' = \zeta'' = 0\}$ and $(\zeta'; z^{*\prime})$ of T^*E, then θ_E is written as $\theta_E(\zeta', z^{*\prime}) =< \zeta', z^{*\prime} >$.

Using θ_E, we have a bundle map over E

$$\text{(2.1.0)} \qquad \begin{array}{ccc} T^*E & \longrightarrow & E\times\mathbb{C} \\ p & \longrightarrow & (\pi_E(p), \theta_E(p)) \end{array}$$

where $\pi_E : T^*E \to E$ is the canonical projection. We denote the bundle map (2.1.0) by the same symbol θ_E. An inclusion $E \hookrightarrow \Lambda$ induces canonical morphisms

$$\text{(2.1.1)} \qquad T^*E \xleftarrow{\varrho_E} E \underset{\Lambda}{\times} T^*\Lambda \xrightarrow{\varpi} T^*\Lambda.$$

Composing the morphisms (2.1.0) and (2.1.1), we have a bundle map over E as

$$\theta_{\varrho_E} : E \underset{\Lambda}{\times} T^*\Lambda \xrightarrow{\varrho_E} T^*E \xrightarrow{\theta_E} E' = E \times \mathbb{C}.$$

We also define subsets E'_+,E'_0 and E'_- in E' by

$$E'_\pm = \{(p, \tau) \in E' = E \times \mathbb{C}_\tau; \pm\Re\tau > 0\}$$

and

$$E'_0 = \{(p, \tau) \in E' = E \times \mathbb{C}_\tau; \Re\tau = 0,\ \tau \ne 0\}.$$

Let $\mathcal{M}$ be a coherent $\mathcal{E}_X$ module in a neighborhood of p. The filtration $Fr_\Lambda^{(\sigma)}(\mathcal{E}_X)$ of $\mathcal{E}_X$ induces a good filtration $Fr_\Lambda^{(\sigma)}(\mathcal{M})$ of $\mathcal{M}$ and its graded module

$$Gr_\Lambda^{(\sigma)}(\mathcal{M}) = \bigoplus_{k\in\mathbb{Z}} \frac{Fr_\Lambda^{(\sigma)}(\mathcal{M})(k+1)}{Fr_\Lambda^{(\sigma)}(\mathcal{M})(k)}.$$

We denote by $\Sigma_\Lambda^{(\sigma)}(\mathcal{M})$ a complex analytic subset of $T^*\Lambda$

$$\Sigma_\Lambda^{(\sigma)}(\mathcal{M}) = \mathrm{supp}(\mathcal{O}_{T^*\Lambda} \underset{\pi_\Lambda^{-1}Gr_\Lambda^{(\sigma)}(\mathcal{E}_X)}{\otimes} \pi_\Lambda^{-1}Gr_\Lambda^{(\sigma)}(\mathcal{M}))$$

and by $V(\mathcal{M})$ a set of irreducible components of $\Sigma_\Lambda^{(\sigma)}(\mathcal{M})$. Remark that $\Sigma_\Lambda^{(\sigma)}(\mathcal{M})$ does not depend on a choice of a good filtration $Fr_\Lambda^{(\sigma)}(\mathcal{M})$ and is not $\mathbb{C}^*$ conic along a fiber of $T^*\Lambda$ in general.

We define a subset $V_\pm(\mathcal{M},p)$ of $V(\mathcal{M})$ by

$$V_\pm(\mathcal{M},p) = \{V \in V(\mathcal{M}); {\pi'_E}^{-1}(p) \cap \theta_{\varrho_E}\varpi^{-1}(V) \subset E'_\pm\},$$

where $\pi'_E : E \times \mathbb{C} \to E$ is the first projection. Then we will introduce a subanalytic subset $St_{\Lambda,E}^{(\sigma)}(\mathcal{M})$ in E which describes Stokes lines.

Definition 2.1.0. *For a holonomic system $\mathcal{M}$ its Stokes set in E along Λ with the irregularity σ is defined by*

$$St_{\Lambda,E}^{(\sigma)}(\mathcal{M}) = \pi'_E(E'_0 \cap \theta_{\varrho_E}\varpi^{-1}(\Sigma_\Lambda^{(\sigma)}(\mathcal{M})).$$

Remark that $St_{\Lambda,E}^{(\sigma)}(\mathcal{M})$ is well behaved under a contact transformation, i.e., for a complex homogeneous canonical transformation ϕ of T^*X and its quantization Φ we have

$$\phi(St_{\Lambda,E}^{(\sigma)}(\mathcal{M})) = St_{\phi(\Lambda),\phi(E)}^{(\sigma)}(\Phi(\mathcal{M})).$$

We also introduce a number $N_{\Lambda,E}^{(\sigma)}(\mathcal{M},p)$ by

$$N_{\Lambda,E}^{(\sigma)}(\mathcal{M},p) = \sum_{V\in V_-(\mathcal{M},p)} \mathrm{mult}_V(\mathcal{O}_{T^*\Lambda} \underset{\pi_\Lambda^{-1}Gr_\Lambda^{(\sigma)}(\mathcal{E}_X)}{\otimes} \pi_\Lambda^{-1}Gr_\Lambda^{(\sigma)}(\mathcal{M})).$$

Here $\mathrm{mult}_\Lambda(\bullet)$ is the multiplicity of an $\mathcal{O}_{T^*\Lambda}$ module along Λ.

3 Microfunction solutions of holonomic systems

In this section, we study the structure of holomorphic microfunction solutions of holonomic systems. The behavior of solutions changes drastically according as the Gevrey order of microfunctions is less or greater than a critical value determined by the irregularity of the system. Thus we consider both cases in the following subsections.

3.0 Regularity theorem.

First we consider the case where the Gevrey order of holomorphic microfunctions is less than or equal to the critical value, that is, the case where the comparison theorem holds.

In this direction we know the following Theorem 3.0.0 that generalizes a result of Kashiwara-Kawai[13] (See also Gérard-Levelt[7], Malgrange[21], Ramis[23] and Komatsu[19] for some related topics for linear ordinary differential equations).

Theorem 3.0.0 (Honda[8]). *Let $U \subset T^*X$ be a $\mathbb{C}^*$ conic open set, $\mathcal{M}$ a holonomic $\mathcal{E}_X$ module in U and $\sigma \geq 1$ a rational number. Then the following conditions (1),(2),(3) and (4) are equivalent.*

(1) *There exists a holonomic $\mathcal{E}_X$ module $\mathcal{M}_{reg}$ with regular singularities satisfying*

$$\mathcal{E}_X^{(s)} \underset{\mathcal{E}_X}{\otimes} \mathcal{M} \simeq \mathcal{E}_X^{(s)} \underset{\mathcal{E}_X}{\otimes} \mathcal{M}_{reg}$$

for all $s \in [1, \frac{\sigma}{\sigma-1}]$.

(2) *For any submanifold $Y \subset X$ and any $s \in [1, \frac{\sigma}{\sigma-1}]$, we have*

$$\mathbb{R}\mathcal{H}om_{\mathcal{E}_X}(\mathcal{M}, \mathcal{C}_{Y|X}^{\mathbf{R},(s)}) \simeq \mathbb{R}\mathcal{H}om_{\mathcal{E}_X}(\mathcal{M}, \mathcal{C}_{Y|X}^{\mathbf{R}}).$$

(3) *For any point $p \in U$ and any submanifold $Y \subset X$, we have*

$$\chi(\mathbb{R}\mathcal{H}om_{\mathcal{E}_X}(\mathcal{M}, \mathcal{C}_{Y|X}^{\mathbf{R},(s)})_p) = \chi(\mathbb{R}\mathcal{H}om_{\mathcal{E}_X}(\mathcal{M}, \mathcal{C}_{Y|X}^{\mathbf{R}})_p)$$

for $s = \frac{\sigma}{\sigma-1}$. Here χ denotes the Euler index of the complex.

(4) *$\mathcal{M}$ has the irregularity at most σ in U.*

Let us note some intuitive meaning of this theorem; the implication "(4) $\Rightarrow$ (1)" means any holonomic module (with irregular singularities) can be transformed into a regular singular holonomic module by a microdifferential operator of Gevrey growth order corresponding to the irregularity of the module, and the implication "(2) $\Rightarrow$ (4)" implies the irregularity of a holonomic module is controlled by the growth order of its solutions. We also note that the existence of $\mathcal{M}_{reg}$ stated in (1) is quite effective in studying the growth order of solutions.

Corollary 3.0.1. *Let M be a real analytic manifold with its complexification X and $\mathcal{M}$ a holonomic $\mathcal{E}_X$ module at $p \in T^*X$ with the irregularity at most σ. Then we have an isomorphism for each $s \in [1, \frac{\sigma}{\sigma-1}]$*

$$\mathbb{R}\mathcal{H}om_{\mathcal{E}_X}(\mathcal{M}, \mathcal{C}_M^{(s)})_p \simeq \mathbb{R}\mathcal{H}om_{\mathcal{E}_X}(\mathcal{M}, \mathcal{C}_M)_p,$$

where $\mathcal{C}_M$ (resp. $\mathcal{C}_M^{(s)}$) is the sheaf of microfunctions (resp. microfunctions of Gevrey class (s)).

If $\mathcal{M}$ is with regular singularities and if the solution sheaf is tempered microfunctions, Corollary 3.0.1 was shown by Andronikof [3].

If we assume the support of $\mathcal{M}$ is a complexification of a real Lagrangian manifold in T_M^*X at its smooth points, we have the same regularity theorem as Theorem 3.0.0 replacing the solution sheaves $\mathcal{C}_{Y|X}^{\mathbb{R}}$ and $\mathcal{C}_{Y|X}^{\mathbb{R},(s)}$ with sheaves of microfunctions $\mathcal{C}_M$ and $\mathcal{C}_M^{(s)}$ respectively.

In particular, we can recover the following result obtained by Komatsu [19].

Theorem 3.0.2 (Komatsu[19]). *Let P be a differential operator and $\sigma > 1$ a rational number. The following conditions (1) and (2) are equivalent.*

(1) *For $s = \frac{\sigma}{\sigma-1}$, we have*

$$\dim \operatorname{Ker}(P : \mathcal{B} \to \mathcal{B}) = \dim \operatorname{Ker}(P : \mathcal{D}^{(s)} \to \mathcal{D}^{(s)}),$$

where $\mathcal{B}$ (resp. $\mathcal{D}^{(s)}$) is the sheaf of hyperfunctions (resp. ultradistributions).

(2) *The irregularity of the differential operator P (in the classical sense) is σ .*

3.1 Example.

As we have seen in Theorem 3.0.0, if a holonomic system $\mathcal{M}$ is with regular singularities, we have

$$\mathbb{R}\mathcal{H}om_{\mathcal{E}_X}(\mathcal{M}, \mathcal{C}_{Y|X}^{\mathbb{R},f}) = \mathbb{R}\mathcal{H}om_{\mathcal{E}_X}(\mathcal{M}, \mathcal{C}_{Y|X}^{\mathbb{R}}). \tag{3.1.0}$$

It is also well known that the complex in the right-hand side of (3.1.0) is cohomologically $\mathbb{C}$ constructible. However, if $\mathcal{M}$ is not with regular singularities and if we replace the solution sheaf $\mathcal{C}_{Y|X}^{\mathbb{R},f}$ of (3.1.0) with $\mathcal{C}_{Y|X}^{\mathbb{R},(s)}$ where Gevrey order (s) is greater than the critical value, the equivalence (3.1.0) dose not hold. Thus the solution complex may not be $\mathbb{C}$ constructible in general even though the support of $\mathcal{M}$ is smooth.

In fact, Example 3.1.0 below indicates that its structure is related to Stokes lines.

Example 3.1.0. Let $X = \mathbb{C}$ with a coordinate (z) and consider the system of the Bessel equation at ∞;

$$\mathcal{M} : \left(D_z - \begin{pmatrix} 0 & 1 \\ -1 + \nu^2 z^{-2} & -z^{-1} \end{pmatrix}\right) u = 0,$$

where we assume $\nu \notin \mathbb{Z}$. The Bessel equation has two linearly independent solutions $H_\nu^{(i)}(z)$ $(i = 1, 2)$ known as Hankel functions. The system $\mathcal{M}$ is equivalent to the following reference system $\mathcal{N}$

$$\mathcal{N} : \left(D_z - \begin{pmatrix} i & 0 \\ 0 & -i \end{pmatrix} \right) v = 0$$

by a transformation $u = H(z)v$ where

$$H(z) = \begin{pmatrix} H_\nu^{(1)}(z) & H_\nu^{(2)}(z) \\ \frac{\partial}{\partial z} H_\nu^{(1)}(z) & \frac{\partial}{\partial z} H_\nu^{(2)}(z) \end{pmatrix} \begin{pmatrix} e^{-iz} & 0 \\ 0 & e^{iz} \end{pmatrix}. \tag{3.1.1}$$

The theory of special functions says:

(P.1) $H_\nu^{(1)}(z)e^{-iz}$ has the following asymptotic expansion near ∞ in the sector $(-\pi, 2\pi)$.

$$H_\nu^{(1)}(z)e^{-iz} = \sqrt{\frac{2}{\pi z}} e^{-i\frac{2\nu\pi+1}{4}}(1 + O(z^{-1})) \quad \arg(z) \in (-\pi, 2\pi). \tag{3.1.2}$$

$H_\nu^{(2)}(z)e^{iz}$ has a similar asymptotic expansion in the sector $(-2\pi, \pi)$. Thus the matrix $H(z)$ and $H(z)^{-1}$ are asymtotically expanded near ∞ in the sector $S = \{z \in \widetilde{\mathbb{C}}^*; \arg(z) \in (-\pi, \pi)\}$ where $p_{\mathbb{C}^*} : \widetilde{\mathbb{C}}^* \to \mathbb{C}^*$ is the universal covering space of $\mathbb{C}^*$. This implies, in particular, $H(z)$ and $H^{-1}(z)$ have polynomial growth when $|z| \to \infty$ in S.

(P.2) On the other hand, by the following well known relations $(m = \pm 1, 2, \cdots)$

$$H_\nu^{(1)}(e^{m\pi i}z) = \frac{\sin(m-1)\nu\pi}{\sin \nu\pi} H_\nu^{(1)}(z) - e^{-\nu\pi i}\frac{\sin m\nu\pi}{\sin \nu\pi} H_\nu^{(2)}(z)$$

and

$$H_\nu^{(2)}(e^{m\pi i}z) = \frac{\sin m\nu\pi}{\sin \nu\pi} H_\nu^{(1)}(z) + e^{-\nu\pi i}\frac{\sin(m+1)\nu\pi}{\sin \nu\pi} H_\nu^{(2)}(z),$$

$||H(z)||$ grows exponentially outside the closure of S near ∞.

Thus the behavior of $H(z)$ changes drastically across the half lines $\arg(z) = \pm\pi$ which are classically called Stokes lines.

These Stokes lines can also be observed in microlocal category as follows. Let $\overset{\circ}{T}{}^*{}_{\{0\}}X = \mathbb{C}_\zeta^*$ be a conormal bundle of the origin (without the zero section). Denote by $\widetilde{\mathcal{E}_X}^{\mathbf{R},(s)}$ the subsheaf of operators with constant coefficients of $\mathcal{E}_X^{\mathbf{R},(s)}$, that is,

$$\widetilde{\mathcal{E}_X}^{\mathbf{R},(s)} = \{P \in \mathcal{E}_X^{\mathbf{R},(s)}; [D_z, P] = 0\}|_{\overset{\circ}{T}{}^*{}_{\{0\}}X}.$$

Consider the following holonomic $\mathcal{E}_X$ modules $\mathcal{M}$ and $\mathcal{N}$:

$$\mathcal{M} = \frac{\mathcal{E}_X^2}{\mathcal{E}_X^2\left(zD_z + \frac{1}{2}\begin{pmatrix} \frac{1}{2} & D_z \\ -1+\nu^2 D_z^{-1} & -\frac{3}{2}\end{pmatrix}\right)},$$

$$\mathcal{N} = \frac{\mathcal{E}_X^2}{\mathcal{E}_X^2\left(zD_z + \frac{1}{4}\begin{pmatrix} 1 & 2D_z \\ -2 & -1\end{pmatrix}\right)}.$$

The system $\mathcal{N}$ is a little more complicated than the corresponding classical one. If we employ a transformation in $\mathcal{E}_X^{\mathbb{R},f}$, however, we have

$$\mathcal{E}_X^{\mathbb{R},f} \underset{\mathcal{E}_X}{\otimes} \mathcal{N} = \frac{\mathcal{E}_X^{\mathbb{R},f^2}}{\mathcal{E}_X^{\mathbb{R},f^2}\left(zD_z + \frac{1}{2}\begin{pmatrix} i & 0 \\ 0 & -i\end{pmatrix} D_z^{\frac{1}{2}}\right)}.$$

Remark that the both systems $\mathcal{M}$ and $\mathcal{N}$ have the irregularity 2 along $\mathring{T}^*{}_{\{0\}}X$. Let us consider a transformation

$$W(D_z) = \begin{pmatrix} D_z^{\frac{1}{4}} & 0 \\ 0 & D_z^{-\frac{1}{4}}\end{pmatrix} H(D_z^{\frac{1}{2}}) \begin{pmatrix} D_z^{-\frac{1}{4}} & -iD_z^{\frac{1}{4}} \\ -iD_z^{-\frac{1}{4}} & D_z^{\frac{1}{4}}\end{pmatrix},$$

where the matrix H was defined in (3.1.1).

The matrix $W(D_z)$ belongs to $\Gamma(\widetilde{\mathbb{C}}^*, p_{\mathbb{C}^*}^{-1}\widetilde{\mathcal{E}_X}^{\mathbb{R},(2)})$ (i.e. a multivalued microdifferential operator of Gevrey class (2)). Since $H(z)$ gives an equivalence of the Bessel equation and its reference system, $W(D_z)$ also gives that of the systems $\mathcal{M}$ and $\mathcal{N}$ by the same calculations. Moreover the properties (P.1) and (P.2) of $H(z)$ imply

(3.1.3) $W(D_z)_\zeta \in p_{\mathbb{C}^*}^{-1}\widetilde{\mathcal{E}_X}^{\mathbb{R},f}$ if $\zeta \in S_1 = \{\zeta \in \widetilde{\mathbb{C}}^*; -2\pi < \arg(\zeta) < 2\pi\}$,

(3.1.4) $W(D_z)_\zeta \notin p_{\mathbb{C}^*}^{-1}\widetilde{\mathcal{E}_X}^{\mathbb{R},f}$ if $\arg(\zeta) = \pm 2\pi$.

Now we identify $(\mathring{T}^*_{\{0\}}X \times \mathring{T}^*_{\{0\}}X) \cap \mathring{T}^*_{\Delta_X}(X \times X) \simeq \mathring{T}^*_{\{0\}}X = \mathbb{C}^*_\zeta$ where Δ_X is the diagonal set of $X \times X$ and denote by $\mathcal{M}^*$ the dual system of $\mathcal{M}$. Set $\mathcal{M}^{\mathbb{R}} = \mathcal{E}_X^{\mathbb{R}} \underset{\mathcal{E}_X}{\otimes} \mathcal{M}$, $\mathcal{M}^{\mathbb{R},f} = \mathcal{E}_X^{\mathbb{R},f} \underset{\mathcal{E}_X}{\otimes} \mathcal{M}$, etc. We obtain a holomorphic microfunction solution $W(D_z)$ of the system $\mathcal{M} \boxtimes \mathcal{N}^*$ since we have

(3.1.5) $$\mathbb{R}\mathcal{H}om_{\mathcal{E}_X^{\mathbb{R}}}(\mathcal{M}^{\mathbb{R}}, \mathcal{N}^{\mathbb{R}}) \simeq \mathbb{R}\mathcal{H}om_{\mathcal{E}_{X\times X}}(\mathcal{M} \boxtimes \mathcal{N}^*, \mathcal{C}^{\mathbb{R}}_{\Delta_X|X\times X})[1].$$

Denote by $\mathcal{L}$ the tempered solution complex of $\mathcal{M} \boxtimes \mathcal{N}^*$

$$\mathbb{R}\mathcal{H}om_{\mathcal{E}_{X\times X}}(\mathcal{M} \boxtimes \mathcal{N}^*, \mathcal{C}^{\mathbb{R},f}_{\Delta_X|X\times X})[1].$$

Remark that the complex $\mathcal{L}$ is concentrated in degree 0. By the properties (3.1.3), (3.1.4) and the uniqueness of the division for $\mathcal{E}_X^{\mathbb{R},f}$, we finally obtain

(3.1.6) $W(D_z) \in \mathrm{H}^0\,\mathbb{R}\Gamma(S_1, p_{\mathbb{C}^*}^{-1}\mathcal{L})$, and

(3.1.7) $W(D_z)_\zeta \notin \mathrm{H}^0\, p_{\mathbb{C}^*}^{-1}\mathcal{L}$ if $\arg(\zeta) = \pm 2\pi$.

Thus we have

$$\overset{\circ}{\pi}_{\overset{\circ}{T}{}^*_{\{0\}}X}(\mathrm{SS}(\mathcal{L})|_{\overset{\circ}{T}{}^*\overset{\circ}{T}{}^*_{\{0\}}X}) \supset \mathbb{R}^+, \tag{3.1.8}$$

where $\overset{\circ}{\pi}_{\overset{\circ}{T}{}^*_{\{0\}}X} : \overset{\circ}{T}{}^*\overset{\circ}{T}{}^*_{\{0\}}X \to \overset{\circ}{T}{}^*_{\{0\}}X$ is the canonical projection.

As a conclusion, the (essential) micro-support of the complex $\mathcal{L}$ should contain the positive real half line which corresponds to the classical Stokes lines of the Bessel system.

3.2 Micro-support of microfunction solutions and Stokes sets.

In this subsection, we show the micro-support of solutions of a holonomic system with irregular singularities is estimated by the Stokes set introduced in section 2.

Let $p \in \overset{\circ}{T}{}^*X$, $\Lambda, \Lambda_0 \subset \overset{\circ}{T}{}^*X$ be $\mathbb{C}^*$ conic complex Lagrangian submanifolds intersecting cleanly at p and $\mathcal{M}$ a holonomic $\mathcal{E}_X$ module in a neighborhood of p with $\mathrm{Supp}(\mathcal{M}) \subset \Lambda$. Let σ be the irregularity of $\mathcal{M}$, $E = \Lambda \cap \Lambda_0$ and d the complex codimension of E in Λ_0.

We introduce the following two conditions (C.1) and (C.2).

(C.1)

$$\pi_E'^{-1}(p) \cap \theta_{\varrho_E}\varpi^{-1}(\Sigma_\Lambda^{(\sigma)}(\mathcal{M})) \cap E \times \{0\} = \phi.$$

(Roughly speaking, this condition guarantees that there is no complex degeneration at p).

(C.2) There exists a good filtration $Fr_\Lambda^{(\sigma)}(\mathcal{M})$ of $\mathcal{M}$ satisfying $(\pi_\Lambda)_*(\mathcal{O}_{T^*\Lambda} \underset{\pi_\Lambda^{-1}Gr_\Lambda^{(\sigma)}(\mathcal{E}_X)}{\otimes} \pi_\Lambda^{-1}Gr_\Lambda^{(\sigma)}(\mathcal{M}))$ is a free $\mathcal{O}_\Lambda$ module.

Let $\mathcal{N}$ be a simple holonomic system along Λ_0. Set

$$\mathcal{S} = \mathbb{R}\mathcal{H}om_{\mathcal{E}_X}(\mathcal{M}, \mathcal{N}^{\mathbb{R},f}) = \mathbb{R}\mathcal{H}om_{\mathcal{E}_X^{\mathbb{R},f}}(\mathcal{M}^{\mathbb{R},f}, \mathcal{N}^{\mathbb{R},f}).$$

Theorem 3.2.0. *Assume that a holonomic system $\mathcal{M}$ is not with regular singularities (i.e. its irregularity $\sigma > 1$) and that it satisfies the conditions (C.1) and (C.2). Then we have:*

(1) *The complex $\mathcal{S}$ is concentrated in degree d and has $\mathbb{R}$ constructible cohomologies.*

(2) *The (essential) micro-support of $\mathcal{S}$ is related to the Stokes set in the following manner:*

$$\overset{\circ}{\pi}_E(\mathrm{SS}(\mathcal{S})|_{\overset{\circ}{T}{}^*E}) = St_{\Lambda,E}^{(\sigma)}(\mathcal{M}).$$

(3) *The Euler index $\chi(\mathcal{S}_p)$ of the complex $\mathcal{S}_p$ is*

$$\chi(\mathcal{S}_p) = (-1)^d \dim_{\mathbb{C}}(\mathrm{H}^d(\mathcal{S}_p)) = (-1)^d(\mathrm{mult}_\Lambda(\mathcal{M}) - N_{\Lambda,E}^{(\sigma)}(\mathcal{M}, p)).$$

Remark. (1) In the one dimensional case, we have more precise results without assuming the conditions (C.1) and (C.2). For example, the statement (2) in Theorem 3.2.0 turns out

$$\mathring{\pi}_E(\mathrm{SS}(\mathcal{S})|_{\mathring{T}^*E}) = \cup_{i=0}^{l} St_{\Lambda,E}^{(\sigma_i)}(\mathcal{M}),$$

where σ_i is the i-th irregularity of the system (i.e. one corresponding to the i-th face of the Newton polygon).
(2) For holomorphic microfunction solutions of Gevrey classes, we can obtain a similar result.

Example 3.2.1. Let us reconsider the Example 3.1.0. By an easy calculation, we have

$$St_{\Lambda,E}^{(\sigma)}(\mathcal{M} \boxtimes \mathcal{N}^*) = \mathbb{R}^+ \subset \mathring{T}^*_{\{0\}}X.$$

Here $\Lambda = \mathring{T}^*_{\{0\}}(X \times X)$ and $E = \Lambda \cap \mathring{T}^*_{\Delta_X}(X \times X) \simeq \mathring{T}^*_{\{0\}}X$. Therefore we have

$$\mathring{\pi}_{\mathring{T}^*_{\{0\}}X}(\mathrm{SS}(\mathbb{R}\mathcal{H}om_{\mathcal{E}_{X\times X}}(\mathcal{M} \boxtimes \mathcal{N}^*, \mathcal{C}^{\mathbb{R},f}_{\Delta_X|X\times X}))|_{\mathring{T}^*\mathring{T}^*_{\{0\}}X}) = \mathbb{R}^+.$$

For the microlocal Bessel equation, the micro-support of a tempered microfunction solution exactly corresponds to the positive real half line, which comes from the classical Stokes lines of the Bessel equation.

Theorem 3.2.0 is only a starting point of our study in this direction. It is an important problem to investigate precisely the relationship between the micro-support of a solution and the Stokes set in a general case. A more important and interesting problem is to extract informations of Stokes connection coefficients from tempered microfunction solutions. These topics will be treated in our forthcoming paper.

References

[1] E. Andronikof, *Microlocalisation tempérée des distributions et des fonctions holomorphes I*, C.R. Acad.Sci **303** (1986), 347-350.

[2] ——, *Microlocalisation tempérée des distributions et des fonctions holomorphes II*, C.R. Acad. Sci **304** (1987), 511-514.

[3] ——, *On the C^∞-singularities of regular holonomic distributions*, Ann. Inst. Fourier **42** (1992), 695-704.

[4] T.Aoki, *Growth order of microdifferential operators of infinite order*, J. Fac. Sci. Univ. Tokyo, Sect IA **29** (1982), 143-159.

[5] ——, *Exponential calculus of microdifferential operators of infinite order I*, Ann. Inst. Fourier **33** (1983), 227-250.

[6] ———, *Exponential calculus of microdifferential operators of infinite order II*, Ann. Inst. Fourier **36** (1986), 143-165.

[7] R.Gérard, A.H.M Levelt, *Invariants mesurant l'irrégularité en un point singulier des systèmes d'équations diférentielles linéaires.*, Ann. Inst. Fourier **23** (1973), 157-195.

[8] N.Honda, *On the reconstruction theorem of holonomic modules in Gevrey classes*, Publ. RIMS, Kyoto Univ. **27** (1991), 923-943.

[9] ———, *Regularity theorems for holonomic modules*, in preparation.

[10] M.Kashiwara, *On the maximally overdetermined systems of linear differential equations I*, Publ. RIMS, Kyoto Univ. **10** (1975), 563-579.

[11] ———, *On the holonomic systems of linear differential equations II*, Inventiones Math. **49** (1978), 121-135.

[12] ———, *The Riemann-Hilbert problem for holonomic systems*, Publ. RIMS Kyoto Univ. **20** (1984), 319-365.

[13] M.Kashiwara, T.Kawai, *On the holonomic systems of microdifferential equations III*, Publ. RIMS, Kyoto Univ. **17** (1981), 813-979.

[14] ———, *Second microlocalization and asymptotic expansions*, Lect. Notes in Phys. **126**, Springer-Verlag, 1980, pp. 21-76.

[15] M.Kashiwara, T.Oshima, *Systems of differential equations with regular singularities and their boundary value problems*, Ann. of Math. **106** (1977), 145-200.

[16] M.Kashiwara, P.Schapira, *Microlocal study of sheaves*, Astérisque **128** (1985).

[17] ———, *Sheaves on manifolds*, Grundlehren der Math. **292**, Springer-Verlag, 1990.

[18] K.Kataoka, *On the theory of Radon transformations of hyperfunctions*, J. Fac. Sci. Univ. Tokyo, Sect IA **28** (1981), 331-413.

[19] H.Komatsu, *On the regularity of hyperfunction solutions of linear ordinary differential equations with real analytic coefficients*, J. Fac. Sci. Univ. Tokyo, Sect IA **20** (1973), 107-119.

[20] Y. Laurent, *Théorie de la deuxième microlocalisation dans le domaine complexe*, Progress in Math. **53**, Birkhäuser, 1985.

[21] B.Malgrange, *Sur les points singuliers des équations différentielles*, Enseignement Math. **20** (1974), 147-176.

[22] J.-P. Ramis, *Devissage Gevrey*, Astérisque **59-60** (1978), 173-204.

[23] M.Sato, T.Kawai, M.Kashiwara, *Microfunctions and pseudodifferential equations*, Lecture Notes in Math. **287**, Springer-Verlag, 1973, pp. 265-529.

[24] P.Schapira, *Microdifferential systems in the complex domain*, Grundlehren der Math. **269**, Springer-Verlag, 1985.

Some algorithmic aspects of the D-module theory

Toshinori Oaku

Department of Mathematical Sciences, Yokohama City University, 22-2 Seto, Kanazawa-ku, Yokohama, Kanagawa 236, Japan

Abstract We consider D-modules from an algorithmic point of view. Our aim is to present algorithms for computing some invariants attached to a D-module, such as the characteristic variety, the multiplicity, the b-function (or exponents), and the induced system. In particular, we obtain algorithms for computing the classical b-function (Bernstein-Sato polynomial) and D-modules associated with an arbitrary polynomial.

1 Introduction—What are computable in the D-module theory?

The theory of D-modules, i.e., modules over the ring of differential operators, was initiated in the complex analytic category by Sato, Kashiwara, Kawai ([K1], [SKK]), and was developed extensively by themselves and mainly by French and Japanese schools (see e.g., [Bj], [S], [K4]). On the other hand, the theory of modules over the Weyl algebra (i.e. the ring of differential operators with polynomial coefficients) was founded by Bernstein ([Be]) and generalized to the theory of algebraic D-modules (cf. [Bj], [Bo]).

The aim of the present paper is to show that the notion and the algorithm of Gröbner basis by Buchberger ([Bu], [CLO], [BW]) for the polynomial ring can be generalized to rings of differential operators to yield algorithms of computing some invariants of D-modules. The application of Gröbner bases to rings of differential operators was initiated by Galligo [G], Castro [C], Noumi [N], Takayama [T1], [T2], [T3]. Especially, Takayama has been developing computer programs for Gröbner basis computation for rings of differential operators (*Kan* [T4] and *Macaulay for D-modules*). There is also a (limited) implementation by Shimoyama-Oaku on a computer algebra system *Risa/Asir* of Noro et al. ([NT]).

We shall use the word *algorithm* in a strict sense; i.e., given a finite set of data (input) an *algorithm* returns an answer (output) as a finite set of data, or else determines that there is no answer, in a finite number of steps using a finite amount of memory in the computation. However, the number of steps and the amount of the memory needed in the computation (i.e. the complexity of the algorithm) may not be estimated in advance. This definition probably coincides with the notion of computation by the Turing machine.

Since we can handle only a finite set as data, it would be reasonable to restrict our attention to modules over the Weyl algebra (or more generally, modules over the ring of algebraic differential operators on a smooth algebraic variery). More precisely, We shall consider only the analytic $\mathcal{D}$-modules of the form $\mathcal{D} \otimes_{A_n} M$ with M being a finitely generated left A_n-module and $\mathcal{D}$ the sheaf of analytic differential operators on $\mathbf{C}^n$.

Let M be a finitely generated (left) module over the Weyl algebra A_n. We assume that a presentation of M is given, i.e., a finite set of generators of a left submodule $N \subset (A_n)^r$ is given so that $M = (A_n)^r/N$. Put $\mathcal{M} := \mathcal{D} \otimes_{A_n} M$. Then there are algorithms of computing

(1) the characteristic variety of $\mathcal{M}$ ([O1]);
(2) the multiplicity of $\mathcal{M}$ at a given point of the characteristic variety (by combining [O1] with Mora [Mo] or Lazard [La]);
(3) a free resolution of M of length $\leq 2n$ and $Ext^i_{A_n}(M, A_n)$. (F.-O. Schreyer (cf. [E]) for the polynomial ring, and Takayama (unpublished) for A_n);
(4) a (global) free resolution of $\mathcal{M}$ of length $\leq 2n$ and $Ext^i_{\mathcal{D}}(\mathcal{M}, \mathcal{D})$ (by (3), cf. also [O3]).

Assume that $\mathcal{M}$ is *specializable* with respect to a hyperplane Y of $\mathbf{C}^n$. Then we have algorithms of computing

(5) the b-function (or the indicial equation) of $\mathcal{M}$ along Y ([O6], [O7]);
(6) the cohomology groups of the induced system (or the inverse image) of $\mathcal{M}$ to Y ([O7];
(7) the micro-characteristic cycle of type $\{\infty\}$ in the sense of Laurent (see [L1], [LS] for the definition);

As applications, given an arbitrary polynomial $f(x)$, we obtain algorithms for computing

(8) the b-function (or the Bernstein-Sato polynomial) of f ([O6]);
(9) the left $\mathcal{D}[s]$-module $\mathcal{D}[s]f^s$ ([O8]);
(10) the algebraic local cohomology group $\mathcal{H}^1_{[Z]}(\mathcal{O})$ with $Z := \{x \in \mathbf{C}^n \mid f(x) = 0\}$ with coefficients in the sheaf of rings of holomorphic functions $\mathcal{O}$ ([O7]).

There is also an algorithm of computing

(11) the characteristic variety of an algebraic D-module on a smooth affine algebraic variety over an algebraically closed field of characteristic zero ([O5]).

In the present paper, we shall discuss algorithms for (1)–(6) and (8)–(10).

2 Gröbner bases for modules over the Weyl algebra and free resolution

We work in the affine space $\mathbf{C}^n$ with coordinate system $x = (x_1, \ldots, x_n)$ and write $\partial = (\partial_1, \ldots, \partial_n)$ with the derivations $\partial_i := \partial/\partial x_i$. We use the following two kinds of rings of differential operators:

(1) *the Weyl algebra* $A_n := \mathbf{C}[x_1, \ldots, x_n]\langle \partial_1, \ldots, \partial_n \rangle$,
(2) *the ring of analytic differential operators* $\mathcal{D}_n := \mathbf{C}\{x_1, \ldots, x_n\}\langle \partial_1, \ldots, \partial_n \rangle$.

Put $X := \mathbf{C}^n$ and let us denote by $\mathcal{D} = \mathcal{D}_X$ the sheaf of rings of analytic differential operators on X. The stalk of $\mathcal{D}$ at the origin $0 \in \mathbf{C}^n$ coincides with $\mathcal{D}_n$, and the stalk $\mathcal{D}_p$ of $\mathcal{D}$ at an aribtrary point $p \in X$ is isomorphic to $\mathcal{D}_n$.

In the sequel, we recall the theory and algorithm of Gröbner bases for a submodule of a finitely generated free module over A_n. The following argument applies without any modification to the Weyl algebra $A_n(K)$ over an arbitrary field K of characteristic zero. The Gröbner basis theory is also possible for $\mathcal{D}_n$ as was shown by Castro [C] (cf. also [O3]).

We fix a total order $\prec$ of $\mathbf{N}^{2n} \times \{1, \ldots, r\}$ with $\mathbf{N} := \{0, 1, 2, \ldots\}$ that satisfies the following conditions:

(A-1) $(\alpha, i) \prec (\beta, j)$ implies $(\alpha + \gamma, i) \prec (\beta + \gamma, j)$ for any $\alpha, \beta, \gamma \in \mathbf{N}^{2n}$ and $i, j \in \{1, \ldots, r\}$;
(A-2) $(0, i) \preceq (\alpha, i)$ for any $\alpha \in \mathbf{N}^{2n}$ and $i \in \{1, \ldots, r\}$.

These conditions guarantee that $\prec$ is a well-order.

An element $\mathbf{P} = (P_1, \ldots, P_r)$ of $(A_n)^r$ is written uniquely as

$$\mathbf{P} = \sum_{i=1}^{r} \sum_{\alpha, \beta \in \mathbf{N}^n} a_{i\alpha\beta} x^\alpha \partial^\beta \mathbf{e}_i$$

with $a_{i\alpha\beta} \in \mathbf{C}$ and $\mathbf{e}_1 = (1, 0, \ldots, 0), \ldots, \mathbf{e}_r = (0, \ldots, 0, 1)$, where we use the notation $x^\alpha = x_1^{\alpha_1} \ldots x_n^{\alpha_n}$, $\partial^\beta = \partial_1{}^{\beta_1} \ldots \partial_n{}^{\beta_n}$ for $\alpha = (\alpha_1, \ldots, \alpha_n)$ and $\beta = (\beta_1, \ldots, \beta_n)$. Then we define the *leading exponent* $\operatorname{lexp}(\mathbf{P})$ and the *leading coeffcient* $\operatorname{lcoef}(\mathbf{P})$ of $\mathbf{P}$ by

$$\begin{aligned} \operatorname{lexp}(\mathbf{P}) &:= \max{}_\prec \{(\alpha, \beta, i) \in \mathbf{N}^{2n} \times \{1, \ldots, r\} \mid a_{i\alpha\beta} \neq 0\}, \\ \operatorname{lcoef}(\mathbf{P}) &:= a_{i\alpha\beta} \quad \text{with } (\alpha, \beta, i) := \operatorname{lexp}(\mathbf{P}), \end{aligned}$$

where $\max_\prec$ denotes the maximum element with respect to the order $\prec$ (we assume $\mathbf{P} \neq 0$). If $\operatorname{lexp}(\mathbf{P}) = (\alpha, i)$, we call $\operatorname{lp}(\mathbf{P}) := i$ the *leading point* of $\mathbf{P}$.

Let N be a left A_n-submodule of $(A_n)^r$. Then the set $E(N)$ of leading exponents of N is defined by

$$E(N) := \{\operatorname{lexp}(\mathbf{P}) \mid \mathbf{P} \in N \setminus \{0\}\} \subset \mathbf{N}^{2n} \times \{1, \ldots, r\}.$$

Definition 2.1 (Gröbner basis) A finite subset $\mathbf{G}$ of a left A_n-submodule N of $(A_n)^r$ is called a *Gröbner basis* of N (with respect to the order $\prec$) if

$$E(N) = \bigcup_{\mathbf{P} \in \mathbf{G}} (\operatorname{lexp}(\mathbf{P}) + \mathbf{N}^{2n})$$

holds, where we put

$$(\alpha, \beta, \nu) + \mathbf{N}^{2n} = \{(\alpha + \alpha', \beta + \beta', \nu) \mid \alpha', \beta' \in \mathbf{N}^n\}.$$

Proposition 2.2 *Let N, M be left A_n-submodules of $(A_n)^r$ such that $N \subset M$. Then $N = M$ if and only if $E(N) = E(M)$.*

In general, for vectors $\alpha = (\alpha_1, \ldots, \alpha_m)$ and $\beta = (\beta_1, \ldots, \beta_m)$ in $\mathbf{N}^m$, we put

$$\alpha \vee \beta := (\max\{\alpha_1, \beta_1\}, \ldots, \max\{\alpha_m, \beta_m\})$$

and $(\alpha, i) \vee (\beta, i) := (\alpha \vee \beta, i)$ for $i \in \{1, \ldots, r\}$.

Definition 2.3 (S-operator) For $\mathbf{P}, \mathbf{Q} \in (A_n)^r$, put $\mathrm{lexp}(\mathbf{P}) = (\alpha, \beta, i) \in \mathbf{N}^n \times \mathbf{N}^n \times \{1, \ldots, r\}$ and $\mathrm{lexp}(\mathbf{Q}) = (\alpha', \beta', j)$. Then the *S-operator* of $\mathbf{P}$ and $\mathbf{Q}$ is defined by

$$\mathrm{sp}(\mathbf{P}, \mathbf{Q}) := \mathrm{lcoef}(\mathbf{Q}) x^{\beta \vee \beta' - \beta} \partial^{\alpha \vee \alpha' - \alpha} \mathbf{P} - \mathrm{lcoef}(\mathbf{P}) x^{\beta \vee \beta' - \beta'} \partial^{\alpha \vee \alpha' - \alpha'} \mathbf{Q}$$

if $i = j$, and $\mathrm{sp}(\mathbf{P}, \mathbf{Q}) := 0$ if $i \neq j$.

Theorem 2.4 ([T1]) *Let $\mathbf{G} = \{\mathbf{P}_1, \ldots, \mathbf{P}_s\}$ be a finite subset of $(A_n)^r$ which generates a left A_n-submodule N of $(A_n)^r$. Then the following two conditions are equivalent:*

(1) $\mathbf{G}$ is a Gröbner basis of N;
(2) For any $i, j \in \{1, \ldots, s\}$ such that $i < j$ and that $\mathrm{lp}(\mathbf{P}_i) = \mathrm{lp}(\mathbf{P}_j)$, there exist $Q_{ij1}, \ldots, Q_{ijs} \in A_n$ so that

$$\mathrm{sp}(\mathbf{P}_i, \mathbf{P}_j) = \sum_{k=1}^{s} Q_{ijk} \mathbf{P}_k$$

and that $Q_{ijk} = 0$ or $\mathrm{lexp}(Q_{ijk}\mathbf{P}_k) \prec \mathrm{lexp}(\mathbf{P}_i) \vee \mathrm{lexp}(\mathbf{P}_j)$ for each k.

The condition (2) of this theorem provides the Buchberger algorithm of computing a Gröbner basis from a given set of generators.

Definition 2.5 For $\mathbf{P}_1, \ldots, \mathbf{P}_s \in (A_n)^r$, their (first) *syzygy module* is defined by

$$S(\mathbf{P}_1, \ldots, \mathbf{P}_s) := \{(Q_1, \ldots, Q_s) \in (A_n)^s \mid \sum_{j=1}^{s} Q_j \mathbf{P}_j = 0\}.$$

This is a left A_n-submodule of $(A_n)^s$.

The following theorem, which was first noticed and implemented in his *Macaulay for D-modules* by Takayama, is a generalization of a theorem of F.-O. Schreyer for the polynomial ring ([E]).

Theorem 2.6 *Let $\mathbf{G} = \{\mathbf{P}_1, \ldots, \mathbf{P}_s\}$ be a Gröbner basis of a left A_n-submodule N of $(A_n)^r$. Take Q_{ijk} satisfying the condition (2) of Theorem 2.4. Put* $\mathrm{lexp}(\mathbf{P}_i) = (\alpha^{(i)}, \beta^{(i)}, \nu_i)$ *and*

$$S_{ji} := \mathrm{lcoef}(\mathbf{P}_i) x^{\beta^{(i)} \vee \beta^{(j)} - \beta^{(i)}} \partial^{\alpha^{(i)} \vee \alpha^{(j)} - \alpha^{(i)}},$$

$$\mathbf{V}_{ij} := (0, \ldots, \overset{(i)}{S_{ji}}, \ldots, \overset{(j)}{-S_{ij}}, \ldots, 0) - (Q_{ij1}, \ldots, Q_{ijs}) \in (A_n)^s$$

if $\nu_i = \nu_j$. *Then the syzygy module* $S(\mathbf{P}_1, \ldots, \mathbf{P}_s)$ *is generated by* $\mathbf{V} := \{\mathbf{V}_{ij} \mid 1 \leq i < j \leq s,\ \nu_i = \nu_j\}$. *Moreover,* $\mathbf{V}$ *is a Gröbner basis of* $S(\mathbf{P}_1, \ldots, \mathbf{P}_s)$ *with respect to an appropriate order on* $\mathbf{N}^{2n} \times \{1, \ldots, s\}$ *satisfying (A-1), (A-2).*

By applying this theorem repeatedly, we can construct a free resolution

$$0 \longleftarrow M \longleftarrow (A_n)^{r_0} \xleftarrow{\psi_1} (A_n)^{r_1} \longleftarrow \cdots \longleftarrow (A_n)^{r_{2n-1}} \xleftarrow{\psi_{2n}} (A_n)^{r_{2n}}.$$

Then we can show that $(A_n)^{r_{2n}}/\mathrm{Ker}\,\psi_{2n}$ is a free A_n-module (we can choose free generators explicitly), and get a free resolution of M of length $\leq 2n$. Note that, since $\mathcal{D}$ is flat over A_n, this also gives a (global) free resolution of $\mathcal{D} \otimes_{A_n} M$ of length $\leq 2n$.

Moreover, via this free resolution, we can compute $Ext^i_{A_n}(M, A_n)$ almost directly. For the computation of $Ext^i_{A_n}(M, \mathcal{D})$, we need a homogenization trick in order to reduce the computation in $\mathcal{D}_n$ to that in A_n (cf. [O3]).

3 The characteristic variety

In general, let $\mathcal{M}$ be a coherent left $\mathcal{D}$-module and $u_1, \ldots, u_r \in \mathcal{M}(\Omega)$ be its local generators with an open set $\Omega \subset X$. For each integer k, let $\mathcal{D}(k)$ be the subsheaf of $\mathcal{D}$ consisting of sections of $\mathcal{D}$ of order at most k. Put

$$\mathcal{M}_k := \mathcal{D}(k)u_1 + \ldots + \mathcal{D}(k)u_r \subset \mathcal{M}.$$

Then $\{\mathcal{M}_k\}_{k \in \mathbf{Z}}$ constitutes *a good filtration* of $\mathcal{M}$. Put $\mathrm{gr}(\mathcal{M}) := \bigoplus_{k \in \mathbf{Z}} \mathcal{M}_k/\mathcal{M}_{k-1}$ and call it the graded module of $\mathcal{M}$ associated with a filtration $\{\mathcal{M}_k\}_k$. In particular, we have an isomorphism

$$\mathrm{gr}(\mathcal{D}) = \bigoplus_{k \geq 0} \mathcal{D}(k)/\mathcal{D}(k-1) \simeq \mathcal{O}[\xi] = \mathcal{O}[\xi_1, \ldots, \xi_n]$$

via the principal symbol, where $\mathcal{O} = \mathcal{O}_X$ denotes the sheaf of holomorphic functions on $X = \mathbf{C}^n$. Then $\mathrm{gr}(\mathcal{M})$ is a coherent $\mathrm{gr}(\mathcal{D})$-module.

Definition 3.1 (characteristic variety) Let $\mathcal{M}$ be a coherent left $\mathcal{D}$-module defined on $\Omega \subset X$ and $\{\mathcal{M}_k\}$ be a good filtration of $\mathcal{M}$. Put

$$\mu(\mathrm{gr}(\mathcal{M})) := \mathcal{O}_{T^*X} \otimes_{\pi^{-1}\mathcal{O}[\xi]} \pi^{-1}\mathrm{gr}(\mathcal{M}),$$

where $T^*X = \{(x, \xi_1 dx_1 + \ldots + \xi_n dx_n)\}$ is the cotangent bundle of $X = \mathbf{C}^n$, and $\pi : T^*X \longrightarrow X$ is the natural projection. Then *the characteristic variety* $\mathrm{Char}(\mathcal{M})$ of $\mathcal{M}$ is defined as the support of the coherent $\mathcal{O}_{T^*X}$-module $\mu(\mathrm{gr}(\mathcal{M}))$. This is independent of the choice of a good filtration of $\mathcal{M}$.

Now let $\mathcal{M}$ be an algebraic $\mathcal{D}$-module given explicitly by

$$\mathcal{M} = \mathcal{D} \otimes_{A_n} ((A_n)^r/N)$$

with

$$N = A_n\mathbf{P}_1 + \ldots + A_n\mathbf{P}_s,$$

where $\mathbf{P}_1, \ldots, P_s \in (A_n)^r$. We assume that the order $\prec$ on $\mathbf{N}^{2n} \times \{1, \ldots, r\}$ satisfies, in addition to (A-1), (A-2),

(A-3) $|\beta| < |\beta'|$ implies $(\alpha, \beta, i) \prec (\alpha', \beta', j)$ for any $\alpha, \alpha', \beta, \beta' \in \mathbf{N}^n$, $i, j \in \{1, \ldots, r\}$;

(A-4) If $|\beta| = |\beta'|$ and $i < j$, then $(\alpha, \beta, i) \prec (\alpha', \beta', j)$ for any $\alpha, \alpha' \in \mathbf{N}^n$, $i, j \in \{1, \ldots, r\}$.

By the Buchberger algorithm compute a Gröbner basis $\mathbf{G}$ of N with respect to the order $\prec$.

Theorem 3.2 **([O1])** *Under the notation above, put* $\mathbf{G}_\nu := \{\mathbf{P} \in \mathbf{G} \mid \mathrm{lp}(\mathbf{P}) = \nu\}$ *for each* $\nu \in \{1, \ldots, r\}$. *Then the characteristic variety of* $\mathcal{M}$ *is given by*

$$\mathrm{Char}(\mathcal{M}) = \bigcup_{\nu=1}^{r} \{(x, \xi) \in T^*X \mid \sigma(\mathbf{P})_\nu(x, \xi) = 0 \quad \textit{for any} \quad \mathbf{P} \in \mathbf{G}_\nu\},$$

where $\sigma(\mathbf{P})_\nu$ *denotes the principal symbol of the* ν*-th component of* $\mathbf{P}$.

Thus the characteristic variety V (which is an algebraic set of T^*X) of an algebraic $\mathcal{D}$-module can be computed in a finite number of steps. The author does not know a correct algorithm of decomposing V into irreducible components as an analytic set. However, there is an algorithm of decomposing the algebraic set V into algebraically irreducible components, and this suffices for many examples.

4 Standard bases —computation of the multiplicity

First let us recall the notion of standard basis of an ideal of the ring of power series introduced by H. Hironaka (1964). Let $\prec_L$ be a lexicographic order on $\mathbf{N}^n$. We define another total order $\prec_P$ of $\mathbf{N}^n$ by

$$\alpha \prec_P \beta \quad \text{if and only if} \quad |\alpha| > |\beta| \quad \text{or} \quad (|\alpha| = |\beta| \text{ and } \alpha \prec_L \beta)$$

for $\alpha = (\alpha_1, \ldots, \alpha_n), \beta = (\beta_1, \ldots, \beta_n) \in \mathbf{N}^n$. Note that the order $\prec_P$ is not a well-order. For an element $f = \sum_\alpha a_\alpha x^\alpha$ of the ring of convergent power series $\mathbf{C}\{x\}$, we define its *leading exponent* $\mathrm{lexp}(f) \in \mathbf{N}^n$ by

$$\mathrm{lexp}(f) := \max{}_P\{\alpha \mid a_\alpha \neq 0\},$$

where $\max_P$ denotes the maximum element in the order $\prec_P$. For an ideal I of $\mathbf{C}\{x\}$, we put

$$E(I) := \{\mathrm{lexp}(f) \mid f \in I \setminus \{0\}\}.$$

Definition 4.1 (standard basis) Let I be an ideal of $\mathbf{C}\{x\}$ and $\mathbf{G}$ be a finite subset of I. Then $\mathbf{G}$ is called *a standard basis* of I if and only if

$$E(I) = \bigcup_{f \in \mathbf{G}} (\mathrm{lexp}(f) + \mathbf{N}^n).$$

There is an abstract algorithm of computing a standard basis from a given set of generators of I, which is similar to the Buchberger algorithm. However, one must employ the so-called Weierstrass-Hironaka division which cannot be correctly computed in finitely many steps in general.

If I is generated by given polynomials, however, there are two algorithms of computing a standard basis of I in finitely many steps: One is the so-called Mora's tangent cone algorithm [Mo], and the other is the trick of homogenization by Lazard [La].

Let $\mathcal{M} = \mathcal{D} \otimes_{A_n} ((A_n)^r/N)$ be as in the preceding section. Let $x^* := (x^0, \xi^0 dx)$ be a non-singular point of $V := \mathrm{Char}(\mathcal{M})$ and assume that x^* belongs to an irreducible component V_i of V. Our purpose is to compute the multiplicity of $\mathcal{M}$ along V_i, or equivalently, the multiplicity of $\mu(\mathrm{gr}(\mathcal{M}))$ at x^*.

Let $\mathbf{G}$ be a Gröbner basis of N with respect to the order $\prec$ used in Section 3. Let $\mathbf{G}_\nu$ $(\nu = 1, \ldots, r)$ be as in Theorem 3.2 and I_ν be the ideal of $(\mathcal{O}_{T^*X})_{x^*} = \mathbf{C}\{x - x^0, \xi - \xi^0\}$ generated by $\sigma(\mathbf{G}_\nu) := \{\sigma(\mathbf{P})_\nu \mid \mathbf{P} \in \mathbf{G}_\nu\}$. Letting J be the maximal ideal of $\mathbf{C}\{x - x^0, \xi - \xi^0\}$, put

$$H(k) := \sum_{\nu=1}^{r} \dim_{\mathbf{C}} \mathbf{C}\{x - x^0, \xi - \xi^0\}/(I_\nu + J^k) \tag{4.1}$$

for $k \in \mathbf{N}$. Then $H(k)$ is a polynomial of k for k sufficiently large and is called *the Hilbert polynomial* of $\mu(\mathrm{gr}(\mathcal{M}))$. In fact, it is easy to see that

$$H(k) = \dim_{\mathbf{C}} \mu(\mathrm{gr}(\mathcal{M}))_{x^*} / J^k \mu(\mathrm{gr}(\mathcal{M}))_{x^*}.$$

It is known that the Hilbert polynomial is written in the form

$$H(k) = \frac{m}{d!} k^d + (\text{lower degree terms with respect to } k)$$

for sufficiently large k. Here d and m coincide with the dimension of V_i and the multiplicity of $\mathcal{M}$ along V_i respectively. Now let $\tilde{\mathbf{G}}_\nu$ be a standard basis of I_ν in $\mathbf{C}\{x - x^0, \xi - \xi^0\}$. Then we have by definition

$$E_\nu := \{\mathrm{lexp}(f) \mid f \in I_\nu \setminus \{0\}\} = \bigcup_{f \in \tilde{\mathbf{G}}_\nu} (\mathrm{lexp}(f) + \mathbf{N}^{2n}).$$

By the Weierstrass-Hironaka division theorem, we have

$$\dim_{\mathbf{C}} \mathbf{C}\{x - x^0, \xi - \xi^0\}/(I_\nu + J^k) = \sharp\{(\alpha, \beta) \in \mathbf{N}^{2n} \mid (\alpha, \beta) \notin E_\nu,\ |\alpha| + |\beta| \leq k - 1\}. \tag{4.2}$$

Thus we can compute the multiplicity $\mathrm{mult}(\mathcal{M}, V_i)$ combining (4.1) and (4.2) through the algorithm of [Mo] or of [La].

5 Gröbner basis and homogenization with respect to a filtration

From now on, we work in $X := \mathbf{C}^{n+1}$. Put

$$A_{n+1} := \mathbf{C}[t, x_1, \ldots, x_n]\langle \partial_t, \partial_1, \ldots, \partial_n \rangle,$$
$$\mathcal{D}_{n+1} := \mathbf{C}\{t, x_1, \ldots, x_n\}\langle \partial_t, \partial_1, \ldots, \partial_n \rangle$$

with $\partial_t := \partial/\partial t$ and $\partial_i = \partial/\partial x_i$. Put $Y = \{(t,x) \in \mathbf{C}^{n+1} \mid t = 0\}$. We use a filtration ($V$-filtration) with respect to Y introduced by Kashiwara and Malgrange ([K5], [M2]) for the study of vanishing cycle sheaves (cf. also [L2], [LS]). An element P of A_{n+1} (or of $\mathcal{D}_{n+1}$) is written in the form

$$P = \sum_{\mu,\nu \geq 0, \alpha,\beta \in \mathbf{N}^n} a_{\mu,\nu,\alpha,\beta} t^\mu x^\alpha {\partial_t}^\nu \partial^\beta. \tag{5.1}$$

For each integer m, define $\mathbf{C}$-subspaces of A_{n+1} and of $\mathcal{D}_{n+1}$ respectively by

$$F_m(A_{n+1}) := \{P = \sum_{\mu,\nu,\alpha,\beta} a_{\mu,\nu,\alpha,\beta} t^\mu x^\alpha \partial_t^\nu \partial^\beta \in A_{n+1} \mid a_{\mu,\nu,\alpha,\beta} = 0 \text{ if } \nu - \mu > m\},$$
$$F_m(\mathcal{D}_{n+1}) := \{P = \sum_{\mu,\nu,\alpha,\beta} a_{\mu,\nu,\alpha,\beta} t^\mu x^\alpha \partial_t^\nu \partial^\beta \in \mathcal{D}_{n+1} \mid a_{\mu,\nu,\alpha,\beta} = 0 \text{ if } \nu - \mu > m\}.$$

For a nonzero element P of $\mathcal{D}_{n+1}$, we define the *F-order* $\mathrm{ord}_F(P)$ of P as the minimum integer m that satisfies $P \in F_m(\mathcal{D}_{n+1})$. When the F-order of P in the form (5.1) is m, we put

$$\hat{\sigma}(P) = \hat{\sigma}_m(P) := \sum_{\nu-\mu=m} a_{\mu,\nu,\alpha,\beta} t^\mu x^\alpha \partial_t^\nu \partial^\beta$$

and call it the *formal symbol* of P along Y (cf. [LS]). We have $\mathrm{ord}_F(PQ) = \mathrm{ord}_F(P) + \mathrm{ord}_F(Q)$ and $\hat{\sigma}(PQ) = \hat{\sigma}(P)\hat{\sigma}(Q)$ for $P, Q \in \mathcal{D}_{n+1}$.

Now let $\prec_F$ be an order on $\mathbf{N}^{2n+2}$ which satisfies (A-1) (with $r = 1$ and n replaced by $n+1$) and

(A-5) if $\nu - \mu > \nu' - \mu'$, then $(\mu, \nu, \alpha, \beta) \succ_F (\mu', \nu', \alpha', \beta')$;
(A-6) $(\mu, \mu, \alpha, \beta) \succeq_F (0, 0, 0, 0)$ for any $\mu \in \mathbf{N}$ and $\alpha, \beta \in \mathbf{N}^n$.

The condition (A-5) implies that $\prec_F$ is not a well-order. However, the definitions in Section 2 apply to this order $\prec_F$. Let us denote by $\mathrm{lexp}_F(P) \in \mathbf{N}^{2n+2}$, $\mathrm{lcoef}_F(P) \in \mathbf{C}$ the leading exponent and the leading coefficient of $P \in A_{n+1} \setminus \{0\}$ with respect to $\prec_F$ respectively. The set of leading exponents $E_F(I) \subset \mathbf{N}^{2n+2}$ is defined in the same way.

Definition 5.1 (FW-Gröbner basis) A finite set $\mathbf{G}$ of generators of a left ideal I of A_{n+1} is called an FW-*Gröbner basis* of I if we have

$$E(I) = \bigcup_{P \in \mathbf{G}} (\mathrm{lexp}(P) + \mathbf{N}^{2n+2}).$$

Since we do not have division algorithm, the Buchberger algorithm does not work directly. In order to bypass this difficulty to obtain an algorithm of computing FW-Gröbner bases, we use the homogenization technique.

Definition 5.2 For $\mu, \nu, \mu', \nu', i, j \in \mathbf{N}$ and $\alpha, \beta, \alpha', \beta' \in \mathbf{N}^n$, an order $\prec_H$ on $\mathbf{N}^{2n+3}$ is defined by $(i, \mu, \nu, \alpha, \beta) \prec_H (j, \mu', \nu', \alpha', \beta'))$ if and only if one of the following conditions holds:

(1) $i < j$;
(2) $i = j$, $\quad (i, \mu + \ell, \nu, \alpha, \beta) \prec_F (j, \mu' + \ell', \nu', \alpha', \beta')$ with $\ell, \ell' \in \mathbf{N}$ such that $\nu - \mu - \ell = \nu' - \mu' - \ell'$;
(3) $(i, \nu, \alpha, \beta) = (j, \nu', \alpha', \beta'), \quad \mu < \mu'$

This definition is independent of the choice of ℓ, ℓ' in view of the condition (A-1).

Lemma 5.3 *(1) $\prec_H$ is a well-order.*
(2) If $\nu - \mu - i = \nu' - \mu' - j$, then $(i, \mu, \nu, \alpha, \beta) \succ_H (j, \mu', \nu', \alpha', \beta')$ if and only if $(\mu, \nu, \alpha, \beta) \succ_F (\mu', \nu', \alpha', \beta')$.

For a nonzero element $P = P(x_0)$ of $A_{n+1}[x_0]$, let us denote by $\operatorname{lexp}_H(P) \in \mathbf{N}^{2n+3}$ and $\operatorname{lcoef}_H(P) \in K$ the leading exponent and the leading coefficient of P with respect to $\prec_H$ respectively. The set $E_H(I)$ of leading exponents of a left ideal I of $A_{n+1}[x_0]$ is also defined.

Definition 5.4 (F-homogeneity) An element P of $A_{n+1}[x_0]$ of the form

$$P = \sum_{i,\mu,\nu,\alpha,\beta} a_{i,\mu,\nu,\alpha,\beta} {x_0}^i t^\mu x^\alpha \partial_t^\nu \partial^\beta$$

is said to be *F-homogeneous* of order m if $a_{i,\mu,\nu,\alpha,\beta} = 0$ whenever $\nu - \mu - i \neq m$.

Definition 5.5 (F-homogenization) For an element P of A_{n+1} of the form

$$P = \sum_{\mu,\nu,\alpha,\beta} a_{\mu,\nu,\alpha,\beta} t^\mu x^\alpha \partial_t^\nu \partial^\beta,$$

put $m = \min\{\nu - \mu \mid a_{\mu,\nu,\alpha,\beta} \neq 0 \text{ for some } \alpha, \beta \in \mathbf{N}^n\}$. Then the *F-homogenization* $P^h \in A_{n+1}[x_0]$ of P is defined by

$$P^h = \sum_{\mu,\nu,\alpha,\beta} a_{\mu,\nu,\alpha,\beta} {x_0}^{\nu-\mu-m} t^\mu x^\alpha \partial_t^\nu \partial^\beta.$$

P^h is F-homogeneous of order m.

Proposition 5.6 *Let I be a left ideal of $A_{n+1}[x_0]$ generated by F-homogeneous operators. Then there exists an H-Gröbner basis (i.e. a Gröbner basis with respect to $\prec_H$) of I consisting of F-homogeneous operators. Moreover, such an H-Gröbner basis can be computed by the Buchberger algorithm.*

Proof: Since $\prec_H$ is a well-order, the arguments in Section 2 apply. Hence we have only to verify that taking the S-operator and computing division both preserve the F-homogeneity. □

Theorem 5.7 *Let I be a left ideal of A_{n+1} generated by $P_1, \ldots, P_d \in A_{n+1}$. Let us denote by I^h the left ideal of $A_{n+1}[x_0]$ generated by $(P_1)^h, \ldots, (P_d)^h$. (Here I^h is not defined uniquely by I.) Let $\mathbf{G} = \{Q_1(x_0), \ldots, Q_k(x_0)\}$ be an H-Gröbner basis of I^h consisting of F-homogeneous operators. Then $\mathbf{G}(1) := \{Q_1(1), \ldots, Q_k(1)\}$ is an FW-Gröbner basis of I.*

Proposition 5.8 *Under the same assumptions as in the preceding theorem, there exist, for any $i, j \in \{1, \ldots, k\}$ with $i < j$, F-homogeneous*

$$U_{ij1}(x_0), \ldots, U_{ijk}(x_0) \in A_{n+1}[x_0]$$

so that

$$\mathrm{sp}(Q_i(x_0), Q_j(x_0)) = S_{ji}(x_0)Q_i(x_0) - S_{ij}(x_0)Q_j(x_0) = \sum_{\ell=1}^{k} U_{ij\ell}Q_k(x_0),$$

where $S_{ji}(x_0)$ is defined in the same way as S_{ji} in Theorem 2.6, and, for each ℓ, we have $\mathrm{lexp}_H(U_{ij\ell}Q_\ell(x_0)) \prec \mathrm{lexp}_F(Q_i(x_0)) \vee \mathrm{lexp}_F(Q_j(x_0))$ if $U_{ij\ell}(x_0) \neq 0$. Put

$$\mathbf{V}_{ij}(x_0) = (0, \ldots, \overset{(i)}{S_{ji}(x_0)}, \ldots, -\overset{(j)}{S_{ij}(x_0)}, \ldots, 0) - (U_{ij1}(x_0), \ldots, U_{ijk}(x_0)) \in (A_{n+1}[x_0])^k.$$

Then the syzygy module

$$S(Q_1(1), \ldots, Q_k(1)) := \{(U_1, \ldots, U_k) \in (A_{n+1})^k \mid \sum_{\ell=1}^{k} U_\ell Q_\ell(1) = 0\}$$

is generated by $\{\mathbf{V}_{ij}(1) \mid 1 \leq i < j \leq k\}$.

Definition 5.9 Let P be a nonzero element of A_{n+1} (resp. $\mathcal{D}_{n+1}$) of F-order m. Then we define $\psi(P)(s) \in A_n[s]$ (resp. $\mathcal{D}_n[s]$) by

$$\hat{\sigma}_0(t^m P) = \psi(P)(t\partial_t) \text{ if } m \geq 0,$$
$$\hat{\sigma}_0(\partial_t^{-m} P) = \psi(P)(t\partial_t) \text{ if } m < 0.$$

Theorem 5.10 *We use the same notation as in Theorem 5.7. Let $\psi(I)$ be the left ideal of $A_n[s]$ generated by the set $\{\psi(P)(s) \mid P \in I \cap (F_0(A_{n+1}) \setminus F_{-1}(A_{n+1}))\}$. Then $\psi(I)$ is generated by $\psi(Q_1(1)), \ldots, \psi(Q_k(1))$.*

The following theorem, which can be proved by using Proposition 5.8, is crucial in the application of FW-Gröbner bases to the D-module theory.

Theorem 5.11 *In the same notation as Theorem 5.10, let $\mathcal{I} = \mathcal{D}_{n+1}I$ be the left ideal of $\mathcal{D}_{n+1}$ generated by I. Let $\psi(\mathcal{I})$ be the left ideal of $\mathcal{D}_n[s]$ generated by the set $\{\psi(P)(s) \mid P \in \mathcal{I} \cap (F_0(\mathcal{D}_{n+1}) \setminus F_{-1}(\mathcal{D}_{n+1}))\}$. Then $\psi(\mathcal{I})$ is generated by $\psi(Q_1(1)), \ldots, \psi(Q_k(1))$.*

6 Computation of the b-function of a D-module

We retain the notation in the preceding section. Let M be a finitely generated left A_{n+1}-module and u a nonzero element of M. In the sequel, we assume that a system of the equations for u is given explicitly; i.e., we assume that a finite set of generators of a left ideal I of A_{n+1} is given so that $A_{n+1}u \simeq A_{n+1}/I$.

More generally, if a presentation of M and a representation of u is known, i.e., if generators of a left A_{n+1}-submodule N of $(A_{n+1})^r$ is given so that $M \simeq (A_{n+1})^r/N$, and also given is an element $\mathbf{U} \in (A_{n+1})^r$ such that u corresponds to the modulo class of $\mathbf{U}$ by the above isomorphism, then there is an algorithm to find generators of the above I by computing syzygies by means of (a generalization of) Theorem 2.6. The following arguments are local, so we may work at the origin without loss of generality.

Put $\mathcal{M} := \mathcal{D}_{n+1} \otimes_{A_{n+1}} M$ and $\mathcal{I} := \mathcal{D}_{n+1}I$. Then we have $\mathcal{D}_{n+1}(1 \otimes u) \simeq \mathcal{D}_{n+1}/\mathcal{I}$. The *b-function* $b_u(s)$ of u is the monic polynomial, if any, $b(s) \in \mathbf{C}[s]$ of the least degree satisfying

$$(b(t\partial_t) + P)(1 \otimes u) = 0 \quad \text{in } \mathcal{M} \tag{6.1}$$

with some $P \in F_{-1}(\mathcal{D}_{n+1})$. Note that (6.1) is equivalent to $b(t\partial_t) + P \in \mathcal{I}$.

The existence of the b-function in this sense was proved by Kashiwara-Kawai [KK], Laurent [L2] when $\mathcal{M}$ is holonomic. Our purpose here is to present an algorithm to determine whether there exists such nonzero $b(s)$, and to find, if it does, $b_u(s)$.

Now let M, u, I be as above and let $\mathbf{G} = \{Q_1(x_0), \ldots, Q_k(x_0)\}$ be an H-Gröbner basis of I^h consisting of F-homogeneous elements as in the preceding section. Let $\psi(I)$ and $\psi(\mathcal{I})$ be left ideals of $A_{n+1}[s]$ and of $\mathcal{D}_{n+1}[s]$ respectively defined in Theorems 5.10 and 5.11. Then $\psi(I)$ and $\psi(\mathcal{I})$ are both generated by $\psi(\mathbf{G}(1)) := \{\psi(Q_1(1)), \ldots, \psi(Q_k(1))\}$.

Let $\prec$ be an order on $\mathbf{N}^{2n+1}$ satisfying (A-1), (A-2) (with $r = 1$ and n replaced by $2n+1$) and

(A-7) if $|\beta| > |\beta'|$, then $(\mu, \alpha, \beta) \succ (\mu', \alpha', \beta')$ for any $\mu, \mu' \in \mathbf{N}$ and $\alpha, \beta, \alpha', \beta' \in \mathbf{N}^n$,

where (μ, α, β) corresponds to $s^\mu x^\alpha \partial^\beta \in A_n[s]$.

For an element P of $A_n[s]$ (resp. $\mathcal{D}_n[s]$), its order $\operatorname{ord}(P)$ is defined to be the usual order with respect to ∂, and the principal symbol $\sigma(P) \in \mathbf{C}[x, \xi, s]$ (resp. $\mathbf{C}\{x\}[\xi, s]$) is defined also in the standard way with $\xi = (\xi_1, \ldots, \xi_n)$.

Theorem 6.1 *Let $\sigma(\psi(I))$ and $\sigma(\psi(\mathcal{I}))$ be ideals of $\mathbf{C}[x, \xi, s]$ and of $\mathbf{C}\{x\}[\xi, s]$ generated by $\{\sigma(P) \mid P \in \psi(I)\}$ and by $\{\sigma(P) \mid P \in \psi(\mathcal{I})\}$ respectively. Let $\mathbf{H}$ be a Gröbner basis of $\psi(I)$ with respect to an order $\prec$ on $\mathbf{N}^{2n+1}$ satisfying (A-1), (A-2), (A-7). Then $\sigma(\psi(I))$ and $\sigma(\psi(\mathcal{I}))$ are both generated by $\sigma(\mathbf{H}) := \{\sigma(P) \mid P \in \mathbf{H}\}$.*

Corollary 6.2 *In the same notation as in Theorem 6.1, put $J := \psi(I) \cap \mathbf{C}[x, s]$ and $\mathcal{J} := \psi(\mathcal{I}) \cap \mathbf{C}\{x\}[s]$. Then J and $\mathcal{J}$ are both generated by $\sigma(\mathbf{H}) \cap \mathbf{C}[x, s]$ as ideals of $\mathbf{C}[x, s]$ and of $\mathbf{C}\{x\}[s]$ respectively.*

Corollary 6.3 *The b-function of u is the monic generator of $\mathcal{J} \cap \mathbf{C}[s]$ if it is not the zero ideal. If $\mathcal{J} \cap \mathbf{C}[s] = 0$, then the b-function does not exist.*

To be general, let J be an ideal of $\mathbf{C}[x, s]$ whose generators are given explicitly and put $\mathcal{J} = \mathbf{C}\{x\}[s]J$. Our purpose is to compute $\mathcal{J} \cap \mathbf{C}[s]$. ($J \cap \mathbf{C}[s]$ is computed easily through the elimination by Gröbner basis.) The following lemma is a consequence of the faithful flatness of $\mathbf{C}\{x\}$ over $\mathbf{C}[x]_0$:

Lemma 6.4 *Let $\mathbf{C}[x]_0$ be the localization of $\mathbf{C}[x]$ with respect to the maximal ideal generated by $x_1, \ldots, x_n$. Put $\mathcal{J}' := \mathbf{C}[x]_0[s]J$. Then we have $\mathcal{J} \cap \mathbf{C}[s] = \mathcal{J}' \cap \mathbf{C}[s]$.*

Thus we can compute $\mathcal{J} \cap \mathbf{C}[s]$ by using the Gröbner basis computation in the polynomial ring and factorization in $\mathbf{C}[s]$ in the following steps:

Algorithm 6.5 Input: a set of generators $f_1(x, s), \ldots, f_k(x, s)$ of J:

(1) Determine whether there exists, and find if any, some $g(x, s) \in J$ such that its leading coefficient with respect to s does not vanish at $x = 0$; This can be done as follows:
   (a) Let $(f_i)^h(x_0, x, s)$ be the homogenization of $f_i(x, s)$ with respect to x; i.e., $(f_i)^h$ is homogeneous with respect to x_0 and x, and $(f_i)^h(1, x, s) = f_i(x, s)$;
   (b) Let $>$ be an order on $\mathbf{N} \times \mathbf{N} \times \mathbf{N}^n \ni (i, \mu, \alpha)$ with (i, μ, α) corresponding to $x_0{}^i s^\mu x^\alpha$. Assume $>$ satisfies (A-1), (A-2) and $(i, \mu, \alpha) > (j, \nu, \beta)$ if $\mu > \nu$ or ($\mu = \nu$ and $i > j$);
   (c) Let $\{g_1(x_0, x, s), \ldots, g_r(x_0, x, s)\}$ be a Gröbner basis of the ideal generated by $(f_1)^h, \ldots, (f_k)^h$ with respect to $>$.
   (d) Let $g(x, s)$ be one of $g_i(1, x, s)$ with the property above; if there is no such $g(x, s)$, then quit (there is no $b(s)$);
(2) Compute the monic generator $f_0(s)$ of the ideal $J(0)$ of $\mathbf{C}[s]$ that is generated by $f_1(0, s), \ldots, f_k(0, s)$ by Gröbner basis or GCD computation; if $f_0(s) = 1$, then put $b(s) := 1$ and exit;
(3) Compute the factorization $f_0(s) = (s - s_1)^{\mu_1} \ldots (s - s_m)^{\mu_m}$ in $\mathbf{C}[s]$;
(4) For each $i = 1, \ldots, m$, determine the least integer $\ell_i = \ell \geq 0$ satisfying $h(x, s)(s - s_i)^\ell \in J$ with some $h(x, s) \in \mathbf{C}[x, s]$ such that $q(0, s_i) \neq 0$, or else determine that there is no such ℓ; This can be done by computing ideal quotient and saturation via Gröbner bases as follows (cf. [BW], [CLO], [E]):
   For $i := 1$ to m do
   (a) Compute a set of generators G_i of the saturation $J : (s - s_i)^\infty$ by means of Gröbner basis;
   (b) Determine whether there is some $h(x, s) \in G_i$ such that $h(0, s_i) \neq 0$; if there is no such h, then put $\ell_i := \infty$ and quit (there is no $b(s)$);
   (c) By computing the ideal quotient $J : (s - s_i)^\ell$ for $\ell = \mu_i, \mu_i + 1, \ldots$ repeatedly, determine the least $\ell \geq \mu_i$ such that $J : (s - s_i)^\ell$ contains an element which does not vanish at $(x, s) = (0, s_i)$. Denote this ℓ by ℓ_i;
(5) Put $b(s) := (s - s_1)^{\ell_1} \ldots (s - s_m)^{\ell_m}$;

Output: $b(s)$ is the monic generator of $\mathcal{J} \cap \mathbf{C}[s]$.

7 Computation of the induced system

Let $M, u, I, \mathcal{M}$ be as in the preceding section. Here we assume that u generates M for the sake of simplicity. The induced system of $\mathcal{M}$ to $Y := \{(t, x) \in X \mid t = 0\}$ at $(0,0) \in Y$ is the complex

$$\mathcal{M}_Y^{\bullet} \quad : \quad 0 \longrightarrow \mathcal{M} \xrightarrow{t} \mathcal{M} \longrightarrow 0$$

of left $\mathcal{D}_n$-modules. (Here we regard $\mathcal{D}_n$ as the stalk at $(0,0) \in Y$ of the sheaf $\mathcal{D}_Y$ of analytic differential operators on Y.) Let us write $\mathcal{M}_Y := \mathcal{M}/t\mathcal{M}$. Our main purpose is to give a sufficient condition for the above map t to be injective and to give an algorithm of computing $\mathcal{M}_Y$ explicitly. Put

$$F_k(\mathcal{M}) := \{P(1 \otimes u) \mid P \in F_k(\mathcal{D}_{n+1})\}, \qquad \mathrm{gr}_k^F(\mathcal{M}) := F_k(\mathcal{M})/F_{k-1}(\mathcal{M})$$

for each integer k. Note that the b-function $b_u(s)$ is the monic polynomial of the least degree that satisfies $b(t\partial_t)\mathrm{gr}_0^F(\mathcal{M}) = 0$.

Lemma 7.1 *Assume that $b(s) \in \mathbf{C}[s]$ satisfies $b(t\partial_t)\mathrm{gr}_0^F(\mathcal{M}) = 0$. Then the homomorphism $t : \mathrm{gr}_{k+1}^F(\mathcal{M}) \longrightarrow \mathrm{gr}_k^F(\mathcal{M})$ is bijective if $b(k) \neq 0$.*

Proposition 7.2 *Assume that $b(s) \in \mathbf{C}[s]$ satisfies $b(t\partial_t)\mathrm{gr}_0^F(\mathcal{M}) = 0$. Put*

$$k_1 := \max\{k \in \mathbf{Z} \mid b(k) = 0\}, \qquad k_0 := \min\{k \in \mathbf{Z} \mid b(k) = 0\}.$$

Then $\mathcal{M}_Y^{\bullet}$ is quasi-isomorphic to the complex

$$0 \longrightarrow F_{k_1+1}(\mathcal{M})/F_{k_0}(\mathcal{M}) \xrightarrow{t} F_{k_1}(\mathcal{M})/F_{k_0-1}(\mathcal{M}) \longrightarrow 0$$

of left $\mathcal{D}_Y$-modules. In particular, $t : \mathcal{M} \longrightarrow \mathcal{M}$ is bijective if $b(k) \neq 0$ for any $k \in \mathbf{Z}$.

Proposition 7.3 *Assume that there exists $b(s) \in \mathbf{C}[s]$ and $m \in \mathbf{N}$ so that*

$$b(t\partial_t)\partial_t{}^m \mathrm{gr}_0^F(\mathcal{M}) = 0.$$

Assume, moreover, $b(k) \neq 0$ for any $k \in \mathbf{Z}$. Then the homomorphism $t : \mathcal{M} \longrightarrow \mathcal{M}$ is injective.

Let P be an element of $F_m(A_{n+1})$. Then we can write P in the form

$$P = \sum_{k=0}^{m} P_k(t\partial_t, x, \partial)\partial_t^k + R$$

uniquely with $P_k \in A_n[t\partial_t]$ and $R \in F_{-1}(A_{n+1})$. Then we put

$$\rho(P, k_0) := \sum_{k=k_0}^{m} P_k(0, x, \partial)\partial_t^k$$

for each integer k_0 with $0 \leq k_0 \leq m$.

Theorem 7.4 *Assume that $b(s) \in \mathbf{C}[s]$ satisfies $b(t\partial_t)\mathrm{gr}_0^F(\mathcal{M}) = 0$. Put*

$$k_1 := \max\{k \in \mathbf{Z} \mid b(k) = 0\}, \qquad k_0 := \max\{0, \min\{k \in \mathbf{Z} \mid b(k) = 0\}\}.$$

(We have $k_1 = m - 1$ and $k_0 = 0$ under the assumption of Proposition 7.3.) Let $\mathbf{G}$ be an FW-Gröbner basis of N. Then we have an isomorphism

$$\mathcal{M}_Y \simeq \bigoplus_{k=k_0}^{k_1} \mathcal{D}_Y \partial_t^k / \mathcal{L},$$

where $\mathcal{L}$ is the left $\mathcal{D}_Y$-module generated by a finite set

$$\mathbf{G}_Y := \{\rho(\partial_t^j P, k_0) \mid P \in \mathbf{G},\ j \in \mathbf{N},\ j + \mathrm{ord}_F(P) \le k_1\}.$$

8 Bernstein-Sato polynomial and D-modules associated with a polynomial

We retain the notation in Sections 5, 6. Let $f(x) \in \mathbf{C}[x]$ be a polynomial with $f(0) = 0$. The following argument is due to Malgrange [M1]. Put $\mathcal{L} = \mathbf{C}\{x\}[f^{-1}, s]f^s$, where we regard f^s as a free generator. Then $\mathcal{L}$ has a structure of left $\mathcal{D}_n[s]$-module defined by

$$\partial_i(g(s)f^{-m}f^s) = \left(\frac{\partial g}{\partial x_i}(s)f^{-m} + (s-m)g(s)\frac{\partial f}{\partial x_i}f^{-m-1}\right)f^s \qquad (i = 1, \ldots, n)$$

for $g(s) \in \mathbf{C}\{x\}[s]$ and $m \in \mathbf{N}$. Moreover, $\mathcal{L}$ has also a structure of left $\mathcal{D}_{n+1}$-module defined by

$$t(g(s)f^s) = g(s+1)f^{s+1}, \qquad \partial_t(g(s)f^s) = -sg(s-1)f^{s-1}$$

for $g(s) \in \mathbf{C}\{x\}[f^{-1}, s]$. We can make an element $a(t) \in \mathbf{C}\{t\}$ operate on $g(s)f^s$ since $f(0) = 0$. It is easy to see that

$$\begin{aligned} &-\partial_t t(g(s)f^s) = sg(s)f^s \quad \text{for any } g(s) \in \mathbf{C}\{x\}[f^{-1}, s], \\ &(t - f(x))f^s = 0, \\ &\left(\partial_i + \frac{\partial f}{\partial x_i}(x)\partial_t\right) f^s = 0 \quad (i = 1, \ldots, n). \end{aligned}$$

Put $\mathcal{N} := \mathcal{D}_n[s]f^s$ and $\mathcal{M} := \mathcal{D}_{n+1}f^s$. Then we have inclusions $\mathcal{N} \subset \mathcal{M} \subset \mathcal{L}$. Set

$$I := A_{n+1}(t - f(x)) + \sum_{i=1}^n A_{n+1}\left(\partial_i + \frac{\partial f}{\partial x_i}\partial_t\right).$$

and $\mathcal{I} := \mathcal{D}_{n+1}I$.

Proposition 8.1 *$\mathcal{M}$ is isomorphic to $\mathcal{D}_{n+1}/\mathcal{I}$.*

Corollary 8.2 *For $P(s) \in \mathcal{D}_n[s]$, we have $P(s)f^s = 0$ in $\mathcal{N}$ if and only if $P(-\partial_t t) \in \mathcal{I}$.*

The (local) b-function (Bernstein-Sato polynomial) $b_f(s)$ of $f(x)$ is the monic polynomial of the least degree $b(s) \in \mathbf{C}[s]$ satisfying

$$P(s,x,\partial)f^{s+1} = b(s)f^s \quad \text{in } \mathcal{N} \tag{8.1}$$

with some $P(s) \in \mathcal{D}_n[s]$. The monic polynomial of the least degree $b(s)$ satisfying (8.1) with some $P(s) \in A_n[s]$ is denoted by $\tilde{b}_f(s)$. Such $\tilde{b}_f(s)$, and hence $b_f(s)$ also exist ([Be], [Bj], [K2]). By definition $b_f(s)$ divides $\tilde{b}_f(s)$.

In view of Corollary 8.2, the equation (8.1) is equivalent to

$$b(-\partial_t t) - P(-\partial_t t, x, \partial)t \in \mathcal{I}.$$

On the other hand, suppose $b(s) \in \mathbf{C}[s]$ and $Q \in F_{-1}(\mathcal{D}_{n+1})$ satisfy

$$(b(t\partial_t) - Q)f^s = 0 \qquad \text{in } \mathcal{M}.$$

Expanding Q in the form

$$Q = \sum_{m=1}^{\infty} Q_m(x, t\partial_t, \partial)t^m,$$

put

$$\rho(Q) := \sum_{m=1}^{\infty} Q_m(x, -s-1, \partial)f^{m-1}.$$

Then we get, in view of Corollary 8.2,

$$(b(-s-1) - \rho(Q)f)f^s = 0 \quad \text{in } \mathcal{N}.$$

In conclusion, the computation of $b_f(s)$ can be done as follows:

Algorithm 8.3 Input: $f(x) \in \mathbf{C}[x]$;

(1) Letting I be the left ideal of A_{n+1} generated by $t - f$ and $\partial_i + (\partial f/\partial x_i)\partial_t$ $(i = 1, \dots, n)$, compute an FW-Gröbner basis $\mathbf{G}$ of I via F-homogenization;
(2) Compute a Gröbner basis $\mathbf{H}$ of the left ideal generated by $\psi(\mathbf{G}) := \{\psi(P) \mid P \in \mathbf{G}\}$ with respect to an order satisfying (A-1), (A-2), (A-7);
(3) Compute the output $b(s) \in \mathbf{C}[s]$ of Algorithm 6.5 with $\mathbf{H} \cap \mathbf{C}[x,s]$ as input;

Output: $b_f(s) := b(-s-1)$.

Remark 8.4 (1) The step (1) of Algorithm 6.5, which is called in the above algorithm, can be skipped since the existence of $\tilde{b}_f(s)$ is assured.
(2) The fact that the roots of $b_f(s)$ are rational (Kashiwara [K2]) makes the steps (3) and (4) of Algorithm 6.5 considerably easier.

Next, let us describe an algorithm of computing $\mathcal{N}$ as a left $\mathcal{D}_n[s]$-module. More precisely, our algorithm computes generators of $\mathcal{I}_f := \{P \in \mathcal{D}[s] \mid Pf^s = 0\}$ for $f \in \mathbf{C}[x]$, where $\mathcal{D}$ denotes the sheaf of analytic differential operators on $\mathbf{C}^n$. Let $\prec_e$ be an order on $\mathbf{N}^{2n+4}$ which satisfies (A-1), (A-2) and

(A-8) $(i,\ell,\mu,\nu,\alpha,\beta) \prec_e (j,\ell',\mu',\nu',\alpha',\beta')$ if $i+\ell < j+\ell'$,

where $(i,\ell,\mu,\nu,\alpha,\beta)$ corresponds to $x_0^i y_0^\ell t^\mu x^\alpha \partial_t^\nu \partial^\beta \in A_{n+1}[x_0,y_0]$.

Algorithm 8.5 Input: $f(x) \in \mathbf{C}[x]$;

(1) Compute a Gröbner basis $\mathbf{G}$ of the left ideal J of $A_{n+1}[x_0,y_0]$ generated by $t - x_0 f$, $\partial_i + x_0(\partial f/\partial x_i)\partial_t$ $(i=1,\dots,n)$, and $1 - x_0 y_0$ with respect to the order $\prec_e$;

(2) $\mathbf{G}_0 := \{\psi(P)(-s-1) \mid P \in \mathbf{G} \cap A_{n+1}\}$;

Output: $\mathbf{G}_0$ is a generator of $\mathcal{I}_f$.

Finally, let us give an algorithm of computing the algebraic local cohomology group with respect to a hypersurface with coefficients in the sheaf $\mathcal{O}_Y$ of holomorphic functions on $Y = \mathbf{C}^n$ as an application of Section 7. Let $f = f(x) \in \mathbf{C}[x]$ be a polynomial and put $Z := \{x \in Y = \mathbf{C}^n \mid f(x) = 0\}$. Then the algebraic local cohomology group $\mathcal{H}^k_{[Z]}(\mathcal{O}_Y)$ has a structure of left $\mathcal{D}_Y$-module and vanishes if $k \neq 1$.

First let us recall the definition (cf. [K3]): Put $\mathcal{J} := \mathcal{O}_Y f$. For a coherent sheaf $\mathcal{F}$ of $\mathcal{O}_Y$-modules, set

$$\Gamma_{[Z]}(\mathcal{F}) := \varinjlim_{\nu} \mathcal{H}om_{\mathcal{O}_Y}(\mathcal{O}_Y/\mathcal{J}^\nu; \mathcal{F}).$$

Then $\mathcal{H}^k_{[Z]}$ is defined as the k-th derived functor of $\Gamma_{[Z]}$.

Lemma 8.6 *We have*

$$\mathcal{H}^k_{[Z]}(\mathcal{O}_Y) \simeq \begin{cases} \mathcal{O}_Y[f^{-1}]/\mathcal{O}_Y & (k=1) \\ 0 & (k \neq 1) \end{cases}$$

Proposition 8.7 *We have a (quasi-) isomorphism $\mathcal{M}^\bullet_Y \simeq \mathbf{R}\Gamma_{[Z]}(\mathcal{O}_Y)$ in the derived category of $\mathcal{D}_Y$-modules.*

Corollary 8.8 *The homomorphism $t : \mathcal{M} \longrightarrow \mathcal{M}$ is injective and we have an isomorphism $\mathcal{H}^1_{[Z]}(\mathcal{O}_Y) \simeq \mathcal{M}_Y$ as $\mathcal{D}_Y$-modules.*

Combining this fact with Theorem 7.4, we obtain immediately an algorithm of computing $\mathcal{H}^1_{[Z]}(\mathcal{O}_Y)$.

9 Implementation and an example of computation

We have implemented our algorithms explained so far in a computer algebra system *Kan* of Takayama [T4]. *Kan* is a system designed especially for Gröbner basis computation in rings of polynomials, differential operators, and (q-) difference operators. Hence we use *Kan* for Gröbner basis computation in the Weyl algebra, while we use a general-purpose computer algebra system *Risa/Asir* [NT] for factorization and Gröbner basis computation in the polynomial ring. We give an example of computation by using our implementation.

Put $f := x^3 + y^2z^2 \in \mathbf{C}[x, y, z]$. Then f has non-isolated singularities. First, we obtain its b-function as

$$b_f(s) = (s+1)\left(s+\frac{5}{6}\right)^2\left(s+\frac{7}{6}\right)^2\left(s+\frac{4}{3}\right)\left(s+\frac{5}{3}\right).$$

A set of involutory generators of $\mathcal{I}_f = \{P(s) \in \mathcal{D}[s] \mid P(s)f^s = 0\}$ is given by

$$\begin{aligned}&\{2x\partial_x + 3y\partial_y - 6s,\ -y\partial_y + z\partial_z,\ 2z^2y\partial_x - 3x^2\partial_y,\\ &2zy^2\partial_x - 3x^2\partial_z,\ (-x^3 - z^2y^2)\partial_z + 2szy^2,\ -x^3\partial_y - z^3y\partial_z + 2sz^2y,\\ &2z^3\partial_z\partial_x - 3x^2\partial_y^2 + 2z^2\partial_x,\ -x^3\partial_y^2 - z^4\partial_z^2 + (2s-2)z^3\partial_z + 2sz^2\}\end{aligned}$$

with $\partial_x = \partial/\partial_x$, and so on. Let us regard s as a complex number satisfying $b_f(s-j) \neq 0$ for any $j = 1, 2, 3, \ldots$ (cf. [K2]). Put $T^*\mathbf{C}^3 := \{(x, y, z; \xi dx + \eta dy + \zeta dz) \mid x, y, z, \xi, \eta, \zeta \in \mathbf{C}\}$ and

$$\begin{aligned}V_f &:= \{(x, y, z; df) \mid f = 0,\ df \neq 0\}\\ &= \{(x, y, z; df) \mid x^3 + y^2z^2 = 0 \text{ and } (x \neq 0 \text{ or } yz \neq 0)\}.\end{aligned}$$

Then the characteristic variety of $\mathcal{D}f^s$ is given by

$$V_f \cup \{x = y = z = 0\} \cup \{x = y = \zeta = 0\} \cup \{x = z = \eta = 0\} \cup \{\xi = \eta = \zeta = 0\},$$

where the component $\{x = y = z = 0\}$ has multiplicity two, while the other components have multiplicity one.

A set of involutory generators of $\mathcal{H}^1_{[Z]}(\mathcal{O}_Y)$ with $Y := \mathbf{C}^3$ and $Z := \{(x, y, z) \in Y \mid f(x, y, z) = 0\}$ is given by

$$\begin{aligned}&\{2x\partial_x + 3y\partial_y + 6,\ -y\partial_y + z\partial_z,\ -x^3 - z^2y^2,\ -3x^2\partial_y + 2z^2y\partial_x,\\ &-3x^2\partial_z + 2zy^2\partial_x,\ -3x^2\partial_y^2 + 2z^3\partial_z\partial_x + 2z^2\partial_x,\\ &x^3\partial_y + z^3y\partial_z + 2z^2y,\ -x^3\partial_y^2 - z^4\partial_z^2 - 4z^3\partial_z - 2z^2\},\end{aligned}$$

and its characteristic variety by

$$V_f \cup \{x = y = z = 0\} \cup \{x = y = \zeta = 0\} \cup \{x = z = \eta = 0\},$$

where the component $\{x = y = z = 0\}$ has multiplicity two, while the other components have multiplicity one.

Acknowledgement: The author is grateful to Professor N. Takayama for stimulus discussions and for the kind assistance in implementing algorithms of the present paper in *Kan*.

References

[BW] Becker, T., Weispfenning, V., *Gröbner Bases*, Springer-Verlag, Berlin, 1993.

[Be] Bernstein, I. N., *Modules over the ring of differential operators*, Functional Anal. Appl. **2** (1971), 1–16.

[Bj] Björk, J.E., Rings of Differential Operators, North-Holland, Amsterdam, 1979.

[Bo] Borel, A. et al., *Algebraic D-Modules*, Academic Press, Boston, 1987.

[BGMM] Briançon, J., Granger, M., Maisonobe, Ph., Miniconi, M., *Algorithme de calcul du polynôme de Bernstein: cas non dégénéré* , Ann. Inst. Fourier **39** (1989), 553–610.

[Bu] Buchberger, B., *Ein algorithmisches Kriterium für die Lösbarkeit eines algebraischen Gleichungssystems*, Aequationes Math. **4** (1970), 374–383.

[C] Castro, F., *Calculs effectifs pour les idéaux d'opérateurs différentiels*, Travaux en Cours **24**, Hermann, Paris, 1987, pp. 1–19.

[CLO] Cox, D., Little, J., O'Shea, D., *Ideals, Varieties, and Algorithms*, Springer, Berlin, 1992.

[E] Eisenbud, D., *Commutative Algebra with a View Toward Algebraic Geometry*, Springer, New York, 1995.

[G] Galligo, A., *Some algorithmic questions on ideals of differential operators*, Lecture Notes in Comput. Sci. **204**, Springer, Berlin, 1985, pp. 413–421.

[K1] Kashiwara, M., *Algebraic study of systems of linear partial differential equations* (in Japanese), Master's thesis, University of Tokyo, 1971.

[K2] Kashiwara, M., *B-functions and holonomic systems –Rationality of roots of b-functions*, Invent. Math. **38**, 1976, 33–53.

[K3] Kashiwara, M., *On the holonomic systems of linear differential equations, II*, Invent. Math. **49** (1978), 121–135.

[K4] Kashiwara, M., *Systems of Microdifferential Equations*, Progress in Math. **34**, Birkhäuser, Boston, 1983.

[K5] Kashiwara, M., *Vanishing cycle sheaves and holonomic systems of differential equations*, Lecture Notes in Math. **1016**, Springer, Berlin, 1983, pp. 134–142.

[KK] Kashiwara, M., Kawai, T., *Second microlocalization and asymptotic expansions*, Lecture Notes in Physics **126**, Springer, Berlin, 1980, pp. 21–76.

[L1] Laurent, Y., *Calcul d'indices et irrégularité pour les systèmes holonomes*, Astérisque **130** (1985), 352–364.

[L2] Laurent, Y., *Polygône de Newton et b-fonctions pour les modules microdifferentiels*, Ann. Sci. Éc. Norm. Sup. **20** (1987), 391–441.

[LS] Laurent, Y., Schapira, P., *Images inverses des modules différentiels*, Compositio Math. **61** (1987), 229–251.

[La] Lazard, D., *Gröbner bases, Gaussian elimination, and resolution of systems of algebraic equations*, Lecture Notes in Comput. Sci. **162**, Springer, Berlin, 1983, pp. 146–156.

[M1] Malgrange, B., *Le polyôme de Bernstein d'une singularité isolée*, Lecture Notes in Math. **459**, Springer, Berlin, 1975, pp. 98–119.

[M2] Malgrange, B., *Polynômes de Bernstein-Sato et cohomologie évanescente*, Astérisque **101–102** (1983), 243–267.

[Mo] Mora, F., *An algorithm to compute the equations of tangent cones*, Lecture Notes in Comput. Sci. **144**, Springer, Berlin, 1982, pp. 158–165.

[N] Noumi, M., *Wronskian determinants and the Gröbner representation of a linear differential equation*, in *Algebraic Analysis* (M. Kashiwara, T. Kawai, eds.), Academic Press, Boston, 1988, pp. 549–569.

[NT] Noro, M, Takeshima T., *Risa/Asir–a computer algebra system*, in *Proceedings of International Symposium on Symbolic and Algebraic Computation* (Paul S. Wnag, ed.), ACM, New York, 1992, pp. 387–396; URL ftp://endeavor.fujitsu.co.jp/pub/isis/asir.

[O1] Oaku, T., *Computation of the characteristic variety and the singular locus of a system of differential equations with polynomial coefficients*, Japan J. Indust. Appl. Math. **11** (1994), 485–497.

[O2] Oaku, T., *Algorithms for finding the structure of solutions of a system of linear partial differential equations*, in *Proceedings of International Symposium on Symbolic and Algebraic Computation* (J. Gathen, M. Giesbrecht, eds.), ACM, New York, 1994, pp. 216–223.

[O3] Oaku, T., *Gröbner Bases and Systems of Linear Partial Differential Equations*, Sophia Kokyuroku in Mathematics, No. **38** (in Japanese), Department of Mathematics, Sophia University, Tokyo, 1994.

[O4] Oaku, T., *Algorithmic methods for Fuchsian systems of linear partial differential equations*, J. Math. Soc. Japan **47** (1995), 297–328.

[O5] Oaku, T., *Gröbner bases for D-modules on a non-singular affine algebraic variety*, to appear in Tôhoku Math. J.

[O6] Oaku, T., *An algorithm of computing b-functions*, (preprint).

[O7] Oaku, T., *Computation of b-functions and induced systems of D-modules*, (in preparation).

[O8] Oaku, T., *Algorithms for the b-function and D-modules associated with an arbitrary polynomial*, (in preparation).

[SKK] Sato, M., Kawai, T., Kashiwara, M., *Microfunctions and pseudo-differential equations*, Lecture Notes in Math. 287, Springer, Berlin, 1973, pp. 265–529.

[SKKO] Sato, M., Kashiwara, M., Kimura, T., Oshima, T., *Micro-local analysis of prehomogeneous vector spaces*, Invent. Math. **62** (1980), 117–179.

[S] Schapira, P., *Microdifferential Systems in the Complex Domain*, Grundlehren der Math. Wiss. vol. **269**, Springer, Berlin, 1985.

[T1] Takayama, N., *Gröbner basis and the problem of contiguous relations*, Japan J. Appl. Math. **6** (1989), 147–160.

[T2] Takayama, N., *An algorithm of constructing the integral of a module —an infinite dimensional analog of Gröbner basis*, in *Proceedings of International Symposium on Symbolic and Algebraic Computation* (S. Watanabe, M. Nagata, eds.), ACM, New York, 1990, pp. 206–211.

[T3] Takayama, N., *An approach to the zero recognition problem by Buchberger algorithm*, J. Symbolic Comput. **14** (1992), 265–282.

[T4] Takayama, N., *Kan: A system for computation in algebraic analysis*, http://www.math.s.kobe-u.ac.jp, 1991—.

[Y] Yano, T., *On the theory of b-functions*, Publ. RIMS, Kyoto Univ. **14** (1978), 111–202.

On higher-codimensional boundary value problems

Kiyoshi Takeuchi

Department of Mathematics, Hiroshima University, 1-3-1, Kagamiyama, Higashi-hiroshima, Hiroshima 739, Japan
(e-mail: `takeuchi@top2.math.sci.hiroshima-u.ac.jp`)

1 Introduction

— Let us consider a real analytic manifold M and its submanifold N of codimension d. We take $Y \subset X$ as a complexification of $N \subset M$. Let $\mathcal{M}$ be a coherent module over the sheaf of ring $\mathcal{D}_X$ of holomorphic differential operators on X and assume that Y is noncharacteristic for $\mathcal{M}$. In this talk, we first report that we have weakened the conditions on $\mathcal{M}$ for the vanishing of the local cohomologies of the hyperfunction solution complex :

$$H^j \mu_N R\mathcal{H}om_{\mathcal{D}_X}(\mathcal{M}, \mathcal{B}_M) \simeq 0 \quad \text{for} \quad j < d, \tag{1}$$

where μ_N denotes Sato's microlocalization functor along N. This kind of problem is very important since many classical results in analysis can be deduced from such vanishing theorems on cohomologies. Holmgren's uniqueness theorem and the famous abstract Edge of the Wedge theorem of Martineau [14], Morimoto [15] and Kashiwara [7] (when $\mathcal{M} = \bar{\partial}$ the Cauchy-Riemann equation) are special cases of it. In particular, the last result was essentially used to construct the microfunction theory by Sato et al. [17] and many extension theorems on holomorphic functions can be obtained from it. For example, we have the Edge of the Wedge theorem of Martineau for holomorphic functions and the local Bochner's tube theorem of Komatsu [13]. After that, Kashiwara-Kawai [6] showed (1) under the additional condition of the ellipticity of $\mathcal{M}$ and extended this result to general systems. In this talk, we shall further extend it into various cases where the solutions are not necessarily real analytic, and we derive several results on the extension of hyperfunction solutions. For this purpose, we make use of the theory of bimicrolocalization developped in [21] and [24] showing in particular that the sheaf $R\mathcal{H}om_{\mathcal{D}_X}(\mathcal{M}, \mathcal{C}_{NM})$ of bimicrofunction solutions of [21] plays the role of the obstruction against the extension of hyperfunction solutions. We also prove (under weaker conditions than that of Kashiwara-Kawai [6]) that every real analytic solutions on an open convex cone $\Omega \subset M = \mathbb{R}^n$ with the edge N automatically extends to a conic open neighborhood of Ω in $X = \mathbb{C}^n$. To investigate this phenomenon, a partial solution to Schapira's conjecture and the result on the injectivity of the microlocal boundary value morphism proved in [23] will be used.

2 Microfunctions for boundary value problems

— In this report, we essentially follow the terminology of [9] and [21]. Let $X \supset M \supset N$ be a sequence of C^∞-manifolds. In [21], we introduced the functor of bispecialization :

$$\nu_{NM} : \mathbf{D}^{\mathrm{b}}(X) \longrightarrow \mathbf{D}^{\mathrm{b}}(T_N M \times_M T_M X), \tag{2}$$

where $\mathbf{D}^{\mathrm{b}}(*)$ is the derived category of sheaves of $\mathbb{C}$-vector spaces on a topological space with bounded cohomologies. We also defined two functors :

$$\begin{cases} \nu\mu_{NM} : \mathbf{D}^{\mathrm{b}}(X) \longrightarrow \mathbf{D}^{\mathrm{b}}(T_N M \times_M T^*_M X) \\ \mu_{NM} : \mathbf{D}^{\mathrm{b}}(X) \longrightarrow \mathbf{D}^{\mathrm{b}}(T^*_N M \times_M T^*_M X) \end{cases} \tag{3}$$

as the Fourier transformations of $\nu_{NM}(*)$. Now let M be a real analytic manifold of dimension n, $N \subset M$ its submanifold of codimension d, and $Y \subset X$ a complexification of $N \subset M$. We set $\mathcal{O}_X$ the sheaf of holomorphic functions on X, $\mathcal{B}_M$ (resp. $\mathcal{C}_M$ and $\mathcal{C}_{N|X}$) the sheaf of Sato's hyperfunctions on M (resp. sheaves of Sato's microfunctions of [17]), and in [21] we introduced the "sheaves" (use Theorem 5.4 of [23]) :

$$\hat{\mathcal{C}}_{NM} := \nu\mu_{NM}(\mathcal{O}_X) \otimes \mathrm{or}_M[n] \quad \text{and} \quad \mathcal{C}_{NM} := \mu_{NM}(\mathcal{O}_X) \otimes \mathrm{or}_N[n]. \tag{4}$$

For coherent $\mathcal{D}_X$-modules $\mathcal{M}$, we have the distinguished triangles :

$$\begin{aligned} &\nu_N R\mathcal{H}om_{\mathcal{D}_X}(\mathcal{M}, \mathcal{O}_X)\,|_{T_N M} \rightarrow \nu_N R\mathcal{H}om_{\mathcal{D}_X}(\mathcal{M}, \mathcal{B}_M) \\ &\rightarrow R\hat{\pi}_{M*} R\mathcal{H}om_{\mathcal{D}_X}(\mathcal{M}, \hat{\mathcal{C}}_{NM}) \rightarrow_{+1} \end{aligned} \tag{5}$$

and

$$\begin{aligned} &R\theta_! R\mathcal{H}om_{\mathcal{D}_X}(\mathcal{M}, \mathcal{C}_{N|X}) \rightarrow \mu_N R\mathcal{H}om_{\mathcal{D}_X}(\mathcal{M}, \mathcal{B}_M) \\ &\rightarrow R\dot{\pi}_{M*} R\mathcal{H}om_{\mathcal{D}_X}(\mathcal{M}, \mathcal{C}_{NM}) \rightarrow_{+1}, \end{aligned} \tag{6}$$

where we used the projections $\hat{\pi}_M : T_N M \times_M \dot{T}^*_M X \rightarrow T_N M$, $\dot{\pi}_M : T^*_N M \times_M \dot{T}^*_M X \rightarrow T^*_N M$, and $\theta : T^*_N X \rightarrow T^*_N M$.

3 Microlocally hyperbolic systems

— First, let us consider the projections $\pi_M : T^*_N M \times_M T^*_M X \rightarrow T^*_N M$ and $\pi_N : T^*_N M \times_M T^*_M X \rightarrow N \times_M T^*_M X$.

Theorem 1. *Let $p \in \dot{T}^*_N M \times_M T^*_M X$ and suppose that $\mathcal{M}$ is micro-hyperbolic (in the sense of [8]) in the direction $\pi_M(p) \in \dot{T}^*_N M$ at $\pi_N(p) \in T^*_M X$. Then we have $R\mathcal{H}om_{\mathcal{D}_X}(\mathcal{M}, \mathcal{C}_{NM})_p \simeq 0$.*

Remark. Funakoshi [4] has already obtained a similar result for second microfunction solutions of Kataoka-Tose [11].

To introduce the notion of microlocal hyperbolicity, we shall use the natural injection

$$i : \dot{T}^*_M X \times_M T^*M \to T^*(T^*_M X) \simeq T_{(T^*_M X)}(T^*X) \tag{7}$$

induced by $\dot{T}^*_M X \to M$ and the natural projection $\dot{\pi} : \dot{T}^*_M X \times_M T^*M \to T^*M$.

Definition 2. We say that the system $\mathcal{M}$ is "microlocally hyperbolic" in the direction $\xi \in \dot{T}^*M$ if the characteristic variety $\mathrm{Ch}\mathcal{M}$ of $\mathcal{M}$ satisfies the condition $i(\dot{\pi}^{-1}(\xi)) \cap C_{T^*_M X}(\mathrm{Ch}\mathcal{M}) = \emptyset$.

The microlocally hyperbolic systems that we introduced here enjoy many applications concerning the extension of hyperfunction (holomorphic) solutions. First of all, we have by (6) and Theorem 1 :

Corollary 3. *Let $\Omega \subset T_N M$ be an open convex proper cone such that the system $\mathcal{M}$ is microlocally hyperbolic for every vector $\xi \in \Omega^{oa} \setminus N \subset \dot{T}^*_N M$. Then we have a natural isomorphism :*

$$\mathrm{R}\Gamma(\Omega; \nu_N R\mathcal{H}om_{\mathcal{D}_X}(\mathcal{M}, \mathcal{O}_X) \mid_{T_N M}) \xrightarrow{\sim} \mathrm{R}\Gamma(\Omega; \nu_N R\mathcal{H}om_{\mathcal{D}_X}(\mathcal{M}, \mathcal{A}_M)), \tag{8}$$

where $\mathcal{A}_M$ denotes the sheaf of real analytic functions on M.

Example 1. The hypothesis of microlocal hyperbolicity is strictly weaker than that of ellipticity or of hyperbolicity, and Corollary 3 generalizes a result of Kashiwara-Kawai [6] for elliptic equations. For example, let us take an elliptic operator E (resp. a hyperbolic operator Q) and set $P = EQ$. Then the $\mathcal{D}_X$-module $\mathcal{M} = \mathcal{D}_X/\mathcal{D}_X P$ satisfies the condition of microlocal hyperbolicity. In fact, to show the isomorphism

$$\Gamma(\Omega; \nu_N \mathcal{H}om_{\mathcal{D}_X}(\mathcal{M}, \mathcal{O}_X) \mid_{T_N M}) \xrightarrow{\sim} \Gamma(\Omega; \nu_N \mathcal{H}om_{\mathcal{D}_X}(\mathcal{M}, \mathcal{A}_M))$$

, we can again weaken the condition on $\mathcal{M}$. In section 5 we shall extend Corollary 3 in various directions.

Theorem 4. *Assume that Y of codimension $d \geq 2$ is noncharacteristic for the system $\mathcal{M}$ and $\mathcal{M}$ is microlocally hyperbolic for $\xi \in \dot{T}^*_N M$.*

(i) For every $j < d$, $H^j \mu_N R\mathcal{H}om_{\mathcal{D}_X}(\mathcal{M}, \mathcal{B}_M)_\xi \simeq 0$.
(ii) (Bochner type theorem of extension) There exists an open subset Ω of M such that the polar set of $C_N(M \setminus \Omega)$ is a closed convex proper cone of $T^*_N M$ which contains ξ in its interior, and every hyperfunction solution $u \in \Gamma(\Omega; \mathcal{H}om_{\mathcal{D}_X}(\mathcal{M}, \mathcal{B}_M))$ on Ω automatically extends to an open neighborhood of N as a solution to $\mathcal{M}$.

Finally we shall give an Edge of the Wedge theorem of Bogoliubov type for microlocally hyperbolic equations.

Definition 5. We say an open subset $\Omega \subset M$ is an open convex cone with the edge N if there exists a local analytic chart $M = \mathbb{R}^d \times \mathbb{R}^{n-d}$ such that $N = \{0\} \times \mathbb{R}^{n-d}$ and $\Omega = \Omega_1 \times \mathbb{R}^{n-d}$ with $\Omega_1 \subset \mathbb{R}^d$ being an open convex proper cone.

Assume $d \geq 2$ and let $M_0 \subset M$ be a submanifold of codimension $d-1$ such that $N \subset M_0$. We divide $M_0 \setminus N$ into two connected components M_{0+} and M_{0-}.

Theorem 6. *Assume that the system $\mathcal{M}$ is noncharacteristic for Y and microlocally hyperbolic for every $\xi \in N \times_{M_0} \dot{T}^*_{M_0} M$. Then there exist two open convex cones $\Omega_\pm$ with the edge N such that $\Omega_\pm \supset M_{0\pm}$, $\Omega_+ \cap \Omega_- = \emptyset$, and the following assertion holds :*

If the boundary values $u_\pm |_N \in \mathcal{H}om_{\mathcal{D}_Y}(\mathcal{M}_Y, \mathcal{B}_N)$ of the hyperfunction solutions $u_\pm \in \Gamma(\Omega_\pm; \mathcal{H}om_{\mathcal{D}_X}(\mathcal{M}, \mathcal{B}_M))$ are the same, $u_\pm$ extend to a neighborhood of N as a solution to $\mathcal{M}$.

Remark. This theorem extends a result of Uchida [29] for elliptic systems when the geometric situation is the same. In fact, he treats also the case when the convex hull of Ω_+ and Ω_- is just a convex cone of M, that is, the Edge of the Wedge theorem of Epstein type.

4 Partially elliptic systems

— In this section, we study the another class of systems which admit the Edge of the Wedge type theorems. First we explain a typical example which inspired us. Let M be a product $M' \times M''$ of two manifolds and $X = X' \times X''$ its complexification. We also consider a closed submanifold $N = N' \times M'' \subset M$ of codimension d and its complexification $Y = Y' \times X'' \subset X$. We take a coherent $\mathcal{D}_{X'}$-module $\mathcal{M}'$ which satisfies the condition of Kashiwara-Kawai [6] :

$$\mathrm{Ch}\mathcal{M}' \cap \dot{T}^*_{Y'} X' = \emptyset \quad \text{and} \quad \mathrm{Ch}\mathcal{M}' \cap \dot{T}^*_{M'} X' = \emptyset$$

and set $\mathcal{M} := \mathcal{M}' \boxtimes \mathcal{D}_{X''}$. Then it seems very natural to expect that the Edge of the Wedge type theorem :

$$H^j \mu_N R\mathcal{H}om_{\mathcal{D}_X}(\mathcal{M}, \mathcal{B}_M) \simeq 0 \quad \text{for} \quad j < d$$

holds, since the case when X'' is just a point coincides with a result of [6]. However the condition imposed on the characteristic variety of $\mathcal{M}$ above is very strong at the zero-section $T^*_X X$ and it is desirable to extend the class of systems so that they can contain elliptic factors in their characteristic varieties. Therefore we shall generalize the Edge of the Wedge type theorem of Kashiwara-Kawai [6] for elliptic systems to the systems which contain partially elliptic factors and elliptic ones at the same time.

From now on we shall formulate the problem more intrinsically. Let $f : M \to M''$ be a smooth morphism of real analytic manifolds, $N \subset M$ a submanifold of codimension d such that $g = f|_N : N \to M''$ is smooth, and $f_{\mathbb{C}} : X \to X''$

a complexification of f. Set $L := f_{\mathbb{C}}^{-1}(M'')$. Then the complex $\mu_{NL}(\mathcal{O}_X)[n]$ for $n = \dim^{\mathbb{R}} M$ is concentrated in degree 0 by the abstract Edge of the Wedge theorem of Kashiwara [7], and we define the sheaf $\mathcal{C}_{NL}$ on $T_N^*L \times_L T_L^*X$ by $\mathcal{C}_{NL} := \mu_{NL}(\mathcal{O}_X) \otimes \mathrm{or}_N[n]$. We think that this sheaf has an essential importance in the theory of second microlocal analysis, especially concerning with boundary value problems. To formulate the partial ellipticity of Bony-Schapira [2] in this relative setting, we use the vector bundles $T^*(M/M'')$ and $T^*(X/X'')$ defined by the exact sequences

$$\begin{cases} 0 \to M \times_{M''} T_{M''}^*X'' \to T_M^*X \to T^*(M/M'') \to 0 \\ 0 \to X \times_{X''} T^*X'' \to T^*X \to T^*(X/X'') \to 0. \end{cases} \tag{9}$$

Set $\Lambda := M \times_L T_L^*X$ and note that it is a closed submanifold of $V := X \times_{X''} T^*X'' \subset T^*X$. Now we have a natural injection :

$$T^*(M/M'') \times_M \Lambda \longrightarrow T^*(X/X'') \times_X V \simeq T_V(T^*X) \tag{10}$$

and a projection $\dot{\pi}_\Lambda : \dot{T}^*(M/M'') \times_M \Lambda \longrightarrow \Lambda$.

Definition Bony-Schapira [B-S] *For $q \in \Lambda$, we say that a system $\mathcal{M}$ is partially elliptic along $V \subset T^*X$ at q if*

$$\dot{\pi}_\Lambda^{-1}(q) \cap C_V(Ch\mathcal{M}) = \emptyset.$$

We can solve the abstract Cauchy problem in the framework of sheaves for our bimicrofunctions and we get a correspondence between the complexes $R\mathcal{H}om_{\mathcal{D}_X}(\mathcal{M}, \mathcal{C}_{NM})$ and
$R\mathcal{H}om_{\mathcal{D}_X}(\mathcal{M}, \mathcal{C}_{NL})$ for partially elliptic systems $\mathcal{M}$. Next we shall consider the condition to vanish the cohomologies of $R\mathcal{H}om_{\mathcal{D}_X}(\mathcal{M}, \mathcal{C}_{NL})$.

Conjecture Let $p \in T_N^*L \times_L T_L^*X$ and $q \in N \times_L T_L^*X \subset \Lambda \subset V$ be its base point. If the system $\mathcal{M}$ is non-microcharacteristic for Y along V at $q \in V$, we have :

$$H^j R\mathcal{H}om_{\mathcal{D}_X}(\mathcal{M}, \mathcal{C}_{NL})_p \simeq 0 \quad \text{for} \quad j < d.$$

To prove this conjecture, we require the theory of quantized contact transformation of our bimicrofunctions which will be developped in the future. Here we only give a partial answer to it. We consider T_Y^*X as a closed submanifold of $T_V(T^*X)$ by the injection

$$T_Y^*X \longrightarrow T_Y^*X \times_X V \longrightarrow T_V(T^*X)$$

obtained by the zero-section of V. Then we have the following theorem which extends a result of Kashiwara-Kawai [6].

Theorem 7. *We assume that Y is noncharacteristic for the system $\mathcal{M}$ and $\mathcal{M}$ satisfies the conditions :*

(i) $N \times_M T^*_M X \cap Ch\mathcal{M} \subset N \times_L T^*_L X$,
(ii) $\mathcal{M}$ *is partially elliptic along* V *on* $N \times_L \dot{T}^*_L X$,
(iii) *For every point* $p \in N \times_L \dot{T}^*_L X$, *there exists a* $\mathcal{D}_X$*-module* $\mathcal{N}$ *such that* $\dot{T}^*_Y X \cap C_V(Ch\mathcal{N}) = \emptyset$ *and* $\mathcal{M} \simeq \mathcal{N}$ *as* $\mathcal{E}_X$*-module in a neighborhood of* p.

Then we have for every $j < d$:

$$H^j \mu_N R\mathcal{H}om_{\mathcal{D}_X}(\mathcal{M}, \mathcal{B}_M) \simeq 0. \tag{11}$$

Remark. When d=1, a similar situation was considered by Tose [To], but his result was microlocal, that is, in the sheaf $\mathcal{B}^2{}_\Lambda$ of second hyperfunction. Also notice that we have used some arguments of Delort [D] to solve the abstract Cauchy problems in the framework of sheaves.

5 Applications to the extension of solutions

— First, we shall give some applications of Theorem 7 and we inherit the notations in it. In particular, we get a Bochner type theorem on the extension of hyperfunction solutions.

Theorem 8. *Suppose that* $Y \subset X$ *of codimension* $d \geq 2$ *is noncharacteristic for the system* $\mathcal{M}$ *and* $\mathcal{M}$ *satisfies the conditions (i)-(iii) of Theorem 7.*

(i) *For every open convex cone* Ω_0 *with the edge* N *in* M, *every hyperfunction solution* $u \in \Gamma_\Omega \mathcal{H}om_{\mathcal{D}_X}(\mathcal{M}, \mathcal{B}_M)|_N$ *to* $\mathcal{M}$ *on* $\Omega := M \setminus \bar{\Omega}_0$ *automatically extends to an open neighborhood of* N *as a solution to* $\mathcal{M}$.
(ii) *Let* $U \subset T_N M$ *be an open cone with connected fibers and* $V \supset U$ *its convex hull. Then we have an isomorphism :*

$$\Gamma(V; \nu_N \mathcal{H}om_{\mathcal{D}_X}(\mathcal{M}, \mathcal{B}_M)) \xrightarrow{\sim} \Gamma(U; \nu_N \mathcal{H}om_{\mathcal{D}_X}(\mathcal{M}, \mathcal{B}_M)).$$

Remark. We used a method of Uchida [29] to show (ii) of Theorem 8. ¿From Theorem 7, we can also deduce an Edge of the Wedge theorem of Epstein type for "hyperfunction" solutions by using an argument of the same article [29].

Example 2. Let $\mathcal{N} = \mathcal{D}_X / \sum_{j=1}^k \mathcal{D}_X P_j$, $P_j \in \mathcal{D}_X$ be a coherent $\mathcal{D}_X$-module which satisfies the conditions (i) and (ii) of Theorem 7 and (iii)′ $\dot{T}^*_Y X \cap C_V(\mathrm{Ch}\mathcal{N}) = \emptyset$. Next take elliptic operators E_j, $j = 1, \ldots, k$ so that $\mathcal{M} = \mathcal{D}_X / \sum_{j=1}^k \mathcal{D}_X E_j P_j$ is noncharacteristic for Y. Then $\mathcal{M}$ satisfies the conditions of Theorem 8. Note that the $\mathcal{D}_X$-modules $\mathcal{M}$ obtained in this way contain partially elliptic factors and elliptic ones at the same time.

Let us mention that many authors ([5], [28] etc.) have treated the extension of real analytic solutions because we can use the results on the propagation of regularities up to the boundaries of codimension one in this case. However in Theorem 8, Theorem 4 etc., we do not assume any regularity of solutions.

Remark. Let $\xi \in \dot{T}^*_N M$ be a fixed direction. In Example 2, even if all E_j's are not elliptic, we can show

$$H^j \mu_N R\mathcal{H}om_{\mathcal{D}_X}(\mathcal{M}, \mathcal{B}_M)_\xi \simeq 0 \quad \text{for} \quad j < d \tag{12}$$

for $\mathcal{M} = \mathcal{D}_X / \sum_{j=1}^k \mathcal{D}_X E_j P_j$ under the condition that all E_j's are microlocally hyperbolic in the direction $\xi \in \dot{T}^*_N M$ and the partially elliptic factor is separated from the others in $N \times_M \dot{T}^*_M X$. Hence in this way, we can construct many examples of $\mathcal{M}$ for which the Bochner type theorem on the extension of hyperfunction solutions in a fixed direction $\xi \in \dot{T}^*_N M$ (like Theorem 4 (ii)) holds.

In the rest of this report, we extend Corollary 3 in various directions. From now on, we always assume $\Omega := \Omega_1 \times \mathbb{R}^{n-d}$ is an open convex cone of $M = \mathbb{R}^n$ with the edge $N := \{0\} \times \mathbb{R}^{n-d} = \{x_1 = \cdots = x_d = 0\}$ and $d \geq 2$, where $\Omega_1 \subset \mathbb{R}^d$ is an open convex proper cone. Applying the functor $\mathrm{R}\Gamma(\Omega; *)$ to the triangle (5), we obtain the exact sequence (we sometimes identify $\Omega \subset M$ with the open convex proper cone $\mathrm{int}C_N(\Omega)$ of $T_N M$) :

$$\begin{aligned} 0 &\longrightarrow \Gamma(\Omega; \nu_N \mathcal{H}om_{\mathcal{D}_X}(\mathcal{M}, \mathcal{O}_X)\,|_{T_N M}) \longrightarrow \Gamma(\Omega; \nu_N \mathcal{H}om_{\mathcal{D}_X}(\mathcal{M}, \mathcal{B}_M)) \\ &\longrightarrow \Gamma(\hat{\pi}_M^{-1}(\Omega); \mathcal{H}om_{\mathcal{D}_X}(\mathcal{M}, \hat{\mathcal{C}}_{NM})) \longrightarrow \cdots . \end{aligned} \tag{13}$$

It implies that every hyperfunction solution $u \in \Gamma(\Omega; \nu_N \mathcal{H}om_{\mathcal{D}_X}(\mathcal{M}, \mathcal{B}_M))$ on Ω extends to a conic neighborhood of Ω in $X = \mathbb{C}^n$ if the image of u in $\Gamma(\hat{\pi}_M^{-1}(\Omega); \mathcal{H}om_{\mathcal{D}_X}(\mathcal{M}, \hat{\mathcal{C}}_{NM}))$ vanishes. This point of view enables us to study the phenomenon in Corollary 3 with the methods of boundary value problems. Let us explain the meaning of the third arrow of (13). Let $p \in \dot{T}_N M \times_M T^*_M X$, and $q_1 \in \dot{T}_N M$ and $q_2 \in N \times_M T^*_M X$ be its base points. Then we proved in section 5 of [23] :

$$\hat{\mathcal{C}}_{NM}\,|_p \simeq \varinjlim_{\Omega} \mathcal{C}_{\Omega|X}\,|_{q_2}, \tag{14}$$

where Ω ranges over the open convex cones with the edge N such that $q_1 \in \mathrm{int}C_N(\Omega)$ and $\mathcal{C}_{\Omega|X}$ is a complex of sheaves on T^*X introduced in [19]. We know from the construction of the functor $\nu\mu_{NM}(*)$ that the third arrow of (13) is compatible with the natural morphism

$$\Gamma(\Omega; \mathcal{H}om_{\mathcal{D}_X}(\mathcal{M}, \mathcal{B}_M)) \longrightarrow \Gamma(N \times_M \dot{T}^*_M X; \mathcal{H}om_{\mathcal{D}_X}(\mathcal{M}, \mathcal{C}_{\Omega|X})). \tag{15}$$

Therefore to show a result like Corollary 3, it is enough to determine the condition of Ω-regularity of [19] for the open convex cones with the edge N.

First let us take a coordinate system $(z; \zeta dz)$, $z = x + \sqrt{-1}y$, $\zeta = \xi + \sqrt{-1}\eta$ of T^*X and the associated coordinate system

$$((x + \sqrt{-1}y; \xi + \sqrt{-1}\eta),\ x^* \frac{\partial}{\partial x} + y^* \frac{\partial}{\partial y} + \xi^* \frac{\partial}{\partial \xi} + \eta^* \frac{\partial}{\partial \eta})$$

of the tangent bundle $T((T^*X)^{\mathbb{R}})$ of the underlying real analytic manifold $(T^*X)^{\mathbb{R}}$ of T^*X. Now take a point $p := (x; \sqrt{-1}\eta)$ of $N \times_M T^*_M X$ and identify

T_x^*M with a closed subspace of $T_p((T^*X)^{\mathbb{R}})$ via Hamiltonian isomorphism as in [22] :

$$-H : T_x^*M \ni (x; x^*dx) \longmapsto ((x; \sqrt{-1}\eta); x^* \frac{\partial}{\partial \xi}) \in T_p((T^*X)^{\mathbb{R}}). \quad (16)$$

Theorem 9. *Suppose that the system $\mathcal{M}$ satisfies the condition :*

$$-H(\theta) \notin C_p(Ch\mathcal{M}, \bar{\Omega} \times_M T_M^*X) \quad (17)$$

for every $p = (x; \sqrt{-1}\eta) \in N \times_M \dot{T}_M^*X$ and $\theta = (x; \xi dx) \in \Omega^{\circ a} \setminus N \subset \dot{T}_N^*M$. Then we have a natural isomorphism :

$$\Gamma(\Omega; \nu_N \mathcal{H}om_{\mathcal{D}_X}(\mathcal{M}, \mathcal{O}_X) \mid_{T_N M}) \xrightarrow{\sim} \Gamma(\Omega; \nu_N \mathcal{H}om_{\mathcal{D}_X}(\mathcal{M}, \mathcal{A}_M)). \quad (18)$$

The hypothesis on $\mathcal{M}$ of this theorem is strictly weaker than that of Corollary 3 and let us give an example.

Example 3. Let $x = (x_1, x')$ and take a semi-hyperbolic operator $Q = D_1^2 - x_1^k D_{x'}^2$ for some $k \geq 2$ and any elliptic one E. Then we can find an open convex cone Ω with the edge $N = \{x_1 = \ldots = x_d = 0\}$ such that $\bar{\Omega} \setminus N \subset \{x_1 > 0\}$ and the system $\mathcal{M} = \mathcal{D}_X / \mathcal{D}_X P$, $P = EQ$ satisfies the condition of Theorem 9. That is, every real analytic solution $u \in \Gamma_\Omega \mathcal{H}om_{\mathcal{D}_X}(\mathcal{M}, \mathcal{A}_M) \mid_N$ on Ω extends to a conic open neighborhood of Ω in $X = \mathbb{C}^n$.

By Theorem 9, we cannot treat the Tricomi operator $Q = D_1^2 - x_1 D_{x'}^2$ in Example 3 nor the operators of Schapira [18] which are non-microcharacteristic along some regular involutive submanifold $V \subset T^*X$. To treat these operators, we have to study the propagation of regularities up to N of higher codimension. The next theorem (Theorem 6.1 of [23]) which extends a result of Oaku [16] allows us to reduce our problem to the results obtained in the case of codimension one.

Theorem 10. *(Microlocal injectivity of boundary value morphism) Let $\mathcal{M}$ be a coherent $\mathcal{D}_X$-module for which Y is noncharacteristic. Then canonical morphism of boundary value :*

$$\varinjlim_{\Omega} H^0 R\mathcal{H}om_{\mathcal{D}_X}(\mathcal{M}, \mathcal{C}_{\Omega|X}) \longrightarrow \mathcal{E}xt^d_{\mathcal{D}_X}(\mathcal{M}, \mathcal{C}_{N|X})$$

is injective on $N \times_X T^*X$, where Ω ranges over the open convex cones with the edge N which contain a fixed vector in $\dot{T}_N M$.

Now let $M_0 = \{x_2 = \ldots = x_d = 0\} \supset N$ be a closed submanifold of M of codimension $d-1$ and X_0 its complexification in X.

Theorem 11. *Let Ω be an open convex cone with the edge N such that $\Omega_0 := \Omega \cap M_0 = \{x_1 > 0\}$ in M_0 and assume that the system $\mathcal{M}$ is noncharacteristic for Y. Suppose that the induced system $\mathcal{M}_{X_0}$ of $\mathcal{M}$ on X_0 is Ω_0-regular on $N \times_{M_0} \dot{T}_{M_0}^* X_0$. Then we have a natural isomorphism :*

$$\Gamma(\Omega; \nu_N \mathcal{H}om_{\mathcal{D}_X}(\mathcal{M}, \mathcal{O}_X) \mid_{T_N M}) \xrightarrow{\sim} \Gamma(\Omega; \nu_N \mathcal{H}om_{\mathcal{D}_X}(\mathcal{M}, \mathcal{A}_M)). \quad (19)$$

Example 4. Assume that the codimension of N in M is two and take an arbitrary operator $P \in \mathcal{D}_X$ for which $X_0 = \{z_2 = 0\} \subset X$ is noncharacteristic. Next set $\mathcal{M} := \mathcal{D}_X/\mathcal{D}_X P + \mathcal{D}_X EQ$ by the Tricomi operator $Q = D_1^2 - x_1 D_{x'}^2 \in \mathcal{D}_{X_0}$, $x' = (x_3, \dots, x_n)$ and an elliptic operator E on X_0, and suppose that $Y = \{z_1 = z_2 = 0\}$ is noncharacteristic for $\mathcal{M}$. Then by Theorem 1.12 of Kataoka [10], the induced system $\mathcal{M}_{X_0}$ on X_0 satisfies the condition of Theorem 11 and we have an isomorphism :

$$\Gamma(\Omega; \nu_N \mathcal{H}om_{\mathcal{D}_X}(\mathcal{M}, \mathcal{O}_X)\,|_{T_N M}) \xrightarrow{\sim} \Gamma(\Omega; \nu_N \mathcal{H}om_{\mathcal{D}_X}(\mathcal{M}, \mathcal{A}_M)) \qquad (20)$$

for every open convex cone Ω with the edge N s.t. $\Omega \cap M_0 = \{x_1 > 0\}$ in M_0.

Remark. In the example above, we may replace EQ by any operator on X_0 which is N_+-regular in the sense of Kataoka [10] (or Ω-regular in the sense of Schapira [19]) on $N \times_{M_0} \dot{T}^*_{M_0} X_0$. To construct more examples, refer to Schapira [18] and Kaneko [5]. Also notice that Uchida [28] gave some criteria of Ω-regularity for systems, that is, for $\mathcal{D}_X$-modules.

References

[1] T. Aoki and S. Tajima, *On a generalization of Bochner's tube theorem for C-R-submanifolds,* Proc. Japan Acad., Ser. A, **63** (1987), 302-303.

[2] J.M. Bony and P. Schapira, *Propagation des singularités analytiques pour les solutions des équations aux derivées partielles,* Ann. Inst. Fourier, Grenoble, **t.26**I (1976), 81-140.

[3] J-M. Delort, *Microlocalisation simultanée et problème de Cauchy ramifié,* to appear in Compositio Math.

[4] S. Funakoshi, , Master thesis presented to the University of Tokyo, 1995.

[5] A. Kaneko, *Singular spectrum of boundary values of solutions of partial differential equations with real analytic coefficients,* Sci. Papers College Gen. Ed. Univ. Tokyo, **25** (1975), 59-68.

[6] M. Kashiwara and T. Kawai, *On the boundary value problem for elliptic systems of linear partial differential equations I-II,* Proc. Japan Acad., **48** (1971), 712-715; ibid., **49** (1972), 164-168.

[7] M. Kashiwara and Y. Laurent, *Théorèmes d'annulation et deuxième microlocalisation,* prépublication d'Orsay, 1983.

[8] M. Kashiwara and P. Schapira, *Micro-hyperbolic systems,* Acta Math. **142** (1979), 1-55.

[9] M. Kashiwara and P. Schapira, *Sheaves on manifolds* Grundlehlen der Math. Wiss. **292**, Springer-Verlag, 1990.

[10] K. Kataoka, *Microlocal theory of boundary value problems I-II,* J. Fac. Sci. Univ. Tokyo **27** (1980), 355-399 ; ibid., **28** (1981), 31-56.

[11] K. Kataoka and N. Tose, *Some remarks in 2nd microlocalization* (in Japanese), RIMS Kokyuroku, Kyoto Univ. **660**(1988), 52-63.

[12] T. Kawai, *Extension of solutions of systems of linear differential equations,* Publ. RIMS, Kyoto Univ. **12** (1976), 215-227.

[13] H. Komatsu, *A local version of Bochner's tube theorem,* J. Fac. Sci. Univ. Tokyo **19** (1972), 201-214.

[14] A. Martineau, *Le 'edge of the wedge theorem' en théorie des hyperfonctions de Sato,* Proc. Intern. Conf. on Functional Analysis and Related Topics, 1969, Univ. Tokyo Press, Tokyo, 1970, pp. 95-106.

[15] M. Morimoto, *Sur la décomposition du faisceau des germes de singularités d'hyperfonctions,* J. Fac. Sci. Univ. Tokyo **17** (1970), 215-239.

[16] T. Oaku, *Higher-codimensional boundary value problem and F-mild hyperfunctions,* in Algebraic Analysis Vol. **II**, (Papers dedicated to Prof. Sato), (M. Kashiwara and T. Kawai (eds.)), Academic Press, 1988, pp. 571-586.

[17] M. Sato, T. Kawai and M. Kashiwara, *Hyperfunctions and pseudodifferential equations* Lecure Notes in Math. **287**, Springer-Verlag, 1973, pp. 265-529.

[18] P. Schapira, *Propagation at the Boundary of Analytic singularities,* Singularities of Boundary Value Problems, Reidel Publ. Co., 1981, pp. 185-212.

[19] P. Schapira, *Front d'onde analytique au bord II,* Séminaire E.D.P., École Polyt., 1986, Exp.13.

[20] P. Schapira, *Microfunctions for boundary value problems,* Algebraic Analysis (Papers dedicated to Prof. Sato) (M. Kashiwara and T.Kawai (eds.)), Academic Press, 1088, pp.809-819.

[21] P. Schapira and K. Takeuchi, *Déformation binormale et bispécialisation,* C.R. Acad. Sc. **t.319**, Série I (1994), 707-712.

[22] P. Schapira and G. Zampieri, *Regularity at the boundary for systems of microdifferential operators,* Pitman Research Notes in Math.**158**, 1987, pp. 186-201.

[23] K. Takeuchi, *Microlocal boundary value problems in higher codimensions,* to appear in Bull. Soc. math. France, t. **124** (1996), 243-276.

[24] K. Takeuchi, *Binormal deformation and bimicrolocalization,* Publ. RIMS, Kyoto Univ., **32** (1996), 115-160.

[25] K. Takeuchi, *Théorèmes de type Edge of the Wedge pour les solutions hyperfonctions,* C.R. Acad. Sc. **t.321**, Série I (1995), 1333-1336.

[26] K. Takeuchi, *Edge of the Wedge type theorems for hyperfunction solutions,* submitting.

[27] N. Tose, *Theory of partially elliptic systems and its applications,* Master thesis presented to the University of Tokyo, 1985.

[28] M. Uchida, *Continuation of analytic solutions of linear differential equations up to convex conical singularities,* Bull. Soc. math. France, t. **121** (1993), 133-152.

[29] M. Uchida *A generalization of Bochner's tube theorem for elliptic boundary value problems,* RIMS Kokyuroku, Kyoto Univ. **845** (1993), 129-138.

Kashiwara's microlocal analysis of the Bergman kernel for domains with corner

Motoo Uchida

Department of Mathematics, Graduate School of Science, Osaka University, Toyonaka, Osaka 560, Japan

Abstarct. We shall show that an analogy of Kashiwara's microlocal analysis works well and will give good information for Bergman kernels of pseudoconvex domains with corner. The strict pseudoconvexity plays an essential role in the case of dimension ≥ 3.

Notations. For a complex manifold X, T^*X denotes the cotangent bundle of X. The antipodal mapping on T^*X is denoted by a, and $p^a = a(p)$ for $p \in T^*X$. $\mathcal{D}_X$ denotes the sheaf of rings of differential operators on X, $\mathcal{E}_X$ the sheaf of rings of microdifferential operators, and $\mathcal{E}_X(k)$ the subsheaf of $\mathcal{E}_X$ of microdifferential operators of order $\leq k$. If Y is a closed complex submanifold of X, $T^*_Y X$ denotes the conormal bundle of Y and $\mathcal{C}_{Y|X}$ the sheaf of holomorphic microfunctions along Y of finite order.

1. Kashiwara's Microlocal Analysis for Strictly Pseudoconvex Domains with Corner

Let X be a complex manifold of dimension $n \geq 2$; z denotes a system of local coordinates $(z_1, \ldots, z_n)$ of X.

Let Ω be an open domain of X. Let $x_0 \in X$. Assume, in a neighborhood of x_0, Ω to be given by

$$f(z, \overline{z}) > 0 \quad \text{and} \quad g(z, \overline{z}) > 0$$

for real-valued analytic functions f and g in $(z, \overline{z})$ with

$$\partial_z f \wedge \partial_z g(x_0) \neq 0.$$

Assume moreover that domains $\{f > 0\}$ and $\{g > 0\}$ are both strictly pseudoconvex in a neighborhood of x_0. We say that such Ω is a strictly pseudoconvex domain with corner.

Let Z denote the conjugate complex manifold of X, w a system of local holomorphic coordinates of Z with $w = \overline{z}$. Let $p_1 : T^*(X \times Z) \to T^*X$, $p_2 : T^*(X \times Z) \to T^*Z$ be the first and the second projection. Let us set :

$$\Lambda = T^*_M(X \times Z),$$

where M is a complex submanifold of $X \times Z$ defined by

$$f(z, w) = g(z, w) = 0.$$

Let Δ be the diagonal of $X \times Z$. For $p \in T^*X$, $p_2^a(p_1^{-1}(p) \cap T^*_\Delta(X \times Z))$, with $p_2^a = a \circ p_2$, consists of one point of T^*Z, which we denote by p^c. If $p = \lambda\partial_z f(x_0) + \mu\partial_z g(x_0)$ with $\lambda, \mu \in \mathbf{R}$, then $p^c = \lambda\partial_{\overline{z}} f(x_0) + \mu\partial_{\overline{z}} g(x_0)$.

Lemma 1.1. *Let* $p = \lambda\partial_z f(x_0) + \mu\partial_z g(x_0)$ *with* $\lambda > 0$, $\mu > 0$. *Then*

$$p_1|_\Lambda : (\Lambda, \widetilde{p}) \longrightarrow (T^*X, p)$$

and

$$p_2|_\Lambda : (\Lambda, \widetilde{p}) \longrightarrow (T^*Z, p^c)$$

are local isomorphisms, where $\widetilde{p} = (p, p^c)$.

Proof. We may assume $df(x_0) = dz_1 + d\overline{z}_1$ and $dg(x_0) = dz_2 + d\overline{z}_2$. Then

$$\det \begin{bmatrix} 0 & 0 & \partial_w f \\ 0 & 0 & \partial_w g \\ \partial_z f & \partial_z g & \lambda\partial_z\partial_w f + \mu\partial_z\partial_w g \end{bmatrix} = \det \left[\lambda\frac{\partial^2 f}{\partial z_i \partial w_j} + \mu\frac{\partial^2 g}{\partial z_i \partial w_j} \right]_{3 \le i, j \le n}$$

at x_0. In virtue of the strict pseudoconvexity of Ω, this determinant is not zero if $\lambda > 0$, $\mu > 0$. QED

Hence Λ defines a homogeneous symplectic transformation

$$\phi : (T^*X, p) \longrightarrow (T^*Z, (p^c)^a)$$

by $\phi = (p_2^a|_\Lambda) \circ (p_1|_\Lambda)^{-1}$ with p_2^a being $p_2^a(z, w, \zeta, \theta) = (w, -\theta)$ and a the antipodal mapping. Now consider a non-degenerate section

$$u = \log f(z, w) \log g(z, w)\, dw$$

of $\mathcal{C}_{M|X\times Z} \otimes_{\mathcal{O}_Z} p_2^{a-1}\Omega_Z^{(n)}$, where $\Omega_Z^{(n)}$ denotes the sheaf of holomorphic n-forms on Z. Quantizing ϕ by this kernel, we get a ring homomorphism $\widetilde{\phi} : \phi^{-1}\mathcal{E}_Z \to \mathcal{E}_X$.

Let $p = \lambda\partial_z f(x_0) + \mu\partial_z g(x_0)$ with $\lambda < 0$ and $\mu < 0$. Using this quantized contact transformation $(\phi, \widetilde{\phi})$ at p^a, we define a quantization $\widetilde{\psi}$ of ϕ at p by

$$\widetilde{\psi}(Q) = \widetilde{\phi}(Q^*)^*, \quad Q \in \phi^{-1}\mathcal{E}_Z,$$

where $*$ denotes the adjoint homomorphism $\mathcal{E}_X \to a^{-1}\mathcal{E}_X$ or $\mathcal{E}_Z \to a^{-1}\mathcal{E}_Z$. This defines a simple holonomic system $\mathcal{E}_{X\times Z}/\mathcal{J}$ with support on Λ, where $\mathcal{J}$ is the ideal generated by microdifferential operators

$$\widetilde{\psi}(Q)(z, D_z) - Q^*(w, D_w), \quad Q \in \mathcal{E}_Z.$$

Then we have :

Theorem 1.2. *Let $p = \lambda\partial_z f(x_0) + \mu\partial_z g(x_0)$ with $\lambda < 0$ and $\mu < 0$. The Bergman kernel $B(z, w)$ of Ω satisfies the holonomic system defined above at $\widetilde{p} = (p, p^c)$: $\mathcal{J}B = 0$.*

Remark. In Theorem 1.2, p is a point of $\mathrm{SS}(\mathbf{C}_\Omega) \setminus \overline{\mathrm{SS}(\mathbf{C}_\Omega) \setminus T^*_{x_0}X}$, where $\mathrm{SS}(\mathbf{C}_\Omega)$ denotes the micro-support of the sheaf $\mathbf{C}_\Omega$ $(= j_! j^{-1}\mathbf{C}_X$ with $j : \Omega \hookrightarrow X)$ over X.

The heuristic or formal reasoning of this observation is given after Kashiwara's idea in the same way as in the smooth boundary case [K2].

Let $K(z, \overline{z}) = Y(f(z, \overline{z}))Y(g(z, \overline{z}))$, and suppose

$$P^*(z, D_z)K(z, \overline{z}) = Q(\overline{z}, D_{\overline{z}})K(z, \overline{z}).$$

Then for any function $u(z)$ holomorphic on Ω, we have

$$\begin{aligned} P(z, D_z)u(z) &= P(z, D_z)\int_\Omega B(z, \overline{w})u(w)dw \wedge d\overline{w} \\ &= P(z, D_z)\int B(z, \overline{w})K(w, \overline{w})u(w)dw \wedge d\overline{w} \\ &= \int P(z, D_z)B(z, \overline{w}) \cdot K(w, \overline{w})u(w)dw \wedge d\overline{w}. \end{aligned}$$

On the other hand, we have

$$\begin{aligned} P(z, D_z)u(z) &= \int_\Omega B(z, \overline{w})Pu(w)dw \wedge d\overline{w} \\ &= \int B(z, \overline{w})Y(f(w, \overline{w}))Y(g(w, \overline{w}))Pu(w)dw \wedge d\overline{w} \\ &= \int B(z, \overline{w}) \cdot P^*(w, D_w)K(w, \overline{w}) \cdot u(w)dw \wedge d\overline{w} \\ &= \int B(z, \overline{w}) \cdot Q(\overline{w}, D_{\overline{w}})K(w, \overline{w}) \cdot u(w)dw \wedge d\overline{w} \\ &= \int Q^*(\overline{w}, D_{\overline{w}})B(z, \overline{w}) \cdot K(w, \overline{w})u(w)dw \wedge d\overline{w}. \end{aligned}$$

Thus we have

$$\int [P(z, D_z) - Q^*(\overline{w}, D_{\overline{w}})]B(z, \overline{w}) \cdot K(w, \overline{w})u(w)dw \wedge d\overline{w} = 0$$

for any u. Hence

$$P(z, D_z)B(z, \overline{w}) = Q^*(\overline{w}, D_{\overline{w}})B(z, \overline{w}).$$

2. The Leading Term of the Bergman Kernel as Microfunction

2.1. Principal symbols of simple microfunctions.

In this subsection, we recall the notion of principal symbols of simple holomorphic microfunctions [SKK, II.3, and SKKO, Sect.3] and prove a lemma which is useful for the purpose of calculating the leading term of the Bergman kernel microfunction.

Let X be a complex manifold of dimension n; x denotes a system of local coordinates $(x^1, \dots, x^n)$ of X. Denote by (x, ξ) the system of associated canonical coordinates of T^*X.

For a microdifferential operator $P = P(x, D_x)$ of order m, we set

$$L_P = H_{P_m} + \left(P_{m-1} - \frac{1}{2} \sum_{\nu} \frac{\partial^2 P_m}{\partial \xi_\nu \partial x^\nu} \right),$$

where H_{P_m} is the Hamiltonian vector field of the principal symbol P_m of P. Then $dx^{-1/2} \otimes L_P \otimes dx^{1/2}$, $dx = dx^1 \wedge \cdots \wedge dx^n$, defines a section of $\Omega_X^{-1/2} \otimes \mathcal{D}_{T^*X}(1) \otimes \Omega_X^{1/2}$.

Let Λ be a Lagrangian submanifold of T^*X and assume $P_m|_\Lambda = 0$. Then $dx^{-1/2} \otimes L_P \otimes dx^{1/2}$ acts on $\Omega_\Lambda^{1/2} \otimes \Omega_X^{-1/2}$ as Lie derivative on Λ, where $\Omega_\Lambda^{1/2}$ signifies a locally free $\mathcal{O}_\Lambda$-module such that $(\Omega_\Lambda^{1/2})^{\otimes 2} \cong \Omega_\Lambda^{(n)}$. Let $\mathcal{M}$ be a simple holonomic $\mathcal{E}_X$-module along Λ, u a non-degenerate section of $\mathcal{M}$. The principal symbol of u is a section s, denoted by $\sigma_\Lambda(u)$, of $\Omega_\Lambda^{1/2} \otimes \Omega_X^{-1/2}$ such that

$$(dx^{-1/2} \otimes L_P \otimes dx^{1/2})s = 0$$

for any $P \in \mathcal{E}_X$ with $Pu = 0$.

Now we consider the following special type of holonomic system at $p \in T^*X$ outside the zero section :

$$\mathcal{E}_X u = \mathcal{E}_X / \mathcal{J} \quad \text{with} \quad \mathcal{J} = \mathcal{E}_X(P_1, \dots, P_r, Q_{r+1}, \dots, Q_n), \quad u = 1 \bmod \mathcal{J}$$

where $P_i \in \mathcal{E}_X(1)$, $1 \le i \le r$, and $Q_j \in \mathcal{E}_X(0)$, $r+1 \le j \le n$. Assume that the principal symbols p_i, q_j of P_i, Q_j resp. satisfy

$$\{p_i, p_{i'}\} = \{q_j, q_{j'}\} = \{p_i, q_j\} = 0$$

for any i, i', j, j', where $\{p, q\}$ signifies the Poisson bracket on T^*X, and

$$dp_1 \wedge \cdots \wedge dp_r \wedge dq_{r+1} \wedge \cdots \wedge dq_n \wedge \omega \ne 0$$

with ω being the canonical 1-form of T^*X. We fix a nowhere vanishing n-form dx on X and set $\Lambda' = a(\Lambda)$ and

$$\mathcal{E}_X u^* = \mathcal{E}_X / \mathcal{J}^* \quad \text{with} \quad \mathcal{J}^* = \mathcal{E}_X(P_1^*, \dots, P_r^*, Q_{r+1}^*, \dots, Q_n^*),$$

$$u^* = 1 \bmod \mathcal{J}^*$$

where R^* denotes the adjoint operator of R; this is a holonomic system defined in a neighborhood of p^a with support on Λ'. Let $(p_1, \ldots, p_n, q_1, \ldots, q_n)$ be a local homogeneous symplectic coordinate system of (T^*X, p); then $(p_1^a, \ldots, p_n^a, q_1^a, \ldots, q_n^a)$ is a local homogeneous symplectic coordinate system of (T^*X, p^a), where $p_i^a = a^*(p_i)$, $q_j^a = a^*(q_j)$. Setting

$$d\lambda = dq_1 \wedge \cdots \wedge dq_r \wedge dp_{r+1} \wedge \cdots \wedge dp_n |_\Lambda \quad \text{and}$$
$$d\lambda^a = dq_1^a \wedge \cdots \wedge dq_r^a \wedge dp_{r+1}^a \wedge \cdots \wedge dp_n^a |_{\Lambda'},$$

we have :

Lemma 2.1. *If* $\sigma_\Lambda(u) = A \cdot d\lambda^{1/2} \otimes dx^{-1/2}$ *with* $A \in \mathcal{O}_\Lambda$, *then*

$$\sigma_{\Lambda'}(u^*) = \frac{1}{a^*A} \cdot (d\lambda^a)^{1/2} \otimes dx^{-1/2}.$$

In particular, $\mathrm{ord}_{\Lambda'}(u^*) = -\,\mathrm{ord}_\Lambda(u) + n - r$.

Proof is straightforward, and we omit it.

2.2. The leading term of the Bergman kernel microfunction.

In this section, we calculate the leading term of the Bergman kernel microfunction by using Lemma 2.1.

Let (z, w, ζ, θ) be the system of homogeneous symplectic coordinates of $T^*(X \times Z)$ associated to the local coordinate (z, w) of $X \times Z$. We may assume :

$$\Delta_{(12)} = \det \begin{bmatrix} \partial f/\partial z_1 & \partial f/\partial z_2 \\ \partial g/\partial z_1 & \partial g/\partial z_2 \end{bmatrix} \neq 0.$$

Let us take λ, μ so that

$$\begin{cases} f(z, w) = g(z, w) = 0, \\ \zeta = \lambda \partial_z f(z, w) + \mu \partial_z g(z, w), \\ \theta = \lambda \partial_w f(z, w) + \mu \partial_w g(z, w) \end{cases}$$

on Λ. Then (λ, μ, z', w), $z' = (z_3, \ldots, z_n)$, is a local coordinate system on Λ. In this system of coordinates, the principal symbol of $\log f(z, w) \log g(z, w)$, which we denote by u_0 in the following, is given by

$$\sigma_\Lambda(u_0) = \frac{1}{\lambda\mu} \frac{1}{\Delta_{(12)}^{1/2}} \cdot \frac{(d\lambda \wedge d\mu \wedge dz' \wedge dw)^{1/2}}{(dz \wedge dw)^{1/2}},$$

where we set $dz' = dz_3 \wedge \cdots \wedge dz_n$, $dw = dw_1 \wedge \cdots \wedge dw_n$.

Now define $p_1, \ldots, p_n, q_1, \ldots, q_n$ by $p_i = -\phi^*\theta_i$, $1 \le i \le n$, and $q_j = \phi^* w_j$, $1 \le j \le n$, and introduce a system of homogeneous symplectic coordinates $(\widetilde{p}_1, \ldots, \widetilde{p}_{2n}, \widetilde{q}_1, \ldots, \widetilde{q}_{2n})$ as

$$\widetilde{p}_i = (\theta_i + p_i)/2, \quad \widetilde{p}_{n+i} = \theta_i - p_i, \quad 1 \le i \le n$$

and

$$\widetilde{q}_j = w_j - q_j, \quad \widetilde{q}_{n+j} = (w_j + q_j)/2, \quad 1 \le j \le n.$$

In these coordinates, Λ is given by $\widetilde{q}_1 = \cdots = \widetilde{q}_n = \widetilde{p}_{n+1} = \cdots = \widetilde{p}_{2n} = 0$.

Let $z_1 = \varphi_1(z', w)$ and $z_2 = \varphi_2(z', w)$ be given by solving the equation $f = g = 0$ in (z_1, z_2). Then

$$\widetilde{p}_i\,|_\Lambda = \theta_i\,|_\Lambda = \lambda \frac{\partial f}{\partial w_i}(\varphi_1, \varphi_2, z', w) + \mu \frac{\partial g}{\partial w_i}(\varphi_1, \varphi_2, z', w) \quad \text{and} \quad \widetilde{q}_{n+j}\,|_\Lambda = w_j\,|_\Lambda.$$

Hence

$$\begin{aligned} &d\widetilde{p}_1 \wedge \cdots \wedge d\widetilde{p}_n \wedge d\widetilde{q}_{n+1} \wedge \cdots \wedge d\widetilde{q}_{2n}\,|_\Lambda \\ &= \det \begin{bmatrix} \partial_{w_1} f & \partial_{w_1} g & \begin{matrix} \lambda\,(\partial_{z_1}\partial_{w_1} f \cdot \partial_{z'}\varphi_1 + \partial_{z_2}\partial_{w_1} f \cdot \partial_{z'}\varphi_2 + \partial_{z'}\partial_{w_1} f) \\ + \mu\,(\partial_{z_1}\partial_{w_1} g \cdot \partial_{z'}\varphi_1 + \partial_{z_2}\partial_{w_1} g \cdot \partial_{z'}\varphi_2 + \partial_{z'}\partial_{w_1} g) \end{matrix} \\ \partial_{w_2} f & \partial_{w_2} g & \cdots \\ \vdots & \vdots & \\ \partial_{w_n} f & \partial_{w_n} g & \cdots \end{bmatrix} \times \\ &\qquad \times d\lambda \wedge d\mu \wedge dz' \wedge dw. \end{aligned}$$

Since

$$\begin{bmatrix} \partial_{z'}\varphi_1 \\ \partial_{z'}\varphi_2 \end{bmatrix} = - \begin{bmatrix} \partial_{z_1} f & \partial_{z_2} f \\ \partial_{z_1} g & \partial_{z_2} g \end{bmatrix}^{-1} \begin{bmatrix} \partial_{z'} f \\ \partial_{z'} g \end{bmatrix},$$

we have

$$d\widetilde{p}_1 \wedge \cdots \wedge d\widetilde{p}_n \wedge d\widetilde{q}_{n+1} \wedge \cdots \wedge d\widetilde{q}_{2n}\,|_\Lambda = \frac{1}{\Delta_{(12)}} J(f, g)\, d\lambda \wedge d\mu \wedge dz' \wedge dw,$$

where

$$\begin{aligned} J(f, g) &= \det \begin{bmatrix} 0 & 0 & \partial_w f \\ 0 & 0 & \partial_w g \\ \partial_z f & \partial_z g & \lambda \partial_z \partial_w f + \mu \partial_z \partial_w g \end{bmatrix} \\ &= \sum_{i<j,\, k<l} (-1)^{i+j+k+l} \frac{\partial(f, g)}{\partial(z_i, z_j)} \cdot \frac{\partial(f, g)}{\partial(w_k, w_l)} \times \\ &\qquad \times \det \left[\lambda \frac{\partial^2 f}{\partial z_p \partial w_q} + \mu \frac{\partial^2 g}{\partial z_p \partial w_q} \right]_{\substack{p \ne i, j \\ q \ne k, l}}. \end{aligned}$$

In summary, we have

$$\sigma_\Lambda(u_0) = \frac{1}{\lambda\mu} J(f,\, g)^{-1/2} \cdot \alpha_\Lambda^{1/2} \otimes (dz \wedge dw)^{-1/2},$$

where

$$\alpha_\Lambda = d\widetilde{p}_1 \wedge \cdots \wedge d\widetilde{p}_n \wedge d\widetilde{q}_{n+1} \wedge \cdots \wedge d\widetilde{q}_{2n}\, |_\Lambda .$$

By Lemma 2.1, we have

$$\sigma_{\Lambda'}(u_0^*) = \lambda\mu\, J(f,\, g)^{1/2} \cdot \alpha_\Lambda^{1/2} \otimes (dz \wedge dw)^{-1/2}. \tag{2.1}$$

Let $N \in \mathbf{Z}$, $N \geq 0$. For a set of holomorphic functions in z and w

$$\{\, C_{\nu_1\nu_2}(z,\, w) \mid \nu_1,\, \nu_2 \in \mathbf{Z},\, \nu_1 \geq 0,\, \nu_2 \geq 0,\, \nu_1 + \nu_2 \leq N \,\},$$

let

$$v = \sum_{\nu_1+\nu_2 \leq N} (-)^{\nu_1+\nu_2} C_{\nu_1\nu_2}(z,\, w) \frac{(\nu_1+1)!}{f^{\nu_1+2}} \frac{(\nu_2+1)!}{g^{\nu_2+2}};$$

then v defines a section of $\mathcal{C}_{M|X\times Z}$ and, for generic choices of $\{C_{\nu_1\nu_2}\}$, is non degenerate (i.e., the symbol ideal of v is reduced and defines $T_M^*(X \times Z)$). In the same way as above, the principal symbol of v is calculated as follows.
(2.2)

$$\sigma_\Lambda(v) = \sum_{\nu_1+\nu_2=N} C_{\nu_1\nu_2}(z,\, w)\lambda^{\nu_1+1}\mu^{\nu_2+1} J(f,\, g)^{-1/2} \cdot \alpha_\Lambda^{1/2} \otimes (dz \wedge dw)^{-1/2}.$$

By the observation of section 1, u_0^* represents the Bergman kernel microfunction B and it follows from (2.1) and (2.2) that

$$\begin{aligned} B(z,\, w) = \mathrm{Ct}\,(1+R)\Bigg\{ \sum_{i<j,\, k<l} (-1)^{i+j+k+l} \frac{\partial(f,\, g)}{\partial(z_i,\, z_j)} \cdot \frac{\partial(f,\, g)}{\partial(w_k,\, w_l)} \times \\ \times \widetilde{\det}\left[\frac{1}{f} \cdot \frac{\partial^2 f}{\partial z_p \partial w_q} + \frac{1}{g} \cdot \frac{\partial^2 g}{\partial z_p \partial w_q}\right]_{\substack{p\neq i,\, j \\ q \neq k,\, l}} \cdot \frac{1}{f^2 g^2} \Bigg\}, \end{aligned}$$

where R is a microdifferential operator in $(z,\, w)$ of order ≤ -1 defined at $\widetilde{p}$, and

$$\begin{aligned} &\widetilde{\det}\left[\frac{1}{f} \cdot \frac{\partial^2 f}{\partial z_p \partial w_q} + \frac{1}{g} \cdot \frac{\partial^2 g}{\partial z_p \partial w_q}\right]_{\substack{p\neq i,\, j \\ q \neq k,\, l}} \\ &\qquad = \left[\det\left[\lambda \frac{\partial^2 f}{\partial z_p \partial w_q} + \mu \frac{\partial^2 g}{\partial z_p \partial w_q}\right]_{\substack{p\neq i,\, j \\ q \neq k,\, l}}\right]_{\substack{\lambda^{\nu_1} = (-)^{\nu_1}(\nu_1+1)!/f^{\nu_1} \\ \mu^{\nu_2} = (-)^{\nu_2}(\nu_2+1)!/g^{\nu_2}}} \end{aligned}$$

(When $n = 2$, $\widetilde{\det}[\cdots]$ reads as 1).

References

[B1] Boutet de Monvel, L., *Complément sur les noyaux de Bergman*, Séminaire E.D.P. 1985-1986, exposé 20, Ecole Polytechnique.

[B2] Boutet de Monvel, L., *Singularity of the Bergman kernel*, Complex Geometry (G. Komatsu and Y. Sakane, eds.), Lecture Notes in Pure and Applied Math. **143**, Dekker, 1992, pp. 13–29.

[BS] Boutet de Monvel, L. and Sjöstrand, J., *Sur les singularités des noyaux de Bergman et de Szegö*, Astérisque, Soc. math. France, **34-35** (1976), 123–164.

[K1] Kashiwara, M., *Micro-local calculus of simple microfunctions*, Complex Analysis and Algebraic Geometry (W. L. Baily and T. Shioda, eds.), Iwanami-Shoten, Tokyo, 1977, pp. 369–374.

[K2] Kashiwara, M., *Analyse microlocale du noyau de Bergman*, Séminaire Goulaouic-Schwartz 1976-77, exposé 8, Ecole Polytechnique.

[SKK] Sato, M., Kawai, T., and Kashiwara, M., *Microfunctions and pseudo-differential equations*, Lecture Notes in Math. **287**, Springer-Verlag, 1973, pp. 265–529.

[SKKO] Sato, M., Kashiwara, M., Kimura, T., and Oshima, T., *Micro-local analysis of prehomogeneous vector spaces*, Invent. Math. **62** (1980), 117–179.

[Y] Yamazaki, S., *Singularity of the Bergman kernel for a two dimensional pseudoconvex tube domain with corners*, J. Math. Sci., Univ. Tokyo **2** (1995), 637–655.